VOLUME FORTY FOUR

Advances in
CHILD DEVELOPMENT
AND BEHAVIOR

VOLUME FORTY FOUR

Advances in
CHILD DEVELOPMENT
AND BEHAVIOR

Embodiment and Epigenesis: Theoretical
and Methodological Issues in Understanding
the Role of Biology within the Relational
Developmental System

Part A: Philosophical, Theoretical, and Biological Dimensions

Volume Editors

RICHARD M. LERNER
Institute for Applied Research in Youth Development,
Tufts University,
Medford, Massachusetts, USA

JANETTE B. BENSON
Department of Psychology,
University of Denver,
Denver, Colorado, USA

ELSEVIER

AMSTERDAM • BOSTON • HEIDELBERG • LONDON
NEW YORK • OXFORD • PARIS • SAN DIEGO
SAN FRANCISCO • SINGAPORE • SYDNEY • TOKYO
Academic Press is an imprint of Elsevier

Academic Press is an imprint of Elsevier
225 Wyman Street, Waltham, MA 02451, USA
525 B Street, Suite 1800, San Diego, CA 92101-4495, USA
Radarweg 29, PO Box 211, 1000 AE Amsterdam, The Netherlands
The Boulevard, Langford Lane, Kidlington, Oxford, OX5 1GB, UK
32 Jamestown Road, London NW1 7BY, UK

First edition 2013

Library of Congress Cataloging-in-Publication Data
A catalogue record for this book is available from the Library of Congress

British Library Cataloguing in Publication Data
A catalog record for this book is available from the British Library

ISBN: 978-0-12-397947-6
ISSN: 0065-2407 (Series)

For information on all Academic Press publications
visit our websie at store.elsevier.com

Printed in the United States of America
13 14   10 9 8 7 6 5 4 3 2 1

Working together
to grow libraries in
developing countries

www.elsevier.com • www.bookaid.org

To Willis F. Overton
Preeminent Developmental Scientist and Esteemed Colleague

CONTENTS (VOLUME 44)

CONTENTS (VOLUME 45)-Available separately

CONTRIBUTORS

Janette B. Benson
Department of Psychology, University of Denver, Denver, CO, USA

Evan Charney
Sanford School of Public Policy, Duke University, Durham, NC, USA

Gary Greenberg
Department of Psychology, Wichita State University, Wichita, KS, USA

Paul E. Griffiths
Department of Philosophy and Charles Perkins Centre, University of Sydney, Sydney, NSW, Australia

Lawrence V. Harper
Department of Human Ecology, University of California, Davis, CA, USA

Shirley Heying
Department of Anthropology, University of New Mexico, Albuquerque, NM, USA

Sheldon Krimsky
Department of Urban and Environmental Policy and Planning, Tufts University, Medford, MA, USA

Richard M. Lerner
Institute for Applied Research in Youth Development, Tufts University, Medford, MA, USA

Robert Lickliter
Department of Psychology, Florida International University, Miami, FL, USA

Megan Kiely Mueller
Eliot-Pearson Department of Child Development, Tufts University, Medford, MA, USA

Willis F. Overton
Department of Psychology, Temple University, Philadelphia, PA, USA

Ken Richardson
Bellevue, Dunblane, UK

Peter T. Saunders
Department of Mathematics, King's College London, London, UK; Institute of Science in Society, London, UK

Kristina Schmid Callina
Eliot-Pearson Department of Child Development, Tufts University, Medford, MA, USA

James Tabery
Department of Philosophy, University of Utah, Salt Lake City, UT, USA

Douglas Wahlsten
Department of Psychology, University of North Carolina Greensboro, Greensboro, NC, USA

David C. Witherington
Department of Psychology, University of New Mexico, Albuquerque, NM, USA

PREFACE

At this writing, cutting-edge theory and research in developmental science is framed by a relational developmental systems perspective as best articulated by Willis F. Overton (e.g., 2003, 2006, 2010, 2011, 2012; Overton & Lerner, 2012; Overton & Müller, 2012). Based on theory and research from multiple disciplines (e.g., evolutionary biology, human genetics, developmental science, sociology, and anthropology), this perspective demonstrates the flaws of split, reductionist conceptions of human development. Scholarship framed by relational developmental systems thinking therefore documents the logical and empirical shortcomings of biological reductionist (genetic or neuronal) models (e.g., sociobiology, evolutionary psychology, or behavioral genetics) and methods (e.g., adoption designs, MZ and DZ twin research, or heritability analysis).

Relational developmental systems theory explains that any facet of individual structure or function (e.g., genes, the brain, personality, cognition, or intelligence) is embodied, or fused, with other features of the individual and with the characteristics of his or her proximal and distal ecology, including culture and history. Embodiment means that biological, psychological, and behavioral attributes of the person, in fusion with history, have a temporal parameter. This integration among the levels of organization within the developmental system has implications across both ontogeny and phylogeny. Thus, embodiment provides a basis for epigenetics across generations, that is, for changes in gene–context relations within one generation being transmitted to succeeding generations. Embodiment also provides the basis for epigenetic change within the life span of an individual, that is, for qualitative discontinuity across ontogeny in relations among biological, psychological, behavioral, and social variables.

Therefore, relational developmental systems conceptions provide an approach to the study of evolutionary and ontogenetic change that capitalizes on the dynamic, mutually influential relation between developing individuals and their complex and changing ecology. Given the singularly important implications of this scholarship for developmental science, the unprecedented step has been taken to devote both volumes of the 2013 publication year of *Advances in Child Development and Behavior* to this work, "Embodiment and Epigenesis: Theoretical and Methodological Issues in Understanding the Role of Biology within the Relational Developmental System."

The purpose of this special two-volume set is to present theory and research pertinent to the role of biological/physiological variables (e.g., in regard to the roles of evolutionary processes and genetic- or brain-related variables) in cognitive, emotional, and behavioral development. From a relational developmental systems perspective, the chapters in the volumes view biology/physiology within the context of the concept of embodiment, and thus eschew the reductionist approach to biology that has plagued developmental science in previous eras. The two volumes bring together key scientists from several disciplines to focus on the role of the active individual, on temporality and culture, and on the mutually influential individual–context relations that both create epigenetic change and that afford a multidimensional understanding of evolution (involving neo-Lamarckian conceptions of evolution) and emphasize ontogenetic plasticity in biological (including genetic) processes. We believe that these two volumes show that cutting-edge theory and research in developmental and biological science demonstrates both the conceptual vacuity and empirical failings of biological reductionism and, in turn, offers a theoretically- and methodologically-rigorous approach to the study of integrated, systematic, and successive change across ontogeny and phylogeny.

There are numerous people to thank in regard to the preparation of this book. First and foremost we are indebted to the authors who contributed chapters to this project. Their scholarship and dedication to excellence enabled this work to be produced. We are also grateful to Jarrett M. Lerner, the Managing Editor at the Institute for Applied Research at Tufts University, for his superb editorial work, his meticulous attention to detail, and gracious and good-natured way of handling all aspects of the publication process. His commitment to quality and productivity, his efficiency and judgment, and his resilience in the face of the tribulations of manuscript production, are greatly admired and deeply appreciated, and contributed mightily to the cohesiveness and clarity of this project. Richard M. Lerner is also grateful to the John Templeton Foundation, the National 4-H Council, and the Thrive Foundation for Youth for supporting his work during the course of working on this project.

Finally, we dedicate these two volumes to Willis F. Overton, the preeminent scholar regarding philosophy and theory in developmental science over the course of the past five decades. His vision, voice, and erudition have been the key intellectual forces involved in establishing and enabling the flourishing of the relational metatheory and relational developmental systems models that, today, frame theory, method, and research about life-span development. We dedicate these two volumes to honoring and celebrating his intellect and scholarly leadership.

R.M. L.
Medford, MA, USA
J.B. B.
Denver, CO, USA

REFERENCES

Overton, W. F. (2003). Development across the life span: philosophy, concepts, theory. In: R. M. Lerner, M. A. Easterbrooks & J. Mistry (Eds.), & Irving B. Weiner (Editor-in-Chief). *Comprehensive Handbook of psychology: Developmental psychology* (Vol. 6, pp. 13–42). New York: Wiley.

Overton, W. F. (2006). Developmental psychology: philosophy, concepts, and methodology. In R. M. Lerner (Ed.), & W. Damon & R. M. Lerner (Editors-in-Chief). *Theoretical models of human development (Handbook of child psychology* (Vol. 1, 6th ed., pp. 18–88). New York: Wiley.

Overton, W. F. (2010). Life-span development: concepts and issues. In W. F Overton (Ed.), & R. M. Lerner (Editor-in-Chief). *Cognition, biology, and methods across the life span. Volume 1 of the handbook of life-span development* (pp. 1–29). Hoboken, NJ: Wiley.

Overton, W. F. (2011). Relational developmental systems and quantitative behavior genetics: Alternative or parallel methodologies? *Research in Human Development, 8*(3–4), 258–63.

Overton, W. F (2012). Evolving scientific paradigms: Retrospective and prospective. In L. L'Abate (Ed.), *The role of paradigms in theory construction.* (pp. 31–65). New York: Springer.

Overton, W. F. & Lerner, R. M (2012). Relational developmental systems: Paradigm for developmental science in the post-genomic era. *Behavioral and Brain Sciences, 35*(5), 375–6.

Overton, W. F. & Müller, U. (2012). Development across the life span: Philosophy, concepts, theory. In R. M Lerner, M. A Easterbrooks & J Mistry (Eds.), & I. B. Weiner (Editor-in-Chief). *Handbook of psychology: Developmental psychology* (Vol. 6, pp. 19–58). New York: Wiley.

CHAPTER ONE

Introduction: Embodiment and Epigenesis: A View of the Issues

Richard M. Lerner*,1, Janette B. Benson†

*Institute for Applied Research in Youth Development, Tufts University, Medford, MA, USA
†Department of Psychology, University of Denver, Denver, CO, USA
1Corresponding author: E-mail: richard.lerner@tufts.edu

Contents

Abstract

Relational developmental systems theories emphasize that any facet of individual structure or function is embodied with other features of the individual and with the characteristics of his or her proximal and distal ecology, including culture and history. Embodiment means that biological, psychological, and behavioral attributes of the person, in fusion with the contexts of human development, have a temporal parameter. Embodiment provides a basis for epigenetics across generations and for epigenetic (qualitative discontinuous) change across ontogeny. We describe how the chapters in this two-volume set present theory and research pertinent to the roles of evolutionary and ontogenetic processes in cognitive, emotional, and behavioral development across the life span. The scholarship presented in these volumes suggests that rigorous, relational developmental theory-predicated research about the plastic, mutually influential relations among individual and ecological processes will enable developmental science to better describe, explain, and optimize the fundamental relational process of human development.

The writing of this chapter was supported in part by grants from the John Templeton Foundation, the National 4-H Council, and the Thrive Foundation for Youth.

1. INTRODUCTION

Developmental science seeks to describe, explain, and optimize intraindividual changes and interindividual differences in intraindividual changes across the life span (Baltes, Reese, & Nesselroade, 1977; Lerner, 2012). Although the goals of description, explanation, and optimization can be found in prior instantiations of the field—in child psychology and then in developmental psychology (Lerner, 2012)—contemporary developmental scientists now approach these three objectives differently than in the past. Whether studying infancy, childhood, adolescence, or the adult and aging portions of the life span, contemporary scholarship in human development attempts to explain how mutually influential relations between individuals and their contexts (i.e., bidirectional, reciprocal, synergistic, or fused relations; e.g., Thelen & Smith, 2006; Tobach & Greenberg, 1984) provide the basis for individual behavior and development.

Today, then, developmental scientists focus on systematic and successive alterations in the course of these relations, and focus on the integration of multiple attributes of the individual (e.g., physiological, cognitive, emotional, motivational, and behavioral characteristics) and multiple levels of the ecology of human development, ranging from the biological level through the sociocultural and historical levels, including the designed and natural environments (Bronfenbrenner & Morris, 2006; Lerner, 2002, 2006). Accordingly, in contemporary developmental science the cutting edge of theory and research aimed at elucidating these relations between individuals and contexts is framed by relational developmental systems theories, models best articulated by Willis F. Overton (2003, 2006, 2010, 2011, 2012; Overton and Lerner 2012; Overton & Müller, 2012).

2. RELATIONAL DEVELOPMENTAL SYSTEMS THEORIES: AN OVERVIEW

The study of human development has evolved from a field dominated by either psychogenic or biogenic approaches to a multidisciplinary approach to the life span that seeks to integrate variables from biological through cultural and historical levels of organization into a synthetic, coactional system (Elder, 1998; Gottlieb, 1997, 1998; Hood, Halpern, Greenberg, & Lerner, 2010). Reductionist accounts of development that adhere to a Cartesian dualism, and that pull apart (split) facets of the integrated developmental system, are rejected by proponents of relational developmental systems theories (Mistry & Wu, 2010; Overton, 2010, 2011; Overton & Müller, 2012). Reductionist views typically raise as key developmental issues such split formulations as nature versus nurture, continuity versus discontinuity, stability versus instability, or basic versus applied science (Lerner, 2002, 2006).

Today, such thinking is eschewed in favor of a relational metamodel that emphasizes that studying the integration of different levels of organization is a means to understand and to study life-span human development (Lerner, 2006; Overton, 2011; Overton & Müller, 2012). Thus, the conceptual emphasis of relational developmental systems theory is placed on the nature of mutually influential relations between individuals and contexts, represented as individual ↔ context relations. As we have noted, all levels of the developmental system are integrated within relational developmental systems theories (Lerner, 2006), ranging from variables involved in biological/physiological processes, through behavioral and social relationship processes, through physical ecological, cultural, and historical processes (e.g., Bronfenbrenner & Morris, 2006; Lerner, 2002). The embeddedness of all levels within history imbues temporality into individual ↔ context relations, and means that there is a potential for plasticity, for organized and systematic change in these relations, across person, time, and place (Elder, 1998; Lerner, 1984, 2002, 2006).

Relational developmental systems theories focus on the "rules," the processes that govern exchanges between individuals and their contexts. Brandtstädter (1998) terms these relations "developmental regulations" and notes that when developmental regulations involve mutually beneficial individual ↔ context relations, they constitute *adaptive* developmental regulations. The possibility of adaptive developmental relations between individuals and their contexts and the potential plasticity of human development are the distinctive features of this approach to human development.

These core features of developmental systems models provide a rationale for making a set of methodological choices that differs in study design, measurement, sampling, and data analytic techniques from selections made by researchers using split, dichotomous or reductionist approaches to developmental science. Moreover, the emphasis on how the individual acts on the context, to contribute to the plastic relations with it, fosters an interest in individual agency (or on intentional self-regulation; Gestsdóttir & Lerner, 2008)—on individuals as producers of their own development (Lerner, 1982; Lerner & Busch-Rossnagel, 1981). This focus is best instantiated by person-centered (as compared to variable-centered) approaches to the study of human development and thus, as well, to individual difference (diversity)-oriented developmental scholarship (Nesselroade & Molenaar, 2010).

In addition, the person-centered focus, as well as the emphases on plasticity and on mutually influential person ↔ context relations, has resulted in relational developmental systems theories being used as a frame for modeling the changing structure of ontogenetic trajectories, and has resulted in the view that developmental science is a nonergodic field (Nesselroade & Molenaar, 2010).The ergodic theorem holds that data sets are marked by (1) homogeneity across individuals in a three-dimensional matrix that involves persons, variables, and time and (2) stationarity of individuals' scores on variables across time (Molenaar, 2007). Framed by relational developmental systems thinking, however, developmental scientists believe that there is variation across people both within time and within people across time in their trajectories of *individual* ↔ *context relations* (i.e., *across time differences*). In other words, people differ in their paths across the life span. As such, the assumptions of homogeneity and stationarity of the ergodic theorem are rejected in contemporary developmental science (Molenaar, 2007, 2010). As a consequence of nonergodicity, developmental scientists place greater importance on not only person-centered research but also, as such, on change-sensitive methodologies for their descriptive and explanatory efforts.

2.1. Summary

The conceptual and associated methodological emphases of scholarship associated with relational developmental systems theories, and with the relational metamodels within which such theories are embedded (Overton, 2011), have led developmental scientists within this perspective to draw on research from multiple disciplines (for instance, evolutionary biology, human genetics, developmental science, sociology, and anthropology) to better understand the integrated changes across the multiple levels of organization within the ecology of human development and, as well, to therefore document the logical and empirical shortcomings of split, biological reductionist (genetic or neuronal) models (e.g., sociobiology, evolutionary psychology (EP), or behavioral genetics) and methods (e.g., adoption designs, MZ and DZ twin research, or heritability analysis). These scholars have used relational developmental systems theory to explain that any facet of individual structure or function (e.g., genes, the brain, personality, cognition, or intelligence) is embodied, that is, is fused, with other features of the individual and with the characteristics of his or her proximal and distal ecology, including culture and history.

Embodiment means that biological, psychological, and behavioral attributes of the person, in fusion with culture, have a temporal (historical) parameter. As such, embodiment, the fusion among the levels of organization within the developmental system, has implications across both ontogeny and phylogeny (Ho, 2010; Jablonka & Lamb, 2005). These implications involve the concept of epigenesis and, as well, the presence of plasticity in phylogeny and ontogeny that occurs because of embodied change; plasticity characterizes the relations between organisms and contexts (Lerner, 1984) that, across time, create epigenetic processes within and across generations.

3. EMBODIMENT, EVOLUTION, AND ONTOGENY

Contemporary scholarship about the character of evolution reflects the concept of embodied change within the developmental system. For instance, Bateson and Gluckman (2011, p. 5) observe that "gene expression is profoundly influenced by factors external to the cell nucleus in which reside the molecules making up the genes: the deoxyribonucleic acid (DNA). A willingness to move between different levels of analysis has become essential for an understanding of development and evolution." Similarly, Keller (2010, pp. 6–7) explains that:

Not only is it a mistake to think of development in terms of separable causes, but it is also a mistake to think of development of traits as a product of causal elements interacting with one another. Indeed, the notion of interaction presupposes the existence of entities that are at least ideally separable – i.e., it presupposes an a priori space between component entities – and this is precisely what the character of developmental dynamics precludes. Everything we know about the processes of inheritance and development teaches us that the entanglement of developmental processes is not only immensely intricate, but it is there from the start. From its very beginning, development depends on the complex orchestration of multiple courses of action that involve interactions among many different kinds of elements – including not only preexisting elements (e.g., molecules) but also new elements (e.g., coding sequences) that are formed out of such interactions, temporal sequences of events, dynamical interactions, etc.

Moreover, Pigliucci and Mueller (2010), in presenting what they term an 'Extended Synthesis' of evolution, note that:

Far from denying the importance of genes in organismal evolution, the extended theory gives less overall weight to genetic variation as a generative force. Rather, [there is a] view of "genes as followers" in the evolutionary process, ensuring the routinization of developmental interactions, the faithfulness of their inheritance, and the progressive fixation of phenotypic traits that were initially mobilized through plastic responses of adaptive developmental systems to changing environmental conditions. In this way, evolution progresses through the capture of emergent interactions into genetic-epigenetic circuits, which are passed to and elaborated on in subsequent generations (p. 14).

In turn, West-Eberhard (2003) argues that 'the universal environmental responsiveness of organisms, alongside genes, influences individual development and organic evolution, and this realization compels us to reexamine the major themes of evolutionary biology in a new light' (p. vii). Linking the presence of plasticity across development with evolution, she makes three major points:

> First, environmental induction is a major initiator of adaptive evolutionary change. The origin and evolution of adaptive novelty do not await mutation; on the contrary, genes are followers not leaders, in evolution. Second, evolutionary novelties result from the reorganization of preexisting phenotypes and the incorporation of environmental elements. Novel traits are not de novo constructions that depend on a series of genetic mutations. Third, phenotypic plasticity can facilitate evolution by the immediate accommodation and exaggeration of change. It should no longer be regarded as a source of noise in a system governed by genes, or as a "merely environmental" phenomenon without evolutionary importance.
>
> **(West-Eberhard, 2003, p. 20).**

Crystallizing the embodiment of variables from all levels of organization within developmental systems that create epigenetic change across generations, Jablonka and Lamb (2005, p. 1) summarize evidence demonstrating that evolution involves four interrelated dimensions:

> Molecular biology has shown that many of the old assumptions about the genetic system, which is the basis of present-day neo-Darwinian theory, are incorrect. It has also shown that cells can transmit information to daughter cells through non-DNA (epigenetic) inheritance. This means that all organisms have at least two systems of heredity. In addition, many animals transmit information to others by behavioral means, which gives them a third hereditary system. And we humans have a fourth, because symbol-based inheritance, particularly language, plays a substantial role in our evolution. It is therefore quite wrong to think about heredity and evolution solely in terms of the genetic system. Epigenetic, behavioral, and symbolic inheritance also provide variation on which natural selection can act.

Accordingly, in a book discussing the transformations of Lamarckian theory that have arisen in relation to the increasingly more active focus on epigenetic processes in the study of both evolution and development (Meaney, 2010), Gissis and Jablonka (2011, p. xiii) note that "*Plasticity* – the capacity of organisms to change in response to varying conditions – is…a large topic, but, just as Lamarck anticipated, an understanding of plasticity is now recognized as being fundamental to an understanding of evolution." In turn, and underscoring the links between plasticity of embodied relations among an organism and the multiple biological through ecological levels of its ecology and epigenetic change, they go on to note that:

> Experimental work now shows that, contrary to the dogmatic assertions of many mid-twentieth-century biologists that it could not occur, even a form of "inheritance of acquired characteristics" does occur and might even be said to be ubiquitous. In particular, new variations induced by stress are sometimes inherited. The molecular

mechanisms that underlie such inheritance – the epigenetic inheritance systems – are now partially understood, and…the existence of various types of [such] soft inheritance affects how we see adaptive evolution and speciation. It also has implications for human health.

(Gissis and Jablonka (2011, p. xiii).

We will return in the concluding section of this chapter to the implications of embodiment and epigenesis for health and positive human development.

Here we may note, however, that the evidence about embodiment, plasticity, and epigenetics that accounts for the character of evolutionary and developmental change understandably elicits skepticism about, indeed the repudiation of, the "extreme nature" (Rose & Rose, 2000) of the claims of some biological reductionists. For instance, EP claims that "everything from children's alleged dislike of spinach to our supposed universal preferences for scenery featuring grassland and water derives from [the] mythic human origin in the African savannah" (Rose & Rose, 2000, p. 2). These claims are predicated on the basis of the assertion that one can explain:

all aspects of human behaviors, and thence culture and society, on the basis of universal features of human nature that found their final evolutionary form during the infancy of our species some 100–600,000 years ago. Thus for EP, what its protagonists describe as the 'architecture of the human mind' which evolved during the Pleistocene is fixed, and insufficient time has elapsed for any significant subsequent change. In this architecture there have been no major repairs, no extensions, no refurbishments, indeed nothing to suggest that micro or macro contextual changes since prehistory have been accompanied by evolutionary adaption.

(Rose & Rose, 2000, p. 1).

Clearly such assertions within EP are inconsistent with the now quite voluminous evidence in support of the epigenetic character of evolution and ontogeny, of the multiple, integrated dimensions of evolution, and of the role of the organism's own agency and of culture in creating change within and across generations.

Nevertheless, such examples of misguided scholarship continue to appear in the literature—making the presentations across the present two volumes both timely and important. An example of the extreme nature of the claims of evolutionary psychologists pointed to by Rose and Rose (2000) occurs in writing about what is termed "paternal investment theory" (Belsky, 2012; Belsky, Steinberg, & Draper, 1991; Draper & Harpending, 1982, 1988). For instance, Ellis, Schlomer, Tilley, and Butler (2012) claim that

paternal investment theory links low male parental investment to more aggressive and hypermasculine behavior in sons and more precocious and risky sexual behavior in daughters (Draper & Harpending, 1982, 1988). The assumption is that natural selection has designed boys' and girls' brains to detect and encode information about their fathers' social behavior and role in the family as the basis for calibrating socio-sexual development in gender-specific ways (p.32).

The purported mechanism for what Ellis et al. (2012) term this evolutionary developmental is that there is

a unique role for fathers in regulating daughters' sexual behavior. The theoretical basis for emphasizing father effects is (a) that the quality and quantity of paternal investment is—and presumably always has been—widely variable across and within human societies; (b) this variation recurrently and uniquely influenced the survival and fitness of children during our evolutionary history…; and (c) variability in paternal investment, much more than maternal investment, was diagnostic of the local mating system (degree of monogamy vs polygyny) and associated levels of male–male competition… The mating system is important because more polygynous cultures and subcultures are characterized by heightened male intrasexual competition, dominance striving, and violence, with concomitant diminution of paternal involvement and investment (Draper & Harpending, 1982, 1988). In turn, female reproductive strategies in this context are biased toward earlier sexual debut, reduced reticence in selecting mates, and devaluation of potential long-term relationships with high-investing males, all of which translate into more RSB [risky sexual behavior] (p. 32).

However, the embodiment of the individual and of his or her plastic developmental biological, psychological, and behavioral processes within the relational developmental systems provides a basis for epigenetics across generations, that is, for changes in gene–context relations within one generation being transmitted to succeeding generations. The "Just So" stories (Gould, 1981) of EP are conceptually flawed, ignore contemporary scholarship about evolutionary processes and their impact on ontogeny (e.g., Gissis & Jablonka, 2011; Ho, 2010; Meaney, 2010), and are therefore empirically counterfactual. Embodiment provides the basis for epigenetic change within the life span of an individual, that is, for qualitative discontinuity across ontogeny in relations among biological, psychological, behavioral, and social variables. Evidence for the plasticity of human development within the integrated levels of the ecology of human development makes biologically reductionist accounts of parenting and offspring development sexuality implausible, at best, and entirely fanciful, at worst.

3.1. Summary

Relational developmental systems conceptions provide an approach to the study of evolutionary and ontogenetic change that capitalizes on the dynamic, mutually influential relation between developing individuals and their complex and changing ecology. These contributions of relational developmental systems theories are in evidence across the chapters presented in these two volumes. We may best explain this evidence by describing the organization and content of these two books.

4. THE PLAN OF THESE TWO VOLUMES

Volume 44 of this work focuses, first, on the philosophical and theoretical bases of relational developmental systems theory and, in this context, on the place of the concepts of embodiment and epigenesis within such thinking. Second, these chapters elucidate the study of biological processes within the embodied, epigenetic system of relations involved in phylogenetic and ontogenetic changes.

In the second chapter, Overton presents a keynote statement for both volumes, explaining that relationism as a metamodel, and relational developmental systems theories in particular, constitute a paradigm for developmental science in the post-Cartesian era. Overton argues that the Cartesian-Split–Mechanistic scientific paradigm that, until recently, functioned as the standard conceptual framework for subfields of developmental science has been progressively failing as a scientific research program. He explains that relational developmental systems are a midrange metatheory that is a more progressive conceptual framework for developmental science than prior conceptions.

Next, Tabery and Griffiths ask what does developmental systems theory explain and how does it explain it? They review major contributions to the origins of developmental systems theory, for example, the idea of probabilistic epigenesis, and contrast developmental systems theory with behavioral genetics and nativist cognitive psychology. They argue that developmental systems theory explains by elucidating mechanisms.

Greenberg, Schmid Callina, and Kiely Mueller frame their chapter by noting that psychology is a biopsychosocial science as well as a developmental science. They discuss how principles and ideas from other sciences play important roles in psychology; for instance, they focus on the concepts from the physics of self-organization and emergence, on the cosmological and evolutionary biology idea of increased complexity over time, on the organizing principle of integrative levels, and on the epigenetic processes that are in part responsible for transgenerational transmission. They emphasize the concepts of embodiment and contextualism structure contemporary thinking about psychological processes.

Richardson aims to understand the relations between the evolution and development of complex cognitive functions by highlighting the context of complex, changeable environments. He notes that what evolves and develops in such contexts cannot be achieved by linear deterministic process based on stable "codes," and argues that what is needed, even in the molecular ensembles of single-cell organisms, are "intelligent" systems with nonlinear dynamic processing, sensitive to informational structures, not just elements, in environments. Noting that this view is emerging in recent research in molecular biology, he explains how this scholarship is also constructing a new "biologic" of both evolution and development, and is providing a clearer rationale for transitions into more complex forms, including epigenetic, physiological, nervous, cognitive, and human sociocognitive forms.

Witherington and Heying point out that relational, systems-oriented approaches are strongly positioned to advance theory and research in developmental science and to cement a process orientation to development at all levels of organization—from the biological to the psychological and sociocultural—despite continued prominence in the field of biologically reductionist explanatory accounts. Accordingly, they highlight the importance of holistically couching interlevel relations—those that obtain vertically between levels of organization, such as between the biological and psychological levels—in terms of wholes and parts and of recognizing the different modes of causal explanation that obtain depending on whether the relations move from parts to whole or whole to parts. They explain that this approach yields an explanatory pluralism under which all living systems, at any level of organization, exist as both subjects and objects.

In his chapter, Lickliter provides evidence from contemporary epigenetic research that indicates that it is not biologically meaningful to discuss genes without reference to the molecular, cellular, organismic, and environmental context within which they are activated and expressed. Genetic and non-genetic factors, including those beyond the organism, constitute a dynamic relational developmental system. Explaining the importance of bringing together genetics, development, and ecology in one explanatory framework for a more complete understanding of the emergence and maintenance of phenotypic stability *and* variability, he presents examples of this integration and explores its implications for developmental and evolutionary science, with a particular emphasis on the origins of phenotypic novelty. He argues that evolutionary explanations cannot be complete without developmental explanations because the process of development generates the phenotypic variation on which natural selection can act.

Charney makes the case that, since the early twentieth century, inheritance was seen as the inheritance of genes and that, along with the acceptance of the genetic theory of inheritance, was the rejection of the idea that the cytoplasm of the oocyte could also play a role in inheritance. By explaining that current evidence underscores that inheritance is a matter of both genetic and cytoplasmic inheritance, Charney points to the growing recognition of the centrality of the cytoplasm to explain both human development and phenotypic variation and notes that this understanding is based on two contemporaneous developments: the continuing elaboration of the molecular mechanisms of epigenetics and the global rise of artificial reproductive technologies.

Saunders notes that, while Darwinism has contributed much to our understanding of the living world, it has not given us an adequate account of why organisms are the way they are and how they came to be that way. He argues that for such an understanding, all the sciences and not just a single discipline should be considered. Nevertheless, EP follows the Darwinian model exclusively, assuming that the brain is largely modular and that human nature is made up of a very large number of functionally specialized psychological mechanisms that have been constructed over time by natural selection. He suggests that to have any confidence in such assertions one must accept the premises of EP. He cautions against such acceptance.

In his chapter, Wahlsten explains that several large-scale searches for genes that influence complex human traits, such as intelligence and personality, in the normal range of variation have failed to identify even one gene that makes a significant difference. All previously published claims for genetic influences of this kind now appear to have been false positives. In turn, the field of molecular genetics has generated evidence affirming principles that show how development is regulated by networks of interacting genes that function in an environmental context. These findings invalidate several key assumptions of statistical genetic analysis that are made when estimating heritability. He draws attention to the need to reform the teaching of genetics and to restrict the funding of further research that attempts to search for elusive genes that account for so little variance in normal behaviors.

Krimsky notes that how we ascertain causes and find agreement about causes depends largely on the methods and tools of science, methods that vary among the disciplines. Accordingly, he discusses causality more generally and then specifically focuses on causality in genetics. Consistent with the argument of Wahlsten in this volume, he argues that for many claims the concept of "genetic cause" does not stand up to critical scrutiny.

In the final chapter of Volume 44, Harper notes that the range of responses made to environmental exigencies by animals, including humans, may be impacted by the experiences of their progenitors. In mammals, pathways have been documented ranging from transactions between mother and developing fetus in the womb through continuity of parenting practices and cultural inheritance. In addition, phenotypic plasticity may be constrained by factors transmitted by the gametes via factors involved in the regulation of gene expression, rather than modifications to the genome itself. Accordingly, he examines possible mediators for this kind of inheritance, and discusses the conditions that might have led to the evolution of such transmission. He points to evidence indicating that physical growth and responses to nutrient availability are domains in which anticipatory, epigenetically inherited adjustments occur. Harper points out that humans have oppressed one another repeatedly and for relatively long periods of time and that behavioral tendencies such as boldness or innovativeness may involve such effects. He discussed the implications of these possibilities in terms of both research and policy.

In Volume 45, ontogenetic dimensions of embodiment and epigenesis are discussed. In the second chapter, Molenaar and Lo discuss a third source of individual differences that is capable of acting independently of genetic and environmental influences but that still generates phenotypic interindividual variation, even if all genetic and environmental factors as well as their interactions would be kept constant: self-organizing epigenetic processes that play important roles during embryogenesis. They present a class of mathematical models of self-organizing biological growth, and demonstrate that these so-called models of biological pattern formation give rise to variation in morphogenetic structures that is caused by self-organizing forces of diffusive origin. They present a proof that the heterogeneity caused by biological pattern formation invalidates standard statistical analysis techniques used in quantitative genetics. The proof is based on general mathematical theorems of ergodic theory, and they argue that application of quantitative genetic techniques to heterogeneous populations requires their adaptation in subject-specific ways. They illustrate this idea through use of the recently developed technique to determine subject-specific heritabilities based on intraindividual variation. Finally, they discuss some relationships between this work and relational developmental systems theory.

Next, Mueller, Baker, and Yeung note that according to recent claims from behavior genetics, executive function (EF) is almost entirely heritable, and that the implications of this claim are significant, given the importance of EF in academic, social, and psychological domains. Accordingly, they examine the behavior genetics approach to explaining individual differences in EF and propose a relational developmental systems model that integrates both biological and social factors in the development of EF and the emergence of individual differences in EF. They review empirical evidence from research on stress, social interaction, and intervention and training and demonstrate that individual differences in EF are experience-dependent.

Ho explains that there are no genes for intelligence in the fluid genome. She indicates that genetic determinists still deny the existence of the fluid genome, since its presence makes identifying genes even for common disease impossible. She argues that the fruitless hunt for intelligence genes serves to expose the poverty of an obsolete paradigm that is obstructing knowledge and preventing fruitful policies from being widely implemented. She explains that environment and maternal effects may account for most correlations among relatives, that identical twins diverge genetically and epigenetically throughout life, and that considerable evidence points to the enormous potential for improving intellectual abilities and health through environmental and social interventions.

In his chapter, Joseph discusses the existence and implications of a 1998 "lost" adoption study of personality that found no genetic relationship between birthparents and their 240 adopted-away biological offspring. He reports that in 1998, Robert Plomin and his Colorado Adoption Project (CAP) colleagues published the results of a longitudinal adoption study of personality, finding an average personality test score correlation of only 0.01 between birthparents and their 240 adopted-away 16-year-old biological offspring, suggesting no genetic effects on personality (Plomin, Corley, Caspi, Fulker, & DeFries, 1998). However, the researchers interpreted their results in the context of previous twin studies that produced an average 14% heritability estimate, and concluded that nonadditive genetic factors underlie personality traits. Joseph challenges these conclusions, and notes that the near-zero correlation stands in contrast to other types of behavioral genetic methods, such as twin studies, that are more vulnerable to environmental confounds and other biases. He also shows that authoritative psychology texts frequently fail to mention the 1998 CAP study and, when it is mentioned, the original researchers' conclusions are usually accepted without critical analysis. Joseph also assesses the results in the context of the failure to discover genes that behavioral geneticists believe underlie personality traits.

Carpendale, Atwood, and Hammond note that morality and cooperation are central to human life and psychological explanations for moral development and cooperative behavior will have biological and evolutionary dimensions. They point to several recent proposals that have argued that aspects of morality are unlearned and innate; they review and critique these claims. In contrast to these nativist assumptions about the role of biology in morality, they present an alternative approach based on a relational developmental systems view of moral development. The role for biology in this approach is in setting up the conditions—the developmental system—in which forms of interaction and later forms of thinking emerge.

Moshman notes that adolescents are commonly seen as irrational, a position supported to varying degrees by many developmentalists, who often appeal to recent research on adolescent brain development. Careful review of relevant evidence, however, shows that (1) adults are less rational than is generally assumed; (2) adolescents (and adults) are categorically different from children with respect to the attainment of advanced levels of rationality and psychological functioning; and (3) adolescents and adults do not differ categorically from each other with respect to any rational competencies, irrational tendencies, brain structures, or neurological functioning. Development often continues in adolescence and beyond but categorical claims about adolescents as distinct from adults cannot be justified. He explains that developmental theory and research suggest that adolescents should be conceptualized as young adults, not immature brains, with important implications for their roles, rights, and responsibilities.

Mascolo explains that, in recent decades, the developmental sciences have undergone a *relational turn*, and points out that concepts such as epigenetic, embodied, relational, and systems approaches are transforming the ways in which the nature and origins of psychological structures are conceptualized. Such concepts view genes and environment, biology and culture, cognition and emotion, self and other as *inseparable* as causal processes in the development of action and experience. By drawing on these ideas, he describes an embodied coactive systems framework for understanding how individual psychological structures develop as a product of socially distributed coactions that occur among elements of the extended *person–environment system*. Mascolo presents a system for the *Developmental Analysis of Joint Action*, and explains that by tracking developmental changes in joint action, the system allows researchers to elucidate the origins of higher order psychological structures through particular sequences of coconstructive activity.

Michel, Nelson, Babik, Campbell, and Marcinowski show that handedness is a product of a multifaceted biosocial developmental process that begins prenatally and continues into adulthood. Although right handedness predominates, handedness varies continuously across the population. Therefore, they discuss both differences in developmental pathways that can lead to *similarities* in handedness as well as similarities in pathways that can lead to *differences* in handedness. They note that the key research task is to identify how, when, and for what actions the trajectory of handedness development can be maintained or changed for an individual. They argue that given the complexity of these developmental pathways, it is likely that the asymmetric sensorimotor activity that occurs during the development of handedness influences other hemispheric variations in neural processing.

Agans, Safvenbom, Bowers, and Lerner argue that although exercise and athletic participation are widely recognized as important aspects of healthy life styles and human development, most of the research on youth athletic participation, exercise, and leisure activity has not yet adopted a theoretical framework useful for understanding the development of individual engagement with these movement contexts. They make the case that in order to gain an adequate understanding of the developmental experiences of involvement in movement contexts, understanding the role of the active individual and the mutually influential relations between individual and context is extremely important. Therefore, they present a new approach to the study of involvement in movement contexts, using relational developmental systems theory and the concept of embodiment to forward the idea of positive movement experiences, a concept that the authors believe may facilitate better understanding of involvement in movement contexts as a fundamental component of human life in general, and of youth development in particular.

In the final chapter in this second volume, Mistry notes that, historically, the focus on the sources of human development was framed as the classic nature versus nurture debate. Noting that, today, much progress has been made in moving beyond such dichotomies, and expectations of single or simple causal factors as explanations for development, she argues that a key, remaining challenge is how to discuss culture and biology in an integrative manner reflecting the mutually constitutive process emphasized in treatments of human development as involving embodied, epigenetic processes. Accordingly, Mistry presents a conceptual frame to represent the integration of biology and culture and, as well, she discusses the nascent field of cultural neuroscience to determine if examples of empirical research exist that illustrate the integration of culture and biology.

5. CONCLUSIONS

As is made clear across both volumes in the present work, relational developmental systems theory, and the evidence from the study of phylogeny and ontogeny framed by such theories, coalesce to indicate that split, biological (e.g., genetic) reductionist ideas are fundamentally flawed. Genes are not the to-be-reduced-to entities that provide any "blueprint" for behavior or development, nor do they function as a "master molecule"; they are not the context-independent governors of the "lumbering robots" (Dawkins, 1976) housing them; and they are not the fixed material basis of the grand synthesis of heredity and Darwinism found in the neo-Darwinian model (e.g., Ho, 2010; Ho & Saunders, 1984). Instead, and consistent with the four-dimensional, and neo-Lamarckian system involved in evolution (e.g., Gissis & Jablonka, 2011), genes are a plastic feature of the four-dimensional, epigenetic, action-oriented, and cultural and historical ontogenetic system that constitutes the fundamental process of human development across the life span.

Given the plasticity of the relational developmental system within which genes are embedded, another split—between basic and applied science—may be overcome. Indeed, across both volumes, authors in several chapters point to the implications of theory and research about the embodied, epigenetic developmental system for applications to policies and to intervention or to health or positive development promotion programs (also see Lerner et al., 2012). We share the optimism of these authors that present and future relational developmental theory-predicated research will be marked by new information about how we can promote epigenetic changes that enhance the probability of more positive development among all individuals across the life course.

Moreover, we are also optimistic about the continued useful role of a relational metatheoretical approach to developmental science, as a superordinate frame for scholars in psychology, sociology, economics, biology, medicine, education, and other fields interested in describing, explaining, and optimizing the course of human life. One bit of evidence in support of this optimism is the forthcoming, seventh edition of the *Handbook of Child Psychology*, with the expanded title of the *Handbook of Child Psychology and Developmental Science* (Lerner, in preparation). This publication will emphasize the use of relational developmental systems perspectives and, as well, will point to the implications of this theoretical frame for both understanding and enhancing individual ↔ context relations.

Indeed, based on the evidence offered across the present two volumes for embodiment and for plasticity in both phylogenetic and ontogenetic processes, we believe that relational developmental theory-predicated research, especially when coupled with change-sensitive, person-centered, and rigorous methodology (Molenaar, 2007; Molenaar & Nesselroade, 2012; Molenaar, Lerner, & Newell, in press; Nesselroade & Molenaar, 2010), will provide increasingly more nuanced information about the mutually influential relations among individual and ecological processes that constitute the fundamental change processes of human development. That is, scholarship about the embodiment of human development within the rich and complex ecology of human life will enable developmental science to be a productive means for promoting more positive, healthier developmental trajectories among all individuals. As such, a developmental science framed by relational developmental systems models has the potential to generate evidence-based actions promoting social justice across the life spans of the diverse people of our world (Fisher, Busch-Rossnagel, Jopp, & Brown, 2012; Lerner & Overton, 2008).

REFERENCES

Baltes, P. B., Reese, H. W., & Nesselroade, J. R. (1977). *Life-span developmental psychology: Introduction to research methods.* Monterey, CA: Brooks/Cole.

Bateson, P., & Gluckman, P. (2011). *Plasticity, development and evolution.* Cambridge, UK: Cambridge University Press.

Belsky, J. (2012). The development of human reproductive strategies: progress and prospects. *Current Directions in Psychological Science, 21*(5), 310–316.

Belsky, J., Steinberg, L., & Draper, P. (1991). Childhood experience, interpersonal development, and reproductive strategy: an evolutionary theory of socialization. *Child Development, 62,* 647–670.

Brandtstädter, J. (1998). Action perspectives on human development. In R. M. Lerner (Ed.), & W. Damon & R. M. Lerner (Editors-in-Chief). *Theoretical models of human development* (5th ed.). *The Handbook of child psychology* (Vol. 1, pp. 807–863). New York: Wiley.

Bronfenbrenner, U., & Morris, P. A. (2006). The bioecological model of human development. In R. M. Lerner (Ed.), Editors-in-chief: Damon, W., & Lerner, R. M. *Theoretical models of human development* (6th ed.). *Handbook of child psychology.* (Vol. 1). Hoboken, NJ: Wiley.

Dawkins, R. (1976). *The selfish gene.* New York: Oxford University.

Draper, P., & Harpending, H. (1982). Father absence and reproductive strategy: an evolutionary perspective. *Journal of Anthropological Research, 38,* 255–273.

Draper, P., & Harpending, H. (1988). A sociobiological perspective on the development of human reproductive strategies. In K. B. MacDonald (Ed.), *Sociobiological perspectives on human development* (pp. 340–372). New York: Springer-Verlag.

Elder, G. H., Jr. (1998). The life course and human development. In R. M. Lerner (Ed.), & W. Damon & R. M. Lerner (Editors-in-Chief). *Handbook of child psychology* (5th ed.). *Theoretical models of human development* (Vol. 1, pp. 939–991). New York: John Wiley.

Ellis, B. J., Schlomer, G. L., Tilley, E. H., & Butler, E. A. (2012). Impact of fathers on risky sexual behavior in daughters: a genetically and environmentally controlled sibling study. *Development and Psychopathology, 24(01),* 317–332.

Fisher, C. B., Busch-Rossnagel, N. B., Jopp, D. S., & Brown, J. L. (2012). Applied developmental science, social justice and socio-political well-being. *Applied Developmental Science*, *16*, 54–64.

Gestsdóttir, G., & Lerner, R. M. (2008). Positive development in adolescence: the development and role of intentional self regulation. *Human Development*, *51*, 202–224.

Gissis, S. B., & Jablonka, E. (2011). Preface. In S. B. Gissis & E. Jablonka (Eds.), *Transformations of Lamarckism: From subtle fluids to molecular biology* (pp. xi–xiv). Cambridge, MA: The MIT Press.

Gottlieb, G. (1997). *Synthesizing nature-nurture: Prenatal roots of instinctive behavior*. Mahwah, NJ: Lawrence Erlbaum Associates, Inc.

Gottlieb, G. (1998). Normally occurring environmental and behavioral influences on gene activity: from central dogma to probabilistic epigenesis. *Psychological Review*, *105*, 792–802.

Gould, S. J. (1981). *The mismeasure of man*. New York: Norton.

Ho, M. W. (2010). Development and evolution revisited. In K. E. Hood, C. T. Halpern, G. Greenberg, & R. M. Lerner (Eds.), *Handbook of developmental systems, behavior and genetics* (pp. 61–109). Malden, MA: Wiley Blackwell.

Hood, K. E., Halpern, C. T., Greenberg, G., & Lerner, R. M. (Eds.) (2010). *The handbook of developmental science, behavior and genetics*. Malden, MA: Wiley Blackwell.

Ho, M. W., & Saunders, P. T. (1984). *Beyond neo-Darwinism: Introduction to the new evolutionary paradigm*. London: Academic Press.

Jablonka, E., & Lamb, M. (2005). *Evolution in Four Dimensions: Genetic, epigenetic, behavioral, and symbolic variation in the history of life*. Cambridge, MA: MIT Press.

Keller, E. F. (2010). *The mirage of a space between nature and nurture*. Durham, NC: Duke University Press.

Lerner, J. V., Bowers, E. P., Minor, K., Lewin-Bizan, S., Boyd, M. J., Mueller, M. K., et al. (2012). Positive youth development: processes, philosophies, and programs. In R. M. Lerner (Ed.), Editor-in-chief: Weiner, I. B. *Handbook of psychology* (2nd ed.). *Developmental psychology* (Vol. 6, pp. 365–392). Hoboken, NJ: Wiley.

Lerner, R. M. (1982). Children and adolescents as producers of their own development. *Developmental Review*, *2*, 342–370.

Lerner, R. M. (1984). *On the nature of human plasticity*. New York: Cambridge University Press.

Lerner, R. M. (2002). *Concepts and theories of human development* (3rd ed.). Mahwah, NJ: Lawrence Erlbaum.

Lerner, R. M. (2006). Developmental science, developmental systems, and contemporary theories of human development. In R. M. Lerner (Ed.), & W. Damon & R. M. Lerner (Editors-in-Chief). *Theoretical models of human development* (6th ed.). *The Handbook of child psychology* (Vol. 1, pp. 1–17). Hoboken, NJ: Wiley.

Lerner, R. M. (2012). Developmental science and the role of genes in development. *Gene-Watch*, *25*(1–2) http://www.councilforresponsiblegenetics.org/genewatch/GeneWatch-Page.aspx?pageId=413.

Lerner, R. M. (Editor-in-chief). *The handbook of child psychology and developmental science* (7th ed.). Hoboken, NJ: Wiley, in preparation.

Lerner, R. M., & Busch-Rossnagel, N. A. (Eds.), (1981). *Individuals as producers of their development: A life-span perspective*. New York: Academic Press.

Lerner, R. M., & Overton, W. F. (2008). Exemplifying the integrations of the relational developmental system: synthesizing theory, research, and application to promote positive development and social justice. *Journal of Adolescent Research*, *23*(3), 245–255.

Meaney, M. (2010). Epigenetics and the biological definition of gene x environment interactions. *Child Development*, *81*(1), 41–79.

Mistry, J., & Wu, J. (2010). Navigating cultural worlds and negotiating identities: a conceptual model. *Human Development*, *53*, 5–25.

Molenaar, P. C. M. (2007). On the implications of the classical ergodic theorems: analysis of developmental processes has to focus on intra-individual variation. *Developmental Psychobiology*, *50*, 60–69.

Molenaar, P. C. M. (2010). On the limits of standard quantitative genetic modeling of inter-individual variation: extensions, ergodic conditions and a new genetic factor model of intro-individual variation. In K. E. Hood, C. T. Halpern, G. Greenberg, & R. M. Lerner (Eds.), *Handbook of developmental systems, behavior and genetics* (pp. 626–648). Malden, MA: Wiley Blackwell.

Molenaar, P. C. M., & Nesselroade, J. R. (2012). Merging the idiographic filter with dynamic factor analysis to model process. *Applied Developmental Science*, *16*(4), 210–219.

Molenaar, P. C. M., Lerner, R. M., & Newell, K. (Eds.). *Handbook of developmental systems theory and methodology*. New York: Guilford, in press.

Nesselroade, J. R., & Molenaar, P. C. M. (2010). Emphasizing intraindividual variability in the study of development over the life span. In W. F. Overton (Ed.), Editor-in-chief: Lerner, R. M. *The handbook of life-span development Cognition, biology, methods* (Vol. 1, pp. 30–54). Hoboken: Wiley.

Overton, W. F. (2003). Development across the life span: philosophy, concepts, theory. In R. M. Lerner, M. A. Easterbrooks & J. Mistry (Eds.), (Editor-in-chief: Irving B. Weiner). *Handbook of psychology: Developmental psychology* (Vol. 6, pp. 13–42). New York: Wiley.

Overton, W. F. (2006). Developmental psychology: philosophy, concepts, methodology. In R. M. Lerner (Ed.), Editors-in-chief: Damon, W., & Lerner, R. M. *Handbook of child psychology* (6th ed.). *Theoretical models of human development* (Vol. 1, pp. 18–88). Hoboken, New Jersey: John Wiley & Sons.

Overton, W. F. (2010). Life-span development: concepts and issues. In W. F. Overton (Ed.), Editor-in-chief: Lerner, R. M. *Cognition, biology, and methods across the lifespan Handbook of life-span development* (Vol. 1, pp. 1–29). Hoboken, NJ: Wiley.

Overton, W. F. (2011). Relational developmental systems and quantitative behavior genetics: alternative of parallel methodologies. *Research in Human Development*, *8*, 258–263.

Overton, W. F. (2012). Evolving scientific paradigms: retrospective and prospective. In L. L'Abate (Ed.), *The role of paradigms in theory construction* (pp. 31–65). New York: Springer.

Overton, W. F., & Lerner, R. M. (2012). Relational developmental systems: paradigm for developmental science in the post-genomic era. *Behavioral and Brain Sciences*, *35*(5), 375–376.

Overton, W. F., & Müller, U. (2012). Development across the life span: philosophy, concepts, theory. In R. M Lerner, M. A Easterbrooks & J Mistry (Eds.), & I. B. Weiner (Editor-in-Chief). *Handbook of psychology: developmental psychology* (Vol. 6, pp. 19–58). New York: Wiley.

Pigliucci, M., & Mueller, G. B. (2010). Elements of an extended evolutionary synthesis. In M. Pigliucci (Ed.), *Evolution – the extended synthesis* (pp. 3–17). Cambridge, MA: MIT Press.

Plomin, R., Corley, R., Caspi, A., Fulker, D. W., & DeFries, J. C. (1998). Adoption results for self-reported personality: evidence for nonadditive genetic effects? *Journal of Personality and Social Psychology*, *75*, 211–218.

Rose, H., & Rose, S. (2000). Introduction. In H. Rose & S. Rose (Eds.), *Alas poor Darwin: Arguments against evolutionary psychology* (pp. 1–13). London: Vintage.

Thelen, E., & Smith, L. B. (2006). Dynamic systems theories. In R. M. Lerner (Ed.), Editors-in-chief: Damon, W., & Lerner, R. M. *Theoretical models of human development* (6th ed.). *Handbook of child psychology* (Vol. 1, pp. 258–312). Hoboken, NJ: John Wiley & Sons.

Tobach, E., & Greenberg, G. (1984). The significance of T. C. Schneirla's contribution to the concept of levels of integration. In G. Greenberg & E. Tobach (Eds.), *Behavioral evolution and integrative levels* (pp. 1–7). Hillsdale, NJ: Lawrence Erlbaum.

West-Eberhard, M. J. (2003). *Developmental plasticity and evolution*. New York: Oxford University Press.

CHAPTER TWO

Relationism and Relational Developmental Systems: A Paradigm for Developmental Science in the Post-Cartesian Era

Willis F. Overton

Department of Psychology, Temple University, Philadelphia, PA, USA
E-mail: overton@temple.edu

Contents

Advances in Child Development and Behavior, Volume 44
ISSN 0065-2407, http://dx.doi.org/10.1016/B978-0-12-397947-6.00002-7

Abstract

This chapter argues that the Cartesian-split-mechanistic scientific paradigm that until recently functioned as the standard conceptual framework for subfields of developmental science (including inheritance, evolution, and organismic—prenatal, cognitive, emotional, motivational, sociocultural—development) has been progressively failing as a scientific research program. An alternative scientific paradigm composed of nested metatheories with relationism at the broadest level and relational developmental systems as a midrange metatheory is offered as a more progressive conceptual framework for developmental science. Termed broadly the *relational developmental systems paradigm*, this framework accounts for the findings that are anomalies for the old paradigm; accounts for the emergence of new findings; and points the way to future scientific productivity.

Discerning trends in any scientific field is a dangerous undertaking, as an individual's trend may well be another's random walk. Nevertheless, between 2001 and 2010 or so, there seems to be clearly identifiable trends in several subfields of developmental science. One such trend involves the most recent recycling of the nature–nurture debate (i.e., the issue of inheritance). Here, advances in epigenetics and a broader understanding of the genome itself have made the route from genotype to phenotype complex to the point that the classic position claiming that who we are and what we become to be a simple additive function of gene × environment interactions has become highly untenable (see, e.g., Charney, 2012; Gottlieb, 1997, 2003; Gottlieb, Walhsten & Lickliter 2006; Greenberg, 2011; Joseph, 2010; Keller, 2010; Lerner, 2012a; Meaney, 2010; Moore, 2001; Partridge, 2005; Wahlsten, 2012).

The second trend entails the relation of evolution and development. Here, the field is rapidly moving from the traditional Modern Synthesis (Pigliucci & Mueller, 2010a)—integrating Mendelian genetics with neo-Darwinian variation and natural selection, and split-off from individual development—to a position in which individual development has become an integral part of the fabric of evolution (e.g., Batson & Gluckman, 2011; Gilbert & Epel, 2009; Gottlieb, 2002; Ho, 2010; Jablonka & Lamb, 2005; Jablonka & Raz, 2009; Lickliter & Schneider, 2006; Laubichler, 2010; Pigliucci & Mueller, 2010b; Robert, 2004; West-Eberhard, 2003). This second trend is intertwined with the first as any analysis of evolution—given the Modern Synthesis—necessarily involves a discussion of population genetics.

The third trend concerns cognition and cognitive development. Here, the standard position that mental processes are exclusively located in the brain is increasingly being challenged by the view that mental processes extend out into the body and into the technological and cultural worlds (e.g., Marshall, 2009; Menary, 2010; Overton, 2006; Rowlands, 2010; Stewart, Gapenne, & Di Paolo, 2010).

Finally, in the area of sociocultural development, there appears to be a clear trend away from the positions that identify individual development and culture as separate and distinct, if interacting, entities, and toward the position that recognizes their coconstruction, codetermination and codevelopment (e.g., Eckensberger, 2003; Mistry, Contreras, & Dutta, 2012).

As these trends advance empirically with increasing frequency, there are also suggestions that earlier conceptual frameworks that have contextualized these fields have proven, at best, inadequate to the task of integrating the new empirical advances, and, at worst, a major obstacle to integration and to scientific advancement. For example, Lickliter and Honeycutt (2010), exploring changes in the understanding of evolution, explicitly capture this sentiment in the very title of their chapter, *Rethinking Epigenesis and Evolution in the Light of Developmental Science.* Similarly, West-Eberhard (2003) addressing the genotype–phenotype issue in both individual development and evolution argues, "The need for a conceptual framework for the study of organization lies at the heart of unsolved problems in both ontogeny and phylogeny (p. 16)." Rowlands (2010), calling for a new science of mind in which embodied processes, the environment, and culture all enter as constitutive features of mind, expresses the need for conceptual reflection in this field when he says that this "new science… is aspirational rather than descriptive… It's premature to say it because the new science, as yet, has *no clear conceptual foundation.*" (p. 25, emphasis added). Charney (2012), in an exceptionally valuable review and analysis of the significant new empirical findings in genetics and epigenetics, expresses the need for conceptual reflection in his argument that although the new evidence creates virtually insurmountable obstacles for the population (quantitative) behavior genetics paradigm, and while the evidence moves genetics into a postgenomic era, it does not itself yet constitute a paradigm because, "the postgenomic perspective has *not yet coalesced around a core set of principles or assumptions* characteristic of a paradigm (2012, p. 332 emphasis added)." Finally, Keller (2010), in trying to bring new light to the nature–nurture debate, finds a "*morass of linguistic and conceptual vegetation grown together*

in ways that seem to defy untangling (p. 9; emphasis added)," and ultimately concludes:

> *Daily, we are discovering new and extraordinarily ingenious ways in which
> noncoding DNA sequences participate in the mammoth projects of regulating
> the spatially and temporally specific transcription of DNA, the construction and
> translation of messenger RNA and the positioning, conformation, and activity
> of proteins. Early concepts of the gene were predicated on the assumption of a
> relatively simple transformation from genotype to phenotype, but now we are
> beginning to understand just how enormously complex that process is.* Such
> findings not only require us to rethink basic assumptions in biology, they also
> create the opportunity for such reconceptualizations *(p. 78; emphasis added).*

In this chapter, I will argue that a good deal of rethinking, conceptual reflection, and reconceptualization has already occurred with respect to these and
other issues in the field of developmental science (Lerner, 2012b). Further,
I will argue that this conceptual work does, in fact, offer a clear conceptual framework entailing a core set of principles or assumptions that taken
together constitute a new scientific paradigm for developmental science,
including each of the subfields (inheritance, evolution, and organismic
development—including prenatal, cognitive, emotional, motivational, and
sociocultural) mentioned above.

1. THE ORIGIN, NATURE, AND FUNCTION OF SCIENTIFIC PARADIGMS

To make this argument concerning the availability of a new scientific
paradigm for developmental science, I need to first describe the nature of
scientific paradigms: what they are composed of, how they come about, and
how they function. Following this, I will briefly describe the broad features of
a specific scientific paradigm that until recently served as the standard for the
physical, biological, and social sciences. This is the Cartesian-Split-Mechanistic
worldview and its midrange metatheories. I will then elaborate on the new
scientific paradigm that serves as a new conceptual framework for developmental science, *the relationism and relational developmental systems paradigm.*

1.1. Nested Concepts

To understand the nature of a scientific paradigm, it is essential to first understand the centrality of nested concepts and how they function in scientific
activity. Contrary to the early twentieth century neopositivist narration of
a bottom-up process, where abstract concepts were ultimately observational

inductions often referred to as "empirical generalizations," scientific epistemology today recognizes that abstract background concepts represent a fundamental and an essential grounding for any scientific research program (Overton, 2006; Overton & Mueller, 2012).

1.1.1. Metatheories and Theories

Background concepts are sometimes called *frameworks*, but more generally they are termed *metatheoretical* or *metatheories*. They transcend (i.e., "meta") theories in the sense that they define the context within which theoretical concepts are constructed. Thus, metatheories ground theories. They also function in the same manner with respect to methods, and then are often called *metamethods*. *Methodology* would also be an appropriate term here if this term was understood in its broad sense as a set of principles that guide empirical inquiry (Asendorpf & Valsiner, 1992; Overton, 2006).

The primary function of metatheories—including metamethods—is to provide a rich coherent source of concepts, out of which theories and methods emerge. As a part of this function, a metatheory provides the meaning context for specific theoretical concepts. For example, "structure" may have a physical meaning in the context of one metatheory, and may refer to a pattern, a form, or an organization in another. Metatheory also provides latent conceptual resources that assist in the advancement of theory. And metatheory provides guidelines that assist in avoiding conceptual confusions and, consequently, assist in avoiding what may ultimately be unproductive ideas and unproductive methods.

Theories are about the empirical phenomena in a specific scientific area of interest, and *methods* are the procedures employed in the empirical analysis of these phenomena. In contrast, *metatheories and metamethods are about* the theories and methods themselves, respectively. More specifically, a *metatheory* is a coherent set of rules, principles, or a narrative, that both describes and prescribes what is acceptable and unacceptable, meaningful and meaningless, as theory—the means of *conceptual* exploration of any scientific domain. A *metamethod* is also a set of rules, principles, or a narrative, but these rules describe and prescribe the nature of acceptable methods—the means of *observational exploration*—in a scientific discipline.

1.1.2. Worldviews and Metatheories of a Midrange

Thus, theories are nested within metatheories (Fig. 2.1). The nested relation is further complicated by the fact that there are levels of metatheories as well as levels of theories. Metatheories operating at the level of broadest generality and abstractness are termed *worldviews*. Worldviews are composed

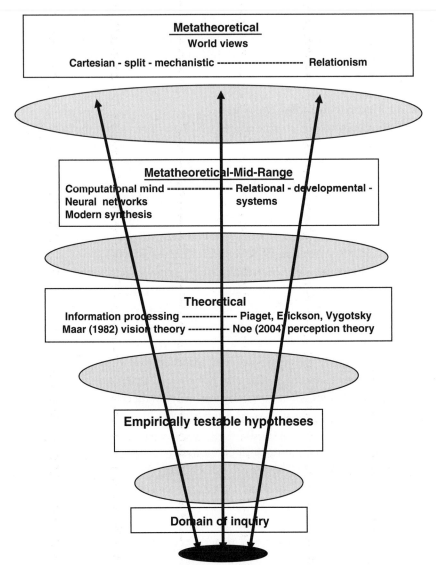

Figure 2.1 Nested concepts of a scientific paradigm.

of coherent sets of *epistemological* principles (i.e., issues pertaining to the sources and justification of knowledge) and *ontological* principles (i.e., issues pertaining to fundamental categories of reality). In essence, a worldview is a framework that presents a vision of the nature of the world and the nature of how we know that world (e.g., Is stasis or change, substance or process, asserted as the fundamental vision of the world? Is knowledge the reflection of a mind-independent reality or do minds actively participate in the constitution of the world as known?).

Nested within worldviews are *metatheories of a middle range*. These concepts are less abstract, are less broad, and entail principles that are identifiably more specific to the observational domains of interest. For example, nested within one particular worldview, a midrange metatheory conceptualizes the human organism as an *active agent* or as a *dynamic system*, whereas nested within another worldview, a midrange metatheory conceptualizes the organism as an *input–output device*.

Any given theory is nested within a coherent set of worldview and midrange metatheories. And specific theories may themselves vary from those seeking to explain a broad field of observational phenomena (e.g., a theory of cognitive development) to those that narrow the breadth of the field of explanation (e.g., dual systems of deductive reasoning). However, whether broad or narrow, the theoretical concepts—*while sensitive to the demands of the observational field of interest*—align themselves with the principles of their affiliated worldviews and midrange metatheories. Failure to do so opens the theories to charges of an eclecticism entailing logical incoherence and inconsistency.

1.2. The Scientific Paradigm

Armed with a background understanding of nested concepts, we may turn to the notion of a scientific paradigm itself (Overton, 2011). The term *paradigm* as used in *scientific paradigm* is generally attributed to Kuhn's (1962) famous text, *The Structure of Scientific Revolutions*. However, paradigm-like concepts have a much longer history, going back at least to the nineteenth century neo-Kantians who argued for the recognition of *cognitive prerequisites of scientific activities*.

1.2.1. Thomas Kuhn and the Nature of Paradigms

In scientific and philosophical circles, the publication of Kuhn's text reignited a long simmering debate about the nature of scientific activity and turned it into a raging firestorm of criticism and countercriticism (see, e.g., Lakatos & Musgrave, 1970). This debate and its history are critical to an understanding of the essential nature of scientific paradigms. The debate was partially about the introduction of sociological matters into the body of scientific theory and method, and partially, it was about the nature of scientific change (e.g., normal vs revolutionary science, scientific crises, anomalies, and gestalt switches). Kuhn's initial use of the term *paradigm* was vague—a sympathetic critic (Masterman, 1970) noted that Kuhn had used the term in 22 different ways in the 1962 edition of the book—but in the second edition, Kuhn (1970) wrote a *Postscript-1969* that defined two distinct meanings, one narrow and one broad (see also Kuhn, 1977). The narrow meaning of scientific paradigm he called *exemplars*. These are examples shared by a community of scientists involving "the concrete problems–solutions that students encounter from the start of their scientific education, whether in laboratories, on examinations, or at the ends of chapters in science texts (p. 187)." For example, the student of physics learns "problems such as the inclined plane, the conical pendulum, and Keplerian orbits, instruments such as the vernier, the calorimeter, and the Wheatstone bridge" (p. 187). Or to take a biological example, the student of population (quantitative) behavior genetics learns issues of genetic kinship relations, computing heritability indices, and employing "genetically informed" research designs.

Kuhn termed the broader meaning of paradigm a *disciplinary matrix*. Two key features of this meaning were (1) shared *metaphysical beliefs*— "from heuristic to ontological models, … [which] supply the group with preferred or permissible analogies and metaphors (p. 184)" and (2) *values*, which are especially important in choosing between incompatible ways of practicing a particular discipline (e.g., a good theory would be coherent, self-consistent, plausible, etc.). While Kuhn, who worked within the natural sciences, was ambivalent about the status of the social sciences, sometimes calling them *preparadigmatic*, he argued that even preparadigmatic sciences were guided by paradigms in the interdisciplinary matrix sense of the term (Kuhn, 1970, p. 179).

1.2.2. The Debate over Paradigms: Neopositivism and Conventionalism

Thus, at the core of Kuhn's argument for paradigms was the notion that worldviews (ontological and epistemological principles) enter scientific activity as a constitutive feature of this activity, and that paradigms entail nested concepts (i.e., ontological to heuristic models and exemplars) (Fig. 2.1). It was the idea that worldviews enter scientific activity as constitutive that reignited the debate concerning the nature of scientific activity, as this idea stood in direct opposition to the prevailing methodologies of neopositivism and conventionalism (see, e.g., Overton & Mueller, 2012).

Neopositivism argued that all scientific concepts, to be scientific, must necessarily be reducible to a neutral observational language, and shown to be pure empirical inductions from that language. Conventionalism, championed by Popper (1959, 1963, 1970), was willing to introduce meta-physical concepts and other nonobservable propositions (e.g., models, like the computer model in cognitive science and information processing) into scientific activity. However, and this is the crux of the matter, for conventionalism, these propositions functioned in a purely heuristic manner. Metaphysical and other nonobservable propositions were encapsulated in a conceptual space, termed the *context of discovery*, which was ancillary to the essential sphere of scientific activity termed the *context of justification*. Further, these propositions functioned *as and only as* convenient and conventional heuristic devices for ordering and organizing hard data (i.e., pristine observations) and for making predictions. Nonobservable propositions could not be allowed to influence the data base itself, and hence, they had no epistemological or cognitive value. Rather, nonobservable propositions operated like pigeonholes or coat racks to classify, arrange, and organize hard data into coherent units. Following this formula, it continued to be possible within conventionalism to retain both the rigid epistemological empiricism of neopositivism and the ontological absolute objectivist truth criterion as norms of science.

In the context of this formula, Popper (1963) argued that science entailed conjecture and refutation. That is, he argued it was scientifically legitimate for the scientist to offer speculative hypotheses drawn from any source (conjecture), as long as those hypotheses were falsifiable (refutation) by pristine observations [hard data] operating in the context of justification. Popper (1959) added a unique dimension to conventionalism through the claim that theories and models should become acceptable in the body of science, *if and only if*, they specify observational results that, if found, would disprove or falsify a theory. Ultimately, it was claimed, *falsification* would lead to the objective truth entailed by scientific realism.

1.2.3. Scientific Paradigm as a Metamethod Revolution

Kuhn's introduction of a scientific paradigm as a disciplinary matrix involving ontological and epistemological propositions as *constitutive features* of any scientific research program was itself a scientific revolution that entailed the rejection of the claims of both neopositivism and conventionalism. This conception of a paradigm was a revolution of metamethod rather than metatheory. Paradigms—conceptual constructions of actively engaged members of the scientific community that function as intrinsic and constitutive features of the scientific process—became the principal vehicles for the rejection of the scientific epistemology of *empiricism*[1] and a scientific ontology of *fixed objectivist foundationalism* claimed by neopositivism and conventionalism. This rejection coupled with the acceptance of a *constructivist* epistemology and an ontology of activity and engagement constituted a paradigm shift in scientific methodology.

1.2.4. Conceptual Antecedents to the Formation of Scientific Paradigms

This metamethod revolution did not emerge without historical precedence. Kant (1871) had argued that the activity of mind is a constitutive feature of all knowing, including scientific knowing. Hegel (1807, 1830) extended this position through his dialectical argument that knower and known constitute a single indissociable complementarity that develops through history. This theme was further elaborated in the contributions of Heidegger (1962) and Merleau-Ponty (1962, 1963) in their phenomenological analysis, leading to the conclusion that all knowledge is the product of engaged human agents coping with the world, and in Ernst Cassirer's (1951) neo-Kantian analysis of the cognitive prerequisites of knowing.

There is also a long list of contemporaries of Kuhn who contributed to the revolution against neopositivism and Popper's conventionalism, of which Kuhn's *paradigm* became a central feature. These scholars included, for example, Anscombe (1957), Bernstein (1983), Dray (1957), Frankel (1957), Putnam (1983, 1987, 1990), Ricoeur (1984, 1991), Taylor (1964), Toulmin (1953, 1961), von Wright (1971), and Winch (1958). Because their works were so central to the paradigm revolution, the specific contributions of the three contemporaries will be briefly detailed here. These include Ludwig Wittgenstein, whose *Philosophical Investigations was published in* 1958, Hans Georg Gadamer, whose *Truth and Method* was first published *in* 1960, and N. R. Hanson, whose *Patterns of Discovery* was published in 1958.

Wittgenstein (1958) and Gadamer (1960) provided the basic scaffolding for the construction of this new methodology. Wittgenstein's fundamental contribution entailed opening the door to the recognition that it is a profound error to treat the activities of science as providing veridical descriptions of a foundational Real. More positively, Wittgenstein's contribution

---

[1]It is important to emphasize that *empiricism* is a philosophical doctrine and *empiricists* are those committed to this doctrine (i.e., knowledge comes from pristine observations and only pristine observations). One can be committed to *empirical* science (i.e., science involving the testing of observational hypotheses) without being an *empiricist*. Thus, the new scientific methodology to emerge from this revolution was an empirical methodology that did not follow the doctrine of epistemological empiricism.

lies in his claim that science is the product of some of the same human actions that underlie the conceptual constructions of our form of life, or *Lebenswelt*. And, in this context, Wittgenstein's concept of *language games* was a direct precursor to Kuhn's paradigms. Gadamer's contribution was a systematic demonstration that this move beyond objectivism and foundationalism did not necessitate a slide into absolute relativism, so feared by those committed to epistemological empiricism and scientific realism.

Hanson's (1958) analysis of the history of the physical sciences was itself significantly influenced by Toulmin (1953) and Wittgenstein (1958), and it provided several necessary prerequisites for Kuhn's introduction of paradigms. On the basis of his analysis, Hanson drew three powerful conclusions about the actual practices of the physical sciences as distinct from the classical language games of neopositivism and conventionalism, which, as suggested above, were at the time being offered as the standard norms of scientific activity. These conclusions themselves provided a blueprint for the new paradigm-based methodology that came to be understood as relational rather than absolutist in character. The conclusions were that (1) there is no absolute demarcation between interpretative theory and observation or between interpretative theory and facts or data; a notion that was captured in his now-famous aphorism, *all data are theory laden*; (2) scientific explanation consists of the discovery of patterns, as well as the discovery of causes (see also, Toulmin, 1953, 1961); and (3) the fundamental logic of science is neither a split-off deductive logic nor a split-off inductive logic, but rather it is an *abductive* (retroductive) logic originally described by the pragmatist philosopher Charles Sanders Pierce (1992). This logic operates by arranging the observation under consideration and all *background ideas* (i.e., the paradigm and theoretical terms) as a relational complementarity. The coordination of the two is explored by asking what, given the background ideas, must necessarily be assumed in order to have that observation. The inference to, or interpretation of, what must necessarily be assumed then constitutes the explanation of the phenomenon. This explanation can then be assessed empirically as a hypothesis to ensure its empirical validity (i.e., its empirical support and scope of application). An important relational feature of this logic is that it assumes the form of the *hermeneutic circle* (Gadamer, 1960; Overton & Mueller, 2012) by moving from the phenomenological level (the commonsense object) to explanation and back in an ever-widening cycle. The difference between this idea and what is often referred to as the hypothetico-deductive explanation of epistemological empiricism is that in abduction, all background ideas (the disciplinary matrix or paradigm and theoretical terms) constitute a necessary feature of the process and empirically supported abductive explanations themselves become a part of the ever-widening corpus of background ideas.

In sum, by the 1980s, the scientific paradigm as disciplinary matrix composed of a core set of assumptions or background ideas – including ontological and epistemological propositions, and nested concept – was relatively well established as a valid scientific methodology. This conclusion is not to argue that neopositivism and conventionalism/instrumentalism completely left the battlefield. As Kuhn himself suggested, old paradigms like old soldiers never die; they fade away as their proponents leave the field and are not replaced (e.g., there are few scientists or philosophers alive today who would defend the central principles of neopositivism).

1.2.5. Refining Scientific Paradigms: Lakatos

While the idea of a scientific paradigm was relatively well established, there were refinements to be made and features of the broader issue of scientific change (e.g., revolutions or evolutions) to be faced. Before describing the central refinement and one of the broader issues, there is a critical point that, although not widely understood, has remained undisputed and requires emphasis. If one accepts the validity of scientific paradigms, then one recognizes that they do not emerge as inductions from data sets, however large and however damaging these data sets may be to currently held beliefs. As Kuhn points out, once the disciplinary matrix and its nested concepts have achieved the status of a paradigm, it "is declared invalid only if an *alternative candidate is available* to take its place....The decision to *reject one paradigm is* always simultaneously *the decision to accept another* (Kuhn, 1970, p. 77, emphasis added)." Thus, regardless of whether there are huge amounts of data standing in opposition to predictions of the classical genetic position concerning inheritance, the classic evolutionary Modern Synthesis, mental processes as brain processes, or the split of person and culture, these positions will never be rejected unless an alternative paradigm is already available to accommodate these new data. It does no good to wait for one to emerge from the data; paradigms must be conceptually constructed. It is the theme of this chapter, to be developed following the present elaboration of paradigms, that such an alternative paradigm is, and has been, available.

The fundamental refinement of Kuhn's concept of scientific paradigm was developed by Imre Lakatos (1970, 1978) and extended by Larry Laudan (1977). Lakatos' refinement entailed more clearly distinguishing among layers of the nested concepts that constitute a paradigm, or what

Lakatos termed a *research program*. For Lakatos, a scientific research program consists of three nested levels ranging from the broadest and most abstract to narrower and less abstract: (1) a level of *hard core* concepts, (2) a level of *positive heuristic* concepts, and (3) a level of concepts called the *belt of auxiliary hypotheses*.

The *hard core* corresponds to Kuhn's disciplinary matrix or worldview (Fig. 2.1). It contains epistemological and ontological concepts (e.g., as to be discussed further later in this chapter, Lakatos referred to "Cartesian metaphysics, that is, the mechanistic theory of the universe—according to which the universe is a huge clockwork (and system of vortices) with push as the only cause of motion (1978, p. 47)" as the hard core of a research program. "It discouraged work on scientific theories—like (the 'essentialist version of) Newton's theory of action at a distance—which were inconsistent with it… On the other hand, it encouraged work on auxiliary hypotheses which might have saved it from apparent counter evidence—like Keplerian ellipses (p. 47, 48)."

The propositions of the hard core do not change. They exert a formative influence on lower levels, and give meaning to the concepts of specific theories. These propositions are also irrefutable in the sense of not being open to empirical falsification. The advantage of this irrefutability is that it allows the next layer to be developed without unnecessary distractions.

The second and lower level—the *positive heuristic*—corresponds to the notion of midrange metatheories discussed earlier (Fig. 2.1). The propositions at this level are influenced by the hard core but they describe the long-term research policy of the program. This level is more flexible than the hard core and consists of a "partially articulated set of suggestions or hints on how to change, and develop the 'refutable variants' of the research program, how to modify, and sophisticate the 'refutable' protective belt of auxiliary hypotheses" (Lakatos, 1978, p. 50).

Lakatos (1978) points out that the hard core and positive heuristic taken together constitute the "conceptual framework" (p. 47) of the program. At this point, it may be helpful to give a brief preview of how these components of scientific paradigms operate with respect to the developmental science subfields described at the beginning of this chapter (i.e., inheritance, evolution, organismic development [including prenatal, cognitive, emotional, motivational, and sociocultural]). The argument here is that each of these subfields has traditionally functioned within a paradigm committed to a Cartesian-split-mechanistic worldview (hard core). Further, it is argued that classic genetics, the evolutionary Modern Synthesis, the computational theory of mind, and traditional cross-cultural approaches in psychology, each entail midrange metatheories (positive heuristics) operating within this hard core. On the other hand, the further argument is that there is and has been an alternative paradigm available that better accommodates the new advances in each of these fields. This paradigm has relationism as its hard core and the broadest midrange metatheory (positive heuristic) within this hard core is relational developmental systems. Other midrange metatheories subsumed under relational developmental systems include developmental, dynamic, dialectical, and transactional systems, and enaction.[2] And each of these midrange metatheories has as core concepts "system," "embodiment," "action," and "epigenesis" (see, however, footnotes[3] and[8]).

The hard core and positive heuristic function as the background ideas that confront the observational domain of scientific interest and operating abductively generate theories—broad or narrow—designed to explain that domain. The theories thus constructed constitute a *family of theories;* theories that assert the same conceptual framework, but differ according to the observational domain they encompass (Fig. 2.1). Thus, for example, Erikson's (1968) theory of socioemotional development, Werner's (1948) theory of comparative development, Piaget's (1967) theory of cognitive development, Bowlby's (1958) theory of emotional development, and Vygotsky's (1978) theory of person-cultural development all represent one family of theories, as each derives from the same set of hard core and positive heuristic assumptions. Similarly, Marcia's (1980) of adolescent identity development is a narrow member of the same family as it operates within an Eriksonian framework. On the other hand, various theories that embrace a Cartesian-split-mechanistic worldview and a midrange computational model of mind metatheory (e.g., Marr's (1982) theory of vision and various information processing theories) belong to another distinct family of theories.

Theories in turn generate Lakatos' third level of a scientific research program, the *belt of auxiliary hypotheses.* This belt is composed of sets of observational hypotheses that, having been formulated abductively, are open to empirical test and constitute the falsifiable or refutable component of the scientific program (Fig. 2.1).

This particular conception then constitutes the contemporary meaning of scientific paradigm: a system composed of at least three layers of nested

---

[2]It has been argued that some system approaches, such as dynamic systems, at times reach the level of theory as they can entail the construction of detailed mathematical models of developmental data (Keller, 2005). Although this is correct, it is not relevant to the current and subsequent argument concerning nested concepts themselves.

concepts: (1) a worldview or hard core constituted by basic ontological and epistemological commitments; (2) midrange metatheories or a positive heuristic whose propositions are influenced by the hard core, but describe the long-term research policy of program; and (3) a theory or theories that generate a belt of auxiliary hypotheses grounded in the conceptual framework of the hard core and positive heuristic, and operating as the refutable component of the paradigm.

Scientific paradigms are necessary and indissociable features of any domain of scientific inquiry. While this judgment was controversial during the middle of the twentieth century, it has now become relatively commonplace. It is found among introductions to the philosophy of science (e.g., Godfrey-Smith, 2003), as well as among the discussions of eminent scientists such as Stephen Hawking:

> There is no picture- or theory-independent concept of reality.... *We will adopt a view that we will call model-dependent realism: The idea that a physical theory or world picture is a model ... and a set of rules that connect the model to observations. This provides a framework with which to interpret modern science.*
> **(Hawking & Mlodinow, 2010; p. 42, 43; emphasis in the original).**

1.2.6. Paradigm Change: Falsification Yields to Anomaly

While these layers of nested concepts constituted the most significant refinement of Kuhn's paradigm, a full understanding of how paradigms function requires a brief examination of one broader issue that Kuhn's work faced. Popper's falsification criterion of scientific progress fell into general disrepute in science through demonstrations by several historians and philosophers of science (e.g., Hanson, 1958; Kuhn, 1962; Lakatos, 1978; Laudan, 1977; Putnam, 1983; Quine, 1953) that, although deductive logic, and hence falsification, is applicable to a specific experimental hypothesis, falsification does not reach to the level of rich theories. As a consequence, Kuhn, Lakatos, Laudan and others, who viewed science not as a scientific realist search for an absolute truth, but as a puzzle or problem solving activity, came to consider any counter evidence or experimental falsification as an *anomaly* or an anomalous incident. These anomalies were to be put aside as the paradigm or program progressed in the sense of successfully proceeding with its work of problem solving.

But, the question remained as to whether and when the buildup of anomalous incidents would actually impact the progress of the scientific program. For Kuhn (1962), anomalies built up until a kind of *gestalt switch* went off and that precipitated a scientific *crisis*, which led to a *scientific revolution* entailing the movement to an alternative available paradigm. While both Lakatos (1978) and Laudan (1977) agreed with the general status of anomalies, each argued that the notion of a *gestalt switch* without a deeper rational for the movement to an alternative paradigm was irrational. As a consequence, each proposed criteria for a program's scientific progress or lack thereof. For Lakatos, the occurrence of anomalies constitutes a local and minor criterion for lack of scientific progress. His criterion of progress of a scientific research program was that it predicted novel or unexpected phenomena with some degree of success. Furthermore, the anticipation of novel events, Lakatos argued, should be guided by a coherent, preplanned positive heuristic rather than via patched-up *ad hoc* auxiliary hypotheses. Laudan (1977) expanded this position by maintaining that science aims "to maximize the scope of solved empirical problems, while minimizing the scope of anomalous and conceptual problems" (p. 66). Thus, scientific progress according to the new methodology is measured ultimately by a pragmatic criterion and not by a scientific realist's truth criterion. With respect to the falsification of observational hypotheses, Lakatos (1978) claimed that the scientist must note them as they occur "but as long as his research programme sustains its momentum, he may freely put them aside ... Only when the driving force of the positive heuristic weakens, may more attention be given to anomalies" (p. 111). Laudan (1977), in his extension of the general strategy, suggested ways in which anomalies can and should be graded also in terms of their cognitive importance (pp. 36–40).

This foray into the place of anomalies and progress in scientific research programs is particularly relevant to the developmental science issues that are the focus of this chapter: the movements (1) beyond classic genetics to a postgenomic world; (2) beyond the evolutionary Modern Synthesis to an evolutionary perspective in which individual development plays a constitutive role in evolution; (3) beyond a cognition and cognitive development perspective that encapsulates mental processes in the brain, and to a position that extends mental processes out into the body and into the technological and cultural worlds; and (4) beyond the sociocultural developmental perspective of the individual and culture as split-off entities and to a view of individual and culture as coconstructed, codetermined, and codeveloped.

To take just the postgenomic movement as an example of this relevance, Charney's (2012) review of new findings in genetics and epigenetics clearly constitutes a set of damaging empirical anomalies for the population (quantitative) behavior genetics metatheory. And when these anomalies are combined with other empirical anomalies and the conceptual flaws that have been described across the decades, including those from biology (e.g., Keller, 2010; Lewontin, 1974, 1991, 2000; Wahlsten, 2012), anthropology (e.g., Ingold, 2000; Sheets-Johnstone, 1990), experimental psychology (e.g., Hirsch, 1967; Kamin, 1974), and developmental science (e.g., Gottlieb, 1997, 2003; Gottlieb, Wahlsten, & Lickliter, 2006; Greenberg, 2011; Greenberg & Partridge, 2010; Joseph, 2010; Lerner, 1978, 2004, 2012a; Meaney, 2010; Moore, 2001; Overton, 1973, 2004, 2011; Partridge, 2005), it is difficult in the extreme not to conclude that population (quantitative) behavior genetics metatheory has been a failed explanation of inheritance, and the body of new empirical data requires an alternative paradigm to ground genetics generally and to guide the further pursuit of new data.

2. ALTERNATIVE PARADIGMS FOR DEVELOPMENTAL SCIENCE

With an understanding of scientific paradigms and their implications as background, we may turn to an examination of the paradigm that has been the standard model across the developmental sciences—the Cartesian-split-mechanistic worldview and its midrange metatheories—and to an examination of the paradigm that Lerner and I have argued (Lerner, 2006, 2011; Lerner & Overton, 2008; Overton & Lerner, 2012; Overton, 2006, 2010, 2012) better accommodates the new data emerging in developmental science—the relationism and relational developmental systems paradigm.

2.1. The Cartesian-Split-Mechanistic Worldview and Split-Mechanistic Midrange Metatheories as Scientific Paradigm

The worldview that constitutes the broad abstract framework for the paradigm to be described here is the Cartesian-split-mechanistic worldview, while metatheories of the midrange subsumed within this worldview vary across subfields of developmental science. Thus, in the field of population genetics, a midrange metatheory has involved the Fisher–Wright ANOVA model; in traditional evolutionary biology, a midrange metatheory is the Modern Synthesis; in cognition, the computational model of mind is a midrange metatheory, and in traditional cultural approaches, it has been the person–culture dichotomy.

The Cartesian worldview as paradigm hard core is, as suggested earlier, the "Cartesian metaphysics, that is, the mechanistic theory of the universe—according to which the universe is a huge clockwork (and system of vortices) with push as the only cause of motion (Lakatos, 1978, p. 47)" and in which fundamental features of this world are split into dichotomous pure forms. A worldview is rarely, if ever, developed by one individual and this point holds for the Cartesian-split-mechanistic worldview. The early protagonists who developed the basic tenets of this broad metatheory were Galileo Galilei, and his physics of a natural world disconnected from mind; Rene Descartes, whose epistemology elevated disconnection or splitting to a first principle and whose ontology began the path to viewing the world in terms of the categories of the machine, later elaborated by Newton's admirers such as John Locke; and Thomas Hobbes who envisioned both mind and nature in terms of an ontology of mechanically operating atomistic materialism. Of the three main figures, Descartes was, perhaps, to have the greatest and the most lasting impact on the text and subtexts of this particular metatheoretical story.

Here I focus primarily on Descartes major epistemological contributions, although the ontological contribution of Hobbs static, fixed materialism cannot be ignored. Descartes epistemological contributions consisted of the introduction of *splitting, foundationalism, and atomism* as key interrelated themes in the story of scientific knowing. *Splitting* is the separation of components of a whole into mutually exclusive pure forms or elements. In splitting, these ostensibly pure forms are cast into an exclusive "either/or" framework that forces them to be understood as contradictions in the sense that one category *absolutely* excludes the other (i.e., follows the logical law of contradiction that it is never the case that A = not A). But, in order to split, one must accept the twin principles of *foundationalism* and *atomism*. These are the metatheoretical axioms that there is ultimately a rock bottom unchanging nature to Reality (i.e., with a capital R, distinguishing this ultimate real from the commonsense reality of everyday objects [Putnam, 1987]). This conception is Descartes' foundationalism, describing a final fixed secure base. It constitutes an absolute, fixed, unchanging bedrock; a final Archimedean point (Descartes, 1969). Further, this rock bottom is composed of elements—pure forms—(the atoms of atomism) that preserve their identity regardless of context. A corollary principle here is the assumption that all complexity is *simple complexity* or *simply complicated* in the sense that any whole is taken to be a purely additive combination of its elements.

Splitting, foundationalism, and atomism are all principles of decomposition; breaking the aggregate down to its smallest pieces, to its bedrock (Overton, 2006). This process also goes by other names, including *reductionism* and *the analytic attitude* (Overton, 2002, 2006). Split metatheory, however, requires another principle to reassemble or recompose the whole. This is the principle of *unidirectional, linear and additive associative or causal sequences*. The elements must be related either according to their contiguous cooccurrence in space and time, or according to simple efficient or material mechanical cause–effect sequences that proceed in a single direction (Bunge, 1962; Overton & Reese, 1973). In fact, split metatheory admits no determination other than individual efficient and material causes, or these individual causes operating in a conjunctive (i.e., additive) plurality. Truly reciprocal causality <-->, or circular causality, are *not permitted* in this system (Bunge, 1962; Overton & Reese, 1973; Witherington, 2011).

2.1.1. Implications of the Cartesian-Split-Mechanistic Worldview for Developmental Science

The standard traditional frame for the issues of developmental science that are the focus of this chapter has been the principles of this Cartesian-split-mechanistic worldview. In classic genetics, the gene was introduced as an analog to the chemical element as the foundational biological element (Keller, 2010); the gene was conceptualized as the "master molecule" that "causes" the production of proteins; a linear additive, unidirectional causal path was asserted to operate from DNA to RNA to protein as defined by the "central dogma of molecular biology" (Gottlieb, 2000). With respect to population (quantitative) genetics, the relation of genes to environment is conceptualized within a completely additive (Lewontin, 1974; Overton & Reese, 1973) model, and statistical gene × environment interactions are themselves completely decomposable into strictly additive elements. As Turkheimer (2011) points out, this assumption of additivity is "the foundation of modern quantitative genetics". And Partridge (2005, 2011), supporting Turkheimer's point, goes on to describe how advances in the Fisher–Wright ANOVA model, such as extensions to multivariate and latent variable models and multilevel models, adhere to the same additive structure as the original Fisher model.

For the evolutionary Modern Synthesis, development and evolution are split (see, e.g., Lickliter & Honeycutt, 2010). Further, internal is split-off from external, yielding an isolated internalism of gene centrism (i.e., gene as the sole unit of variation), and an isolated externalism of change (i.e., emphasis on natural selection as the virtually sole vehicle of change) (see, e.g., Pigliucci & Mueller, 2010a). The Modern Synthesis also entails commitment to evolutionary "gradualism" (i.e., additive continuity) that derives from the same mathematical formalism identified above that Pigliucci and Mueller (2010a) refer to as the "backbone" of population genetics, and, hence the backbone of the Modern Synthesis.

The field of cognition and cognitive development, more than the biological or cultural fields of developmental science, has most frequently given explicit recognition to the fact that, at least until recently, the discipline has been framed by the Cartesian metatheory (e.g., Marshall, 2009; Mueller & Newman, 2008; Rowlands, 2010; Varela, Thompson, & Rosch, 1991). Early in this field's emergence—known as the Cognitive Revolution—what became termed "cognitivism" was based on the midrange metatheory of the computational model of mind. Not only did this midrange metatheory strictly follow the dictates of the Cartesian worldview of splitting mind from body, it also explicitly framed itself within the worldview's basic category system, the machine. As Marshall (2009) describes it, "The ascent of cognitivism depended on making the mind more transparent by using computers to model mental processes.... The Cartesian foundation of this approach has inspired a rising tide of criticism over the last three decades, mainly centered around the problem that the computational mind of cognitivism lacks a brain, a body, and a culture (p. 120)." Later, as Rowlands (2010) pointed out, "from the mid-1980s on, this [computer model of mind] emphasis gradually gave way to a renewed emphasis on 'hardware' in the form of connectionist or neural network approaches (p. 2)." These models attempt a mechanical modeling of cognition that, while perhaps more neurobiologically realistic (see, however, Edelman, 1992), continues to adhere to Cartesian dictates, thereby leaving mental process locked in the brain, split-off from the full functioning of the body and from culture. This Cartesian position is clearly articulated by Adams and Aizawa (2010) in their argument that "There are processes that (1) are recognizably cognitive, (2) take place in the brain, (3) do not take place outside of the brain, and (4) do not cross from the brain into the external world (p. 69)." In the end, as Goode (2007) notes, "on the cognitivist view... the starting point is the solitary Cartesian subject detached from the world and its objects (including other people). Thus, the cognitivist has to account for the way the knower 'hooks on to' the world and to other people in it (p. 272)."

In the area of culture and development, Cartesian tracks are found in a number of areas where individual and culture are viewed as split-off pure forms. For example, in their analysis of the culture and development literature, Mistry et al. (2012) point to the fact that in both cross-cultural and ecological approaches, culture is treated as an antecedent variable that influences, but is not constitutive of, the individual and development:

> On the issue of how culture should be conceptualized, cross-cultural psychologists tended to view culture as an 'independent variable' that influenced human behavior. Some cross-cultural psychologists suggested that culture should be operationalized as a set of conditions … For example Poortinga (1997) defined the cross-cultural approach as: '… a tendency to take cultural context, including ecological as well as sociocultural variables, as a set of antecedent conditions, while behavior phenomena, including attitudes and meanings as well as observed behaviors as outcomes or consequents' (p. 350). In the ecological model (Bronfenbrenner, 1979, 1986), which had been particularly influential in developmental psychology, culture was operationalized as the macro (pervasive) influence on the developing person.

Although still an influential worldview, the Cartesian metatheory and the midrange metatheories it subsumes have come under increased criticisms as an adequate scientific paradigm for developmental science. These criticisms are cross-disciplinary, coming both from those cited above and, more generally, from biology and neuroscience (e.g., Damasio, 1994; Gallese & Lakoff, 2005), philosophy (Gallagher, 2005; Taylor, 1995), anthropology (Ingold, 2000; Sheets-Johnstone, 1990), and psychology (Barsalou, Simmons, Barbey, & Wilson, 2003; Colombetti & Thompson, 2008; Hobson, 2002; Mueller & Newman, 2008; Smith, 2005).

2.2. Relationism and Relational Developmental Systems: A Paradigm for Developmental Science

The question arises as to whether these conceptual criticisms raised against the Cartesian–split–mechanistic worldview and its midrange metatheories; the empirical anomalies faced by the paradigm; and failures of the paradigm to generate new ideas and new data are sufficient to conclude that this paradigm should be rejected as invalid? The answer is no. While these may be necessary conditions for dismissing the Cartesian–split–mechanistic metatheory and its subsumed midrange metatheories, only an available alternative paradigm constitutes a sufficient condition for rejection. It is argued here that there is such a paradigm available; one that (1) better accommodates the new data from several fields; (2) overcomes the conceptual problems of the Cartesian–split–mechanistic metatheory and its subsumed midrange metatheories, and (3) generates novel and empirically productive predictions for the field of developmental science. This is the relationism (as worldview) and relational developmental systems (as midrange metatheory) paradigm, which I next describe.

2.2.1. Relationism

Relationism finds its historical origins in Aristotle's insistence that form and matter cannot be separated into two discrete elements, and later in Kant's attempt to reconcile empiricism and rationalism and in Hegel's elaboration of dialectical logic. One broad effect of adopting relationism as a paradigm worldview is that it leads to the healing of the classic fundamental antimonies (e.g., subject–object, mind–body, nature–nurture, culture–individual, culture–biology, self–other) (Table 2.1) and provides concepts that are inclusive and concepts that adequately ground science generally and developmental science specifically. In an analysis of the historical failures of classical split metatheory, as well as the emptiness of its seeming rival—postmodern thought—Bruno Latour (1993, 2004) proposed a move away from the extremes of Cartesian splits to a center or *middle kingdom* position where entities and ideas are represented, not as pure forms, but as forms that flow across fuzzy boundaries. This movement is one toward what Latour terms *relationism,* a metatheoretical space where foundations are groundings, not bedrocks of certainty, and analysis is about creating categories, not about cutting nature at its joints. The present version of relationism builds on Latour's proposal.

Relationism is a worldview formed as a principled synthesis of Pepper's (1942) organicism and contextualism (for details, see Overton, 2007a; Overton & Ennis, 2006a, 2006b).[3] As a worldview, it is composed of a coherent set of intertwined ontological and epistemological principles. The ontology of relationism offers a Real based on *process-substance* rather than a split-off substance (Bickhard, 2008). This ontology is what Gadamer (1989) argues to be the movement of *to and fro* and what has been sometimes defined as an ontology of *Becoming* (Allport, 1955; Overton, 1991). It includes process, activity, dialectic change, emergence, and necessary organization as fundamental defining categories, but it does not exclude categories of substance, stability, fixity, additivity, and contingent organization.

The epistemology of relationism is, first and foremost, a *relatively inclusive* epistemology, involving both knowing and known as equal and indissociable complementary processes in the construction, acquisition, and growth of knowledge. It is *relatively* inclusive, because *inclusion* itself—much like Hegel's master–slave dialectic—can be grasped only in relation to its

[3]Witherington (2007, 2011) employs a similar organicism–contextualism integration to distinguish between dynamic systems metatheories based on a strict contextualist worldview (e.g., Thelen & Smith, 1994; Spencer, Perone, & Johnson, 2009) from those based on an organicist–contextualist integration (e.g., van Geert, 2003; van der Maas & Molenaar, 1992; Lewis, 2011).

Table 2.1 Fundamental antimonies/false dichotomies

Brain	Body
Mind	Body
Biology	Culture
Person	Biology
Culture	Person
Nature	Nurture
Subject	Object
Form	Matter
Stability	Change
Continuous	Discontinuous
Quantitative	Qualitative
Internal	External
Universal	Particular
Transcendent	Immanent
Analysis	Synthesis
Unity	Diversity
Interpretation	Observation
Absolute	Relative
Variation	Transformation
Reason	Emotion

complement *exclusion*. Thus, just as *freedom* must be identified in the context of *constraint, inclusion* must be identified in the context of *exclusion*. Relational epistemology specifically excludes Cartesian dualistic ways of knowing because Cartesian epistemology trades on absolute exclusivity; it constitutes a *nothing but* epistemology. For the same reason, relationalism rejects both the mechanistic worldview and a strict contextualist interpretation of the contextualist worldview (Overton, 2007a; Witherington, 2007, 2011). Epistemologically, relationism begins by clearing the "nothing but" *splitting, foundationalism, atomism,* and *objectivism* from the field of play and in so doing, it moves toward transforming antinomies into coequal, indissociable complementarities. In the relational frame, fixed absolute elements are replaced by contextually defined parts.

In place of the rejected splitting, foundationalism and atomism, relationism installs *holism* as the overarching epistemological first principle. Building from the base of holism, relationism moves to specific principles that define the relations among parts and the relations of parts to wholes. In other words, relational metatheory articulates principles of analysis and synthesis necessary for any scientific inquiry. These principles are (1) *The Identity of Opposites,* (2) *The Opposites of Identity*, and (3) *The Synthesis of Wholes.*

2.2.2. Holism

Holism is the principle that the identities of objects and events derive from the relational context in which they are embedded. Wholes define parts and parts define wholes. The classic example is the relation of components of a sentence. Patterns of letters form words and particular organizations of words form sentences. Clearly, the meaning of the sentence depends on its individual words (parts define whole). At the same time, the meaning of words is often defined by the meaning of the sentence (wholes define parts). Consider the word meanings in the following sentences: (1) The *party leaders* were *split* on the *platform*; (2) The *disc jockey* discovered a *black rock star*; and (3) The *pitcher* was *driven home* on a *sacrifice fly*. The meaning of the sentence is obviously determined by the meaning of the words, but the meaning of each *italicized* word is determined by context of the sentence it is in. Parts determine wholes, and wholes determine their parts (Gilbert & Sarkar, 2000).

Holistically, the whole is not an aggregate of discrete elements but an organized system of parts, each part being defined by its relations to other parts and to the whole. Complexity in this context is *organized complexity* (Luhmann, 1995; von Bertalanffy, 1968a, 1968b), in that the whole is not decomposable into elements arranged in additive sequences of mechanistic cause–effect relations (Overton & Reese, 1973). In the context of holism, principles of splitting, foundationalism, and atomism are, by definition, rejected as meaningless approaches to analysis, and fundamental antimonies are similarly rejected as false dichotomies (Table 2.1). In an effort to avoid *standard* (i.e., neopositivistic and conventionalist) misunderstandings here, it must be strongly emphasized that *nondecomposability does not mean that analysis itself is rejected*. It means that *analysis of parts must occur in the context of the parts' functioning in the whole*. The *context-free specifications* of any object, event, or process—whether it be a DNA, cell, neuron, evolution, the architecture of mind, or culture—are illegitimate within a holistic system (see, e.g., Ingold, 2000). Bunge (2003) well[4] captures both the problem of reductionism and the issue of holism in the following:

> At first sight, the discovery that genetic material is composed of DNA molecules proves that genetics has been reduced to chemistry .... However, chemistry only accounts for DNA chemistry: it tells us nothing about the biological functions of DNA – for instance that it controls morphogenesis and protein synthesis. In other words, DNA does not perform any such functions

[4] I am indebted to Gary Greenberg for pointing me to this quote.

when outside a cell, anymore than a stray screw holds anything together. Besides, DNA does nothing by itself: it is at the mercy of the enzymes and RNAs that determine which genes are to be expressed or silenced. In other words, the genetic code is not the prime motor it was once believed to be. This is what epigenesis is all about (p. 138).

Although holism is central to relationism, holism does not in itself offer a detailed program for resolving many dualisms that have framed scientific knowing and knowledge. A complete relational program requires principles according to which the individual identity of each concept of a formerly dichotomous pair is maintained, while simultaneously it is affirmed that each concept constitutes, and is constituted by, the other. This understanding is accomplished by considering identity and differences as two *moments of analysis*. The first moment is based on the principle of the *identity* of opposites; the second moment is based on the principle of the *opposites* of identity.

2.2.3. The Identity of Opposites

The principle of the identity of opposites establishes the *identity among parts* of a whole by casting them, not as exclusive contradictions as in the split epistemology but, as differentiated polarities (i.e., coequals) of a unified (i.e., indissociable) inclusive matrix—as a *relation*. As differentiations, each pole is defined recursively; each pole defines and is defined by its opposite. In this identity moment of analysis, the law of contradiction is suspended and each category contains and, in fact, *is* its opposite. Further—and centrally—as a differentiation, this moment pertains to character, origin, and outcomes. The character of any contemporary behavior, for example, is 100% nature because it is 100% nurture; 100% biology because it is 100% culture. There is no origin to this behavior that was some other percentage—regardless of whether we climb back into the womb, back into the cell or back into the DNA—nor can there be a later behavior that will be a different percentage.

There are a number of ways to illustrate this principle; one particularly clear illustration is found in the famous ink sketch by M. C. Escher titled *Drawing Hands* (Overton, 2006). In this sketch, a left and a right hand assume a relational posture according to which each is simultaneously drawing and being drawn by the other (Fig. 2.2(a) is a schematic illustration). In this matrix, there is a sense in which each hand is different (opposite left and right hand) and a sense in which the hands are identical (each is drawing and being drawn). In the latter analytic Identity of Opposites moment, the hands are identical (i.e., A = Not A), thus coequal and indissociable.

This moment of analysis is one in which the law of contradiction (i.e., not the case that A = not A) is relaxed and identity (i.e., A = Not A) reigns. In this identity moment of analysis, pure forms or the notion of "natural kinds" collapse and categories flow into each other. Here, each category contains, and is, its opposite. As a consequence, there is a broad inclusivity established among categories.

Within the identity moment of analysis, it is often a useful exercise to write on each hand (or the arrows of the schematic) one of the bipolar terms of an often split dualisms (e.g., genotype and phenotype, development and evolution, encapsulated and extended mental processes, person, and culture) and to explore the resulting effect (see, for example, Fig. 2.2(b), (c)). This exercise is quite different than an illustration of a familiar bidirectionality of mechanical cause and effects. This exercise makes tangible a central tenet of the relational metatheory; seemingly dichotomous ideas often thought of as competing alternatives (Table 2.1) can, in fact, enter

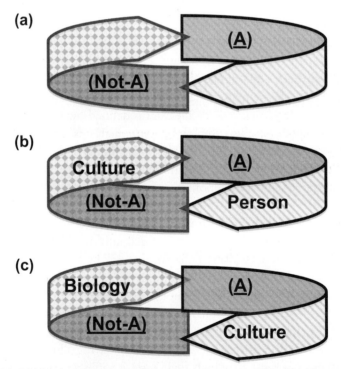

Figure 2.2 (a) Identity of opposites (b) culture–person as an identity of opposites (c) person–biology as an identity of opposites (d) culture–biology as an identity of opposites (not shown). For color version of this figure, the reader is referred to the online version of this book.

into inquiry as coequal and indissociable. This exercise also concretizes the meaning of "causality" within relationism. In this framework, the concepts *reciprocal determination* (Overton & Reese, 1973), *coaction* (Gottlieb, Wahlsten, & Lickliter, 2006), *fusion*, (Greenberg, 2011; Partridge, 2011) as well as *relational bidirectional* (**<-->**) *causality* (Lerner, 2006), *relational causality* (Gottlieb, 2003), and *circular causality* (Witherington, 2011) are relatively similar terms used to differentiate the positive and negative feedback loops of relationism from additive (even bidirectionally additive) causality of the Cartesian-split-mechanistic worldview.

The principle of the identity of opposites imposes theoretical and methodological constraints on any field of inquiry—biological, evolutionary, individual, and cultural—just as other metatheories impose constraints on any field of inquiry. The primary constraints within relationism are that (1) splits are not permitted (e.g., the split of genotype and phenotype in genetics, the split of internalism and externalism in the Modern Synthesis, the split of brain, body, and culture in cognitivism, and the cultural split of individual and culture) and (2) phenomena cannot be thought of as being decomposable into independent and additive pure forms (e.g., the Fisher–Wright AVOVA model in genetics and in the Modern Synthesis).[5]

If the principle of the identity of opposites introduces constraints, it also opens possibilities. One of these is the recognition that—to paraphrase Searle (1992)—the fact that a behavior implicates activity of the biological system does not imply that it does not implicate activity of the cultural system, and the fact that the behavior implicates activity of the cultural system does not imply that it does not implicate activity of the biological system. In other words, the identity of opposites establishes the metatheoretical rationale for the theoretical position that biology, person, and culture operate in a truly *interpenetrating relational* manner.

2.2.4. The Opposites of Identity

Although the identity of opposites sets constraints and opens possibilities, it does not in itself set a positive agenda for empirical scientific inquiry. The limitation of the identity moment of analysis is that, in establishing a flow of categories of one into the other, a stable base for inquiry that was

---

[5]West-Eberhard's (2003) evolutionary work provides a biological example of the identity of opposites in her resolution of the conflict between the quantitative genetics of continuous variation and the developmental biology of the discrete traits. This resolution is "a theory of the phenotype based on the *complementarity* of continuous and discrete variation (p. 13 emphasis added)." All antinomies are best viewed as complementaries. Relationism articulates the meaning of *complementarity*.

provided by bedrock material *atoms* of the split metatheory is eliminated. In the split approach, no relativity entered the picture; all was absolute. Reestablishing a *stable base*—not an absolute fixity, nor an absolute relativity, but a *relative relativity* (Latour, 1993)—within relational metatheory requires moving to a second moment of analysis. This is the oppositional moment, where the figure of identity and ground of opposites reverses and opposites become figure. This moment becomes dominated by a *relational exclusivity*. Thus, in this opposite moment of analysis, it becomes clear that despite the earlier identity, the schematic of Escher's sketch does illustrate both a *right* hand and a *left* hand (see Fig. 2.3 for *culture* and person). In this moment of opposition, the law of contradiction is reasserted and categories again exclude each other. As a consequence of this exclusion, parts exhibit *unique* identities that differentiate each from the other. These unique differential qualities are stable within any holistic system and, thus, may form relatively stable platforms for empirical inquiry. The platforms created according to the principle of the opposites of identity become *standpoints, points-of-view,* or *lines-of-sight,* in recognition that they do not reflect absolute foundations (Latour, 1993, 2004) but perspectives in a multiperspective world. They may also be considered under the common rubric *levels of analysis* when these are not understood as bedrock foundations.

Again, thinking of the Escher sketch (or the schematic of Fig. 2.3), when left hand as left hand (A) and right as right (Not-A) are separately the focus of attention, it then becomes quite clear that, were they large enough, one could stand at either hand and examine the structures and functions of that

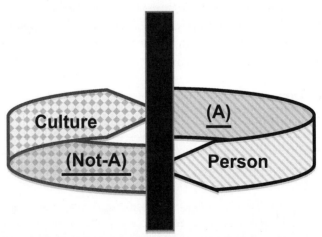

Figure 2.3 Opposites of identity. For color version of this figure, the reader is referred to the online version of this book.

location, as well as its relation to the other location (i.e., the *coactions* of parts). Thus, to return to the example of nature–nurture, although explicitly recognizing that any behavior is both 100% biology and 100% culture, alternative points of view permit the scientist to analyze the acts of the person from a *biological* or from a *cultural standpoint*. Biology and culture no longer constitute competing alternative explanations; rather, they are two points of view on an object of inquiry that has been created by, and will be fully understood only through, multiple viewpoints. More generally, *the unity that constitutes the organism and its development becomes discovered only in the diversity of multiple interrelated lines of sight.*

2.2.5. The Synthesis of Wholes

Engaging fundamental bipolar concepts as relatively stable standpoints opens the way, and takes an important first step toward establishing a broad stable base for empirical inquiry within relational metatheory. However, this solution is incomplete as it omits a key relational component, the relation of parts to the whole. The oppositional quality of the bipolar pairs reminds us that their contradictory nature still remains, and still requires a resolution. Further, the resolution of this tension cannot be found in the split approach of reduction to a bedrock absolute reality. Rather, the relational approach to a resolution is to move away from the extremes to the center and above the conflict, and there discover a novel system that will coordinate the two conflicting systems. This principle is the *synthesis of wholes*, and the synthesis itself is another standpoint.

The synthesis of interest for the general metatheory would be a system that is a coordination of the most universal bipolarity that can be imagined. Arguably, there are several candidates for this level of generality, but the polarity between matter or nature, on the one hand, and society, on the other, is sufficient for present purposes (Latour, 1993). Matter and society represent systems that stand in an identity of opposites. To say that an object is a social or cultural object in no way denies that it is matter; to say that an object is matter in no way denies that it is social or cultural. And further, the object can be analyzed from either a social–cultural or a physical standpoint. The question for synthesis becomes the question of what system will coordinate these two systems. Arguably, the answer is that it is *life* or living systems that represent the coordination of matter and society. Because our specific focus of inquiry is the psychological subject, we can reframe this matter–society polarity back into a nature–nurture polarity of *biology* (matter) and *culture* (society). In the context of psychology, then, as an illustration, if we again write *biology* on one and *culture* on the other Escher hand (or schematic figure), and question what system represents the coordination of these systems, it is life, the human organism, the *person* (Fig. 2.4). That is, the person is the relational synthesis of biological and sociocultural processes.

At the synthesis, then, a standpoint coordinates and resolves the tension between the other two components of the relation. This synthesis provides a particularly broad and stable base for launching empirical inquiry. A *person standpoint* opens the way for the empirical investigation of universal dimensions of psychological structure–function relations (e.g., processes of perception, thought, emotions, values), the particular variations associated with these wholes, their individual differences, and their development across the life span. Because universal and particular are themselves relational concepts, no question can arise here about whether the focus on universal processes excludes the particular; it clearly does not as we already know from the earlier discussion of relations. The fact that a process is viewed from a universal standpoint in no way suggests that it is not situated and contextualized; the fact that it is viewed from an individual standpoint in no way denies its universality.

It is important to recognize that one standpoint of synthesis is relative to other synthesis standpoints. *Life and Society* are coordinated by *Matter*. As a consequence, if we are broadly considering the scientific field of psychology, *biology* represents a standpoint as the synthesis of person and culture (Fig. 2.4). The implication of this idea is that a relational biological approach to psychological processes investigates the biological conditions and settings of psychological structure–function relations and the actions they express. This exploration is quite different from split foundationalist Cartesian-split-mechanistic approaches to biological inquiry that assumes an atomistic and

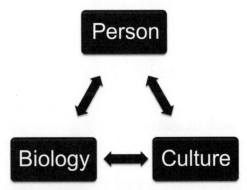

Figure 2.4 The synthesis of wholes: person–culture–biology become points of synthesis.

reductionistic stance toward the object of study. Neurobiologist Antonio Damasio's (1994, 1999) work on the brain–body basis of a psychological self and emotions is an excellent illustration of this biological relational standpoint. In the context of this standpoint, Damasio (1994) is emphatic that:

> A task that faces neuroscientists today is to consider the neurobiology supporting *adaptive supraregulations [e.g., the psychological subjective experience of self]* ... I am not attempting to reduce social phenomena to biological phenomena, but rather to discuss the powerful connection between them. ... Realizing that there are biological mechanisms behind the most sublime human behavior does not imply a simplistic reduction to the nuts and bolts of neurobiology [emphasis added] (pp. 124–125).

A similar biological example comes from the Nobel laureate neurobiologist Gerald Edelman's (1992; 2006) work on the brain–body base of consciousness:

> I hope to show that the kind of reductionism that doomed the thinkers of the Enlightenment is confuted by evidence that has emerged both from modern neuroscience and from modern physics. ... To reduce a theory of an individual's behavior to a theory of molecular interactions is simply silly, a point made clear when one considers how many different levels of physical, biological, and social interactions must be put into place before higher order consciousness emerges.
>
> **(Edelman, 1992, p. 166).**

And finally, Gilbert and Epel (2009) in presenting ecological developmental biology (eco-devo) describe several "revolutions" occurring in biology, including a new relational orientation: "Rather than analyzing independent 'things' a new focus of developmental biology concerns *'relationships'.* Nothing , it seems, exists except as part of a network of interactions (p. xiii, emphasis added)."

A third synthesis standpoint recognizes that *Person and Matter* are coordinated by *Society,* and again granting that our domain of scientific interest is psychological inquiry about psychological processes, then *culture* or *sociocultural* represents a standpoint as the synthesis of *person* and *biology* (Fig. 2.4). Thus, a relational cultural approach to psychological processes explores the cultural conditions and settings of psychological structure–function relations. From this *cultural standpoint,* the focus is on cultural differences in the context of psychological functions as complementary to the person standpoint's focus on psychological functions in the context of cultural differences.

Valsiner (1998) gives one illustration of a relational, developmentally oriented *cultural standpoint* in his examination of the "social nature of human psychology". Focusing on the "social nature" of the person, Valsiner stresses the importance of avoiding the temptation of trying to reduce person processes to social processes. To this end, he explicitly distinguishes between the *dualisms* of split foundationalist metatheory and *dualities* of the relational stance he advocates.

When the three points of synthesis—biology, person, and socioculture—are cast as a unity of interpenetrating coacting parts, there emerges what Greenberg and Partridge (2010) describe as a *biopsychosocial* model of the organism. In their tripartite relational approach, each part interpenetrates and *coconstructs* the other or *coevolves* with the other. Development of the biological organism begins from a relatively undifferentiated biosocial action matrix, and through coconstructive interpenetrating coactions, the biological, the cultural, and the psychological or person part systems emerge, differentiate, and continue their interpenetrating coconstruction, moving through levels of increased complexity toward developmental ends.

2.2.6. Relational Developmental Systems

Taken as a whole—including both its ontological and epistemological assumptions—relationism operates as the contextual frame for the construction of midrange metatheories. These latter metatheories are less broad in scope, more specific to particular domains of inquiry, and together with relationism, constitute a *conceptual framework* for a scientific paradigm. Relational developmental system is itself the broadest of these midrange metatheories, all of which incorporate systems concepts, including developmental, dynamic, dialectical, transactional systems, and enaction.[2] Relational developmental systems represents an extension (Lerner, 2006, 2011; Lerner & Overton, 2008; Overton & Lerner, 2012; Overton, 2006, 2010, 2012) of the original developmental systems "theory" described by Ford and Lerner (1992) and Gottlieb (1996) (see also Lerner, 2002). This extension was motivated by an increasing recognition of relationism as a central feature of the conceptual framework of an alternative scientific paradigm to that formulated within the Cartesian-split-mechanistic worldview.

Relational developmental systems is a perspective on developmental science (i.e., development [behavioral, cognitive, motivational, emotional, and sociocultural], inheritance, and evolution). The relational nature of the system emphasizes causality as reciprocal bi- or multidirectional (<-->) or circular (positive and negative feedback loops).[6] All facets of the individual and the context exist in mutually influential relations (Elder, 1998; Molenaar, 2007). Accordingly, the potential for *plasticity* (Batson & Gluckman, 2011; Charney, 2012; West-Eberhard, 2003) of *intraindividual change* is a hallmark of Relational developmental systems.

This metatheory conceptualizes living organisms as *active agents*, (Overton, 1976) that is, as relational, spontaneously active, complex adaptive systems, that are self-creating (i.e., enactive; autopoetic), self-organizing (i.e., process according to which higher level system organization arises solely from the coaction of lower level components of the system), and self-regulating. Further, the development process—including embryogenesis, ontogenesis, and phylogenesis—is conceptualized as entailing, five defining features: (1) *nonlinearity* (i.e., inputs are not proportional to outputs), (2) *order* and *sequence*, (3) *direction*, (4) *relative permanence* and *relative irreversibility*, and (5) *epigenesis* and *emergence*. Epigenesis is conceptualized as "probabilistic epigenesis" (Gottlieb, 1992), which designates a *holistic* approach to understanding developmental complexity. Probabilistic epigenesis is the principle that the role played by any part of a relational developmental system—DNA, cell, tissue, organ, organism, physical environment, and culture—is a function of all of the interpenetrating and coacting parts of the system. It is through complex *reciprocal* bidirectional and *circular reciprocal interpenetrating* actions among the coacting parts that the system moves to levels of increasingly organized complexity. Thus, epigenesis identifies the system as being completely *contextualized* and *situated*.

Epigenesis entails the closely related feature of *emergence of system novelty*.[7] As systems change, they become increasingly complex. This increased complexity is a complexity of form rather than an additive complexity of elements. The butterfly emerges from the caterpillar through the differentiation and reintegration of organization, the frog from the tadpole, the plant from the seed, and the organism from the zygote. In an identical manner, higher order psychological structures emerge from lower order structures; also in

[6]In order to avoid serious conceptual confusion, it is essential to differentiate this type of causality from "mechanical" or "mechanistic" causality. An example of the failure to make this distinction appears in the writings of Pigliucci and Mueller (2010a) and Mueller (2010). In discussions of new trends in evolution, these authors acknowledge the centrality of systems concepts, but simultaneously describe this as "a shift towards a causal-mechanistic approach," "a shift ... to a causal-mechanistic theory (Pigliucci & Mueller, 2010a, p. 12)," and a "turn towards the mechanistic explanation of phenotypic change (Mueller, 2010, p. 309)". There is a profound difference between the claim that there has been a trend away from correlational approaches to causal approaches, and the claim that there has been a trend away from correlational approaches to *mechanistic* causal approaches.

[7]For an extensive, in-depth analysis of meanings of "emergence," which are and are not compatible with an integrated organism–contextualism (i.e., relationism) worldview, see Witherington (2011).

an identical manner, new forms of organization exhibit novel features that cannot be reduced to (i.e., completely explained by) or predicted from earlier forms. The novel features are termed *systemic*, indicating that they are properties of the whole system and not properties of any individual part. This emergence of novelty is commonly referred to as *qualitative* change in the sense that it is the change that cannot be represented as purely additive. Similarly, reference to "discontinuity" in development is simply the recognition of emergent novelty and qualitative change of a system.

System constitutes the core concept of this metatheory and this concept has been defined in various ways. For example, van Geert (2003) offers "any collection of phenomena, components, variables" (p. 655). However, this conception and other "collection" or aggregate-like definitions are inconsistent with holism and, consequently, inconsistent with relational developmental systems. A more adequate relational definition of system is "a whole which functions as a whole by virtue of the interdependence of its parts" (Overton, 1975). Thus, a system is by its nature organized and organized holistically. Further, the relational system is an adaptive system. Here, *adaptation* refers to how the system responds to changing environments—"perturbations" in systems language—so as to increase its probability of survival, not in the sense of *adjusting* to an environment. Adaptive systems are defined in contrast with "determined" systems. In determined systems, the relation between inputs and outputs are exactly and reproducibly connected. For example, an automobile is a determined system. When the driver presses the accelerator or turns the steering wheel, both driver and passenger expect the auto to speed up or turn. All components of the auto must be fully determined to achieve this collective response. And determined systems are linear—small inputs resulting in small outputs; large inputs in large outputs—thus, outputs are predictable. In adaptive systems, the parts follow simple rules, whereas the behavior of the whole system is not determined.

The second core concept of relational developmental systems is *action*. The relational developmental system is the source of action. At *subpersonal levels,* where it is not necessary to limit a definition to organismic development or even to living systems, action is defined as the *characteristic functioning of any complex adaptive self-creating and system-organizing system.* For example, weather systems form high-and low-pressure areas and move from west to east. Living systems, on the other hand, organize, and adapt to, their biological, sociocultural, and environmental worlds. At the *person level,* organismic development action is defined as *intentional activity* (i.e., meaning giving activity). Intentionality, however, is not to be identified solely with consciousness.

While all acts are intentional, only some intentions are conscious or self-conscious. In a similar manner, intention is not to be identified solely with a symbolic or reflective level of knowing. Following Brentano (1973), all acts, even those occurring at early sensorimotor levels of functioning, intend some object.

The primary function of action is that at the microscopic level, it represents *the general mechanism for all development*. It is through the coconstituting actions of any target system of interest (e.g., genetic, epigenetic, cell, zygote, embryo, fetus, infant ... species) with its environments, as well as the resistances (perturbations) the target system encounters that the system changes and, hence, becomes differentiated and reintegrated at increasingly complex and novel levels of organization.

The final core concept is that of *embodiment* (Overton, 1994, 2008). All acts are embodied acts and, consequently, the general case is that *embodied action is the general mechanism for all development*. Embodiment represents the interpenetrating relations between person, biology, and culture. It is the claim that perception, thinking, feelings, and desires—*the way we behave, experience, and live the world*—are contextualized by our being *active agents* with this *particular kind of body* (Taylor, 1995). The kind of body we have is a constitutive precondition for having the kind of behaviors, experiences, and meanings that we have. Embodiment includes not merely the physical structures of the body but *the body as a form of lived experience, actively engaged with the world of sociocultural and physical objects*. The *body as form* references the biological point of view, the *body as lived experience* references the psychological subject standpoint, and the *body actively engaged with the world* represents the sociocultural point of view. Within a relational context, embodiment is a concept that bridges and joins in a unified whole these several research points of synthesis without any appeal to splits, foundationalism, elements, atomism, and reductionism.[8]

3. CONCLUSIONS

Relationism as worldview and relational developmental systems as metatheory of a midrange constitute a paradigmatic framework. To simplify the cumbersomeness of this phraseology, these may be joined simply as the *relational developmental systems paradigm*, recognizing both the ever-present relationism as background and the necessity of developing a belt of auxiliary hypotheses for any specific domain of research.

[8]See Withherington (this volume) for a discussion of alternative interpretations of "embodiment" within a strictly contextualist and within an organismic–contextualist (relationism) worldview.

It has been argued in this chapter that the Cartesian-split-mechanistic paradigm has historically represented the conceptual framework for developmental science's subfields, including inheritance, evolution, and organismic—prenatal, cognitive, emotional, motivational, and sociocultural—development. Further, it has been argued that there is currently overwhelming evidence, both conceptual and empirical, that according to any reasonable criteria of scientific productivity and positive growth, this paradigm has been progressively failing.

The relational developmental systems paradigm is offered both as an alternative conceptual framework to the Cartesian-split-mechanistic paradigm and as a broad beginning for further reconceptualizations of the research domains encompassed by developmental science. Relational developmental systems accommodates the new findings in genetics, evolutionary biology, and organismic development (e.g., Gissis & Jablonka, 2011; Ho, 2012; Jablonka & Lamb, 2005) and it points the way to new findings in each of these domains. There emerges genetics with reciprocal causal relations to other developing parts of the total system; an evolutionary perspective with no split between genetics and development; a cognitive approach which, from the beginning, recognizes embodiment as a constitutive feature of mental process; and a sociocultural approach in which culture is not a mere antecedent condition to the individual, but a constitutive feature of individual - culture. All emerge from the relational developmental systems paradigm.

Relational developmental systems is also providing the grounding for a number of areas of developmental science not focused on in this chapter. These areas include (1) the development of fluid intelligence and brain functioning (Blair, 2010); (2) the development of thriving (Bundick, Yeager, King, & Damon, 2010; J. Lerner et al., 2012); (3) the development of social understanding (Carpendale & Lewis, 2010); positive youth development (Kurtines et al., 2008; Lerner, 2007); (4) the development of consciousness and emotions (Lewis, 2010); (5) the development of self-regulation (Geldhof, Little, & Colombo, 2010; McClelland, Ponitz, Messersmith, & Tominey, 2010); (6) language development (MacWhinney, 2010); (7) the relational development of thinking, feeling and acting (Mascalo & Fischer, 2010); (8) the development of representation and concepts (Mueller & Racine, 2010); (9) methods in the study of intraindividual variability across the life span (Molenaar & Nesselroade, 2012; Nesselroade & Molenaar, 2010); (10) the development of reasoning (Ricco & Overton, 2011); (11) developmental psychopathology (Marshall, in press; Santostefano, 2010); and (12) moral development (Turiel, 2010). And, finally, it should be recognized that there are major contributions being made to developmental science from within several of the family of metatheories that relational developmental system represents. This includes work entailing *enaction* (e.g., Stewart et al., 2010), *dynamic systems* (e.g., van Geert & Steenbeck, 2005),[3] *dialectic* and other forms of *transactionalism* (e.g., Kuczynski & Parkin, 2009), and *social mutualism* (e.g., Goode, 2007), among others.

In short, a new scientific paradigm exists for developmental science. It is already providing a rich, productive frame for innovation and theory-predicated research about the human life span.

REFERENCES

Adams, F., & Aizawa, K. (2010). Defending the bounds of cognition. In R. Menary (Ed.), *The extended mind* (pp. 67–85). Cambridge, MA: The MIT Press.

Allport, G. (1955). *Becoming*. New Haven: Yale University Press.

Anscombe, G. E. M. (1957). *Intention*. Oxford: Basil Blackwell.

Asendorpf, J. B., & Valsiner, J. (1992). Introduction: Three dimensions of developmental perspectives. In J. B. Asendorpf & J. Valsiner (Eds.), *Stability and change in development: A study of methodological reasoning*. (pp. 9–22). London: Sage Publications.

Barsalou, L. W., Simmons, W. K., Barbey, A. K., & Wilson, C. D. (2003). Grounding conceptual knowledge in modality-specific systems. *Trends in Cognitive Sciences, 7*, 84–91.

Batson, P., & Gluckman, P. (2011). *Plasticity, robustness, development and evolution*. New York: Cambridge University Press.

Bernstein, R. J. (1983). *Beyond objectivism and relativism: Science, hermeneutics, and praxis*. Philadelphia: University of Pennsylvania Press.

Bickhard, M. H. (2008). Are you social? the ontological and developmental emergence of the person. In U. Mueller, J. I. M. Carpendale, N. Budwig & B. Sokol (Eds.), *Social life and social knowledge: Toward a process account of development* (pp. 17–42). New York: Taylor & Francis Group.

Blair, C. (2010). Fluid cognitive abilities and general intelligence: a life-span neuroscience perspective. In W. F. Overton (Ed.), & Richard M. Lerner (Editor-in-Chief). *Handbook of life-span development Cognition, biology, and methods across the lifespan* (Vol. 1, pp. 226–258). Hoboken, NJ: Wiley.

Bowlby, J. (1958). The nature of the child's tie to his mother. *International Journal of Psycho-analysis, 39*, 350–373.

Brentano, F. (1973). *Psychology from an empirical standpoint*. In A. C. Rancurello, D. B. Terrell, & L. McAlister, Trans. (Eds.), London: Routledge.

Bronfenbrenner, U. (1979). *The ecology of human development*. Cambridge, MA: Harvard University Press.

Bronfenbrenner, U. (1986). Ecology of the family as a context for human development. *Developmental Psychology, 22*(6), 723–742.

Bundick, M. J., Yeager, D. S., King, P. E., & Damon, W. (2010). Thriving across the life span. In W. F. Overton (Ed.), & Richard M. Lerner (Editor-in-Chief). *Handbook of life-span development Cognition, biology, and methods across the lifespan* (Vol. 1, pp. 882–926). Hoboken, NJ: Wiley.

Bunge, M. (1962). *Causality: The place of the causal principle in modern science*. New York: The World Publishing.

Bunge, M. A. (2003). *Emergence and convergence: Qualitative novelty and the unity of knowledge*. Toronto: University of Toronto Press.

Carpendale, J. I. M., & Lewis, C. (2010). The development of social understanding: a relational perspective. In W. F. Overton (Ed.), & Richard M. Lerner (Editor-in-Chief). *Handbook of life-span development Cognition, biology, and methods across the lifespan* (Vol. 1, pp. 584–627). Hoboken, NJ: Wiley.

Cassirer, E. (1951). *The philosophy of the enlightenment.* Boston: Beacon.

Charney, E. (2012). Behavior genetics and postgenomics. *Behavioral and brain sciences, 35,* 331–358.

Colombetti, G., & Thompson, E. (2008). The feeling body: towards an enactive approach to emotion. In Overton, W. F., Müller, U., & Newman, J. L. (Eds.), *Developmental perspectives on embodiment and consciousness* (pp. 313–342). New York: Taylor & Francis.

Damasio, A. (1994). *Descartes' error: Emotion, reason, and the human brain.* New York: Grosset/ Putnam.

Damasio, A. (1999). *The feeling of what happens: Body and emotion in the making of consciousness.* New York: Harcourt Brace.

Descartes, R. (1969). (Haldane, E. S., Ross, G. R. T., Trans.), *The philosophical works of Descartes.* (Vol. 2). Cambridge, England: Cambridge University Press.

Dray, W. H. (1957). *Laws and explanation in history.* Oxford University Press.

Eckensberger, L. H. (2003). Wanted: a contextualized psychology: plea for a cultural psychology based on action theory. In T. S. Saraswathi (Ed.), *Cross-cultural perspectives in human development* (pp. 70–101). New Delhi, India: Sage Publications.

Edelman, G. M. (1992). *Bright air, brilliant fire: On the matter of the mind.* New York: Basic Books.

Edelman, G. M. (2006). *Second nature: Brain science and human knowledge.* New Haven: Yale University Press.

Elder, G. H., Jr. (1998). The life course and human development. In R. M. Lerner (Ed.), & W. Damon (Editor-in-Chief). *Handbook of child psychology Theoretical models of human development* (Vol. 1, 5th ed., pp. 939–991). New York: Wiley.

Erikson, E. H. (1968). *Identity youth and crisis.* New York: W. W. Norton & Company.

Ford, D. H., & Lerner, R. M. (1992). *Developmental systems theory: An integrative approach.* Newbury Park, CA: Sage.

Frankel, C. (1957). Explanation and interpretation in history. *Philosophy of Science, 24,* 137–155.

Gadamer, H. G. (1960/1989). *Truth and method* (Weinsheimer, J., Marshall, D., Trans.), (2nd Revised ed.). New York: The Crossroad Publishing Corp.

Gallagher, S. (2005). *How the body shapes the mind.* Oxford: Clarendon Press.

Gallese, V., & Lakoff, G. F. (2005). The brain's concepts: the role of the sensory-motor system in conceptual knowledge. *Cognitive Neuropsychology, 22,* 455–479.

Geldhof, G. J., Little, T. D., & Colombo, J. (2010). Self-regulation across the lifespan. In M. E. Lamb & A. M. Freund (Eds.), & Richard M. Lerner (Editor-in-Chief). *Social and emotional development. The handbook of lifespan development.* (Vol. 2). Hoboken, NJ: Wiley.

Gilbert, S. F., & Epel, D. (2009). *Ecological developmental biology: Integrating epigenetics, medicine and evolution.* Sunderland, MA: Sinauer Associates.

Gilbert, S. F., & Sarkar, S. (2000). Embracing complexity: organicism for the 21st century. *Developmental Dynamics, 219,* 1–9.

Gissis, S. B., & Jablonka, E. (2011). *Transformations of Lamarckism: From subtle fluids to molecular biology.* Cambridge, MA: The MIT Press.

Godfrey-Smith, P. (2003). *Theory and reality: An introduction to the philosophy of science.* Chicago: The University of Chicago Press.

Goode, J. M. M. (2007). The affordances for social psychology of the ecological approach to social knowing. *Theory & Psychology, 17*(2), 265–295.

Gottlieb, G. (1992). *Individual development and evolution: The genesis of novel behavior.* New York: Oxford University Press.

Gottlieb, G. (1996). A systems view of psychobiological development. In D. Magnusson (Ed.), *Individual development over the lifespan: Biological and psychosocial perspectives* (pp. 76–103). New York: Cambridge University Press.

Gottlieb, G. (1997). *Synthesizing nature-nurture.* Erlbaum.

Gottlieb, G. (2000). Environmental and behavioral influences on gene activity. *Current Directions in Psychological Science, 9*(3), 93–97.

Gottlieb, G. (2002). Developmental-behavioral initiation of evolutionary change. *Psychological Review, 109*(2), 211–218.

Gottlieb, G. (2003). On making behavioral genetics truly developmental. *Human Development, 46*, 337–355.

Gottlieb, G., Wahlsten, D., & Lickliter, R. (2006). The significance of biology for human development: a developmental psychobiological systems view. In R. M. Lerner, W. Damon & R. M. Lerner (Eds.), *Handbook of child psychology. Theoretical models of human development* (Vol. 1, 6th ed., pp. 210–257). Wiley.

Greenberg, G. (2011). The failure of biogenetic analysis in psychology: why psychology is not a biological science. *Research in Human Development, 8*(3–4), 173–191.

Greenberg, G., & Partridge, T. (2010). Biology, evolution, and psychological development. In W. F. Overton (Ed.), & Richard M. Lerner (Editor-in-Chief). *Handbook of life-span development Cognition, biology, and methods across the lifespan* (Vol. 1, pp. 115–148). Hoboken, NJ: Wiley.

Hanson, N. R. (1958). *Patterns of discovery*. New York: Cambridge University Press.

Hawking, S., & Mlodinow, L. (2010). *The grand design*. New York: Bantham Books.

Hegel, G. W.F. (1807). *Phenomenology of spirit*. New York: Oxford University Press. (Miller, A. V., Trans).

Hegel, G. W.F. (1830). *Hegel's logic: Being part one of the encyclopedia of the philosophical sciences*. New York: Oxford University Press. (Wallace, W., Trans).

Heidegger, M. (1962). *Being and time*. New York: Harper and Row. (Macquarrie, J., Robinson, E., Trans).

Hirsch, J. (1967). *Behavior-genetic analysis*. New York: McGraw-Hill.

Ho, M. W. (2010). Development and evolution revisited. In K. Hood, C. Tucker Halpern, G. Greenberg & R. Lerner (Eds.), *Handbook of developmental science, behavior, and genetics* (pp. 61–109). Hoboken, NJ: Wiley.

Ho, M. W. (2012). No genes for intelligence in the fluid genome. *Advances in Child Development and Behavior*.

Hobson, R. P. (2002). *The cradle of thought*. London: Macmillan.

Ingold, T. (2000). Evolving skills. In H. Rose & S. Rose (Eds.), *Alas, poor Darwin: Arguments against evolutionary psychology* (pp. 273–297). Harmony Books.

Jablonka, E., & Lamb, M. W. (2005). *Evolution in four dimensions: Genetic, epigenetic, behavioral, and symbolic variation in the history of life*. MIT Press.

Jablonka, E., & Raz, G. (2009). Transgenerational epigenetic inheritance: prevalence, mechanisms, and implications for the study of heredity and evolution. *The Quarterly Review of Biology, 84*, 131–176.

Joseph, J. (2010). Genetic research in psychiatry and psychology: a critical overview. In K. Hood, C. Tucker Halpern, G. Greenberg & R. Lerner (Eds.), *Handbook of developmental science, behavior, and genetics* (pp. 557–625). Hoboken: NJ: Wiley.

Kamin, L. J. (1974). *The science and politics of I. Q.* Erlbaum.

Kant, I. (1871). *Critique of pure reason*. (Max Muller, F., Trans), New York: Anchor. Books edition, 1966.

Keller, E. F. (2005). DDS: dynamics of developmental systems. *Biology and Philosophy, 20*, 409–416.

Keller, E. F. (2010). *The mirage of a space between nature and nurture*. Durham, NC: Duke University Press.

Kuczynski, L., & Parkin, C. M. (2009). Pursuing a dialectical perspective on transaction: a social relational theory of micro family processes. In A. Sameroff (Ed.), *The transactional model of development: How children and contexts shape each other*. Washington, DC: American Psychological Association.

Kuhn, T. S. (1962). *The structure of scientific revolutions*. Chicago, IL: University of Chicago Press.

Kuhn, T. S. (1970). *The structure of scientific revolutions* (2nd ed.). University of Chicago Press.

Kuhn, T. S. (1977). *The essential tension: Selected studies in the scientific tradition and change*. Chicago, IL: University of Chicago Press.

Kurtines,W. M., Ferrer-Wreder, L., Berman, S. L., Lorente, C. C., Silverman,W. K., & Montgomery, M. J. (2008). Promoting positive youth development: new directions in developmental theory, methods, and research. *Journal of Adolescent Research, 23*(Whole Issue), 233–378.

Lakatos, I. (1970). Falsification and the methodology of scientific research programmes. In I. Lakatos & A. Musgrave (Eds.), *Criticism and the growth of knowledge* (pp. 91–196). New York: Cambridge University Press.

Lakatos, I. (1978). *The methodology of scientific research programmes: Philosophical papers.* (Vol. 1). New York: Cambridge University Press.

Lakatos, I., & Musgrave, A. (1970). *Criticism and the growth of knowledge.* New York: Cambridge university press.

Latour, B. (1993). *We have never been modern.* Cambridge, MA: Harvard University Press.

Latour, B. (2004). *Politics of nature.* Cambridge, MS: Harvard University Press.

Laubichler, M. D. (2010). Evolutionary developmental biology offers a significant challenge to the neo-Darwinian paradigm. In F. J. Ayala & R. Arp (Eds.), *Contemporary debates in philosophy of biology* (pp. 199–212). Malden, MA: Wiley-Blackwell.

Laudan, L. (1977). *Progress and its problems: Towards a theory of scientific growth.* Berkeley: University of California Press.

Lerner, R. M. (1978). Nature, nurture, and dynamic interactionism. *Human Development, 21,* 1–20.

Lerner, R. M. (2002). *Concepts and theories of human development* (3rd ed.). Mahwah, NJ: Erlbaum.

Lerner, R. M. (2004). Genes and the promotion of positive human development: hereditarian versus developmental systems perspectives. In C. Garcia Coll, E. Bearer & R. Lerner (Eds.), *Nature and nurture: The complex interplay of genetic and environmental influences on human behavior and development* (pp. 1–34). Erlbaum.

Lerner, R. M. (2006). Developmental science, developmental systems, and contemporary theories of human development. In R. M. Lerner, W. Damon & R. M. Lerner (Eds.), *Handbook of child psychology. Theoretical models of human development* (Vol. 1, 6th ed., pp. 1–17). Wiley.

Lerner, R. M. (2007). *The good teen: Rescuing adolescents from the myths of the storm and stress years.* New York, NY: The Crown Publishing Group.

Lerner. (2011). Structure and Process in Relational, Developmental Systems Theories: A Commentary on Contemporary Changes in the Understanding of Developmental Change across the Life Span. *Human Development, 54,* 34–43.

Lerner, R. M. (2012a). Developmental science and the role of genes in development. *GeneWatch, 25*(1–2).http://www.councilforresponsiblegenetics.org/genewatch/GeneWatchPage.aspx?pageId=413.

Lerner, R. M. (2012b). Developmental science: past, present and future. *International Journal of Developmental Science, 6,* 29–36.

Lerner, J., Bowers, E., Minor, K., Boyd, M. J., Mueller, M. K., Schmid, K. L., et al. (2012). Positive youth development: processes, philosophies, and programs. In R. M. Lerner, M. A. Easterbrooks & J. Mistry (Eds.), & Irving B. Weiner (Editor-in-Chief). *Comprehensive handbook of psychology: Developmental psychology* (Vol. 6, pp. 365–392). New York: Wiley.

Lerner, R. M., & Overton, W. F. (2008). Exemplifying the integrations of the relational developmental system: synthesizing theory, research, and application to promote positive development and social justice. *Journal of Adolescent Research, 23,* 245–255.

Lewis, M. (2010). The emergence of consciousness and its role in human development. In W. F. Overton (Ed.), & Richard M. Lerner (Editor-in-Chief). *Handbook of life-span development Cognition, biology, and methods across the lifespan* (Vol. 1, pp. 628–670). Hoboken, NJ: Wiley.

Lewis, M. D. (2011). The slippery slope of downward causation. *Human Development, 54,* 101–105.

Lewontin, R. C. (1974). The analysis of variance and the analysis of causes. *American Journal of Human Genetics, 26,* 400–411.

Lewontin, R. C. (1991). Biology as ideology: the doctrine of DNA. *Harper Perennial.*

Lewontin, R. C. (2000). *The triple helix: Gene, organism and environment.* Cambridge, MA: Harvard University Press.

Lickliter, R., & Honeycutt, H. (2010). Rethinking epigenesis and evolution in light of developmental science. In M. S. Blumberg, J. H. Freeman & S. R. Robinson (Eds.), *Oxford handbook of developmental behavioral neuroscience* (pp. 30–47). Oxford University Press.

Lickliter, R., & Schneider, S. M. (2006). The role of development in evolutionary change: a view from comparative psychology. *International Journal of Comparative Psychology* (19), 150–167.

Luhmann, N. (1995). *Social systems*. Stanford, CA: Stanford University Press.

MacWhinney, B. (2010). Language development. In W. F. Overton (Ed.), & Richard M. Lerner (Editor-in-Chief). *Handbook of life-span development Cognition, biology, and methods across the lifespan* (Vol. 1, pp. 467–508). Hoboken, NJ: Wiley.

Marcia, J. (1980). Adolescent identity. In J. Adison (Ed.), *Handbook of adolescent psychology*. New York: John Wiley & Sons.

Marr, D. (1982). *Vision*. San Francisco: W. H. Freeman.

Marshall, P. J. (2009). Relating psychology and neuroscience. *Perspectives on Psychological Science, 4*, 113–125.

Marshall, P. J. (in press). Coping with complexity: Developmental systems and multilevel analyses in developmental psychopathology. Development and Psychopathology.

Mascalo, M. F., & Fischer, K. W. (2010). The dynamic development of thinking, feeling, and acting over the life span. In W. F. Overton (Ed.), *Cognition, biology, and methods across the lifespan (Handbook of life-span development)* (Vol. 1, pp. 149–194). Hoboken, NJ: Wiley.

Masterman, M. (1970). The nature of a paradigm. In I. Lakatos & A. Musgrave (Eds.), *Criticism and the growth of knowledge* (pp. 59–90). New York: Cambridge University Press.

McClelland, M. M., Ponitz, C. C., Messersmith, E. E., & Tominey, S. (2010). Self-Regulation: Integration of cognition and emotion. In W. F. Overton (Ed.), & Richard M. Lerner (Editor-in-Chief). *Handbook of life-span development Cognition, biology, and methods across the lifespan* (Vol. 1, pp. 509–553). Hoboken, NJ: Wiley.

Meaney, M. J. (2010). Epigenetics and the biological definition of gene × environment interactions. *Child Development, 81*, 41–79.

Menary, R. (2010). *The extended mind*. Cambridge, MA: The MIT Press.

Merleau-Ponty, M. (1962). *Phenomenology of perception*. (Colin Smith, Trans), London: Routledge and Kegan Paul.

Merleau-Ponty, M. (1963). *The structure of behavior*. (Alden Fisher, Trans), Boston: Beacon.

Mistry, J., Contreras, M., & Dutta, R. (2012). Culture and child development. In Lerner, R. M., Easterbrooks, M. A., & Mistry, J. (Eds.), & Irving B. Weiner (Editor-in-Chief). *Comprehensive handbook of psychology: Developmental psychology* (Vol. 6, pp. 265–285). New York: Wiley, *16*(4), 210–219.

Molenaar, P. C.M. (2007). On the implications of the classical ergodic theorems: analysis of developmental processes has to focus on intra-individual variation. *Developmental Psychobiology, 50*, 60–69.

Molenaar, P. C. M., & Nesselroade, J. R. (2012). Merging the idiographic filter with dynamic factor analysis to model process. *Applied Developmental Science*.

Moore, D. S. (2001). *The dependent gene: The fallacy of "nature vs. nurture"*. New York: Henry Holt.

Mueller, G. B. (2010). Epigenetic innovation. In M. Pigliucci & G. B. Mueller (Eds.), *Evolution: The extended synthesis* (pp. 3070332). Cambridge, MA: The MIT Press.

Mueller, U., & Newman, J. L. (2008). The body in action: perspectives on embodiment and development. In W. F. Overton, U. Müller & J. L. Newman (Eds.), *Developmental perspectives on embodiment and consciousness* (pp. 313–342). New York, NY: Taylor & Francis.

Mueller, U., & Racine, T. P. (2010). The development of representation and concepts. In W. F. Overton (Ed.), & Richard M. Lerner (Editor-in-Chief). *Handbook of life-span development Cognition, biology, and methods across the lifespan* (Vol. 1, pp. 346–390). Hoboken, NJ: Wiley.

Nesselroade, J. R., & Molenaar, P. C. M. (2010). Emphasizing intraindividual variability in the study of development over the life span: concepts and issues. In W. F. Overton (Ed.), & Richard M. Lerner (Editor-in-Chief). *Handbook of life-span development Cognition, biology, and methods across the lifespan* (Vol. 1, pp. 30–54). Hoboken, NJ: Wiley.

Noe, A. (2004). *Action in perception*. Cambridge, MA: MIT Press.

Overton, W. F. (1973). On the assumptive base of the nature–nurture controversy: additive versus interactive conceptions. *Human Development, 16*, 74–89.

Overton, W. F. (1975). General systems, structure and development. In K. Riegel & G. Rosenwald (Eds.), *Structure and transformation: Developmental aspects* (pp. 61–81). New York: Wiley InterScience.

Overton, W. F. (1976). The active organism in structuralism. *Human Development, 19*, 71–86.

Overton, W. F. (1991). The structure of developmental theory. In H. W. Reese (Ed.), *Advances in child development and behavior* (Vol. 23, pp. 1–37). New York: Academic Press.

Overton, W. F. (1994). The arrow of time and cycles of time: concepts of change, cognition, and embodiment. *Psychological Inquiry, 5*, 215–237.

Overton, W. F. (2002). Understanding, explanation, and reductionism: finding a cure for Cartesian anxiety. In L. Smith & T. Brown (Eds.), *Reductionism* (pp. 29–51). Mahwah, NJ: Lawrence Erlbaum Associates.

Overton, W. F. (2004). Embodied development: ending the nativism–empiricism debate. In C. Garcia-Coll, E. Bearer & R. Lerner (Eds.), *Nature and nurture: The complex interplay of genetic and environmental influences on human behavior and development* (pp. 201–223). Erlbaum.

Overton, W. F. (2006). Developmental psychology: philosophy, concepts, and methodology. In R. M. Lerner (Ed.), & William Damon & Richard M. Lerner (Editors-in-Chief). *Theoretical models of human development The handbook of child psychology* (Vol. 1, pp. 18–88). New York: Wiley.

Overton, W. F. (2007). A coherent metatheory for dynamic systems: relational organicism–contextualism. *Human Development, 50*, 154–159.

Overton, W. F. (2008). Embodiment from a relational perspective. In W. F. Overton, U. Mueller & J. L. Newman (Eds.), *Developmental perspective on embodiment and consciousness* (pp. 1–18). Hillsdale, NJ: Erbaum Associates.

Overton, W. F. (2010). Life-span development: concepts and issues. In W. F. Overton (Ed.), & Richard M. Lerner (Editor-in-Chief). *Handbook of life-span development Cognition, biology, and methods across the lifespan* (Vol. 1, pp. 1–29). Hoboken, NJ: Wiley.

Overton, W. F. (2011). Relational-developmental-systems and quantitative behavior genetics: alternative or parallel methodologies? *Research in Human Development, 8*(3–4), 258–263.

Overton, W. F. (2012). Evolving scientific paradigms: retrospective and prospective. In L. L'Abate (Ed.), *The role of paradigms in theory construction* (pp. 31–65). New York: Springer.

Overton, W. F., & Ennis, M. (2006a). Cognitive-developmental and behavior-analytic theories: evolving into complementarity. *Human Development, 49*, 143–172.

Overton, W. F., & Ennis, M. (2006b). Relationism, ontology, and other concerns. *Human Development, 49*, 180–183.

Overton, W. F., & Lerner, R. M. (2012). Relational-developmental-systems: paradigm for developmental science in the postgenomic era. *Brain and Behavioral Science, 35*, 375–376.

Overton, W. F., & Mueller, U. (2012). Meta theories, theories, and concepts in the study of development. In: R. M. Lerner, M A. Easterbrooks & J. Mistry (Eds.), & Irving B. Weiner (Editor-in-Chief). *Comprehensive handbook of psychology: Developmental psychology* (Vol. 6, pp. 19–58). New York: Wiley.

Overton, W. F., & Reese, H. W. (1973). Models of development: methodological implications. In J. R. Nesselroade & H. W. Reese (Eds.), *Life-span developmental psychology; methodological issues* (pp. 65–86). New York: Academic Press.

Partridge, T. (2005). Are genetically informed designs genetically informative? *Developmental Psychology, 41*(6), 985–988.

Partridge, T. (2011). Methodological advances toward a dynamic developmental behavioral genetics: bridging the gap. *Research in Human Development, 8*(3–4), 242–257.

Pepper, S. (1942). *World hypotheses.* Los Angeles: University of California Press.

Piaget, J. (1967). *Six psychological studies.* New York: Random House.

Pierce, C. S. (1992). *Reasoning and the logic of things: The Cambridge conference lectures of 1898.* Cambridge, MA: Harvard University Press.

Pigliucci, M., & Mueller, G. B. (2010a). Elements of an extended evolutionary synthesis. In M. Pigliucci & G. B. Mueller (Eds.), *Evolution: The extended synthesis* (pp. 1–17). Cambridge, MA: The MIT Press.

Pigliucci, M., & Mueller, G. B. (Eds.), (2010b). *Evolution: The extended synthesis.* Cambridge, MA: The MIT Press.

Poortinga, Y. H. (1997). Towards convergence. In J. W. Berry, Y. H. Poortinga, J. Pandey, P. R. Dasen, T. S. Saraswathi, M. H. Segall, C. Kagitcibasi, J. W. Berry, Y. H. Poortinga & J. Pandey (Eds.), *Handbook of cross-cultural psychology. Theory and method* (Vol. 1, 2nd ed., pp. 347–387). Needham Heights, MA: Allyn & Bacon.

Popper, K. (1959). *The logic of scientific discovery.* London: Hutchinson.

Popper, K. (1963). *Conjectures and refutations.* London: Routledge & Kegan Paul.

Popper, K. (1970). Normal science and its dangers. In I. Lakatos & A. Musgrave (Eds.), *Criticism and the growth of knowledge* (pp. 51–58). New York: Cambridge University press.

Putnam, H. (1983). *Realism and reason: Philosophical papers.* (Vol. 3). New York: Cambridge University Press.

Putnam, H. (1987). *The many faces of realism.* Cambridge, England: Cambridge University Press.

Putnam, H. (1990). *Realism with a human face.* Cambridge, MA: Harvard University Press.

Quine, W. V. (1953). *From a logical point of view.* Cambridge, MA: Harvard University Press.

Ricco, R., & Overton, W. F. (2011). Dual systems competence ↔ procedural processing: a relational-developmental-systems approach to reasoning. *Developmental Review 31,* 119–150.

Ricoeur, P. (1984). (McLalughlin, K., Pellauer, D., Trans.), *Time and narrative.* (Vol. 1). Chicago: The University of Chicago Press.

Ricoeur, P. (1991). *From text to action: Essays in hermeneutics, II.* (Blamey, K., Thompson, J. B., Trans.), Evanston, Ill: Northwestern University Press.

Robert, J. S. (2004). *Embryology, epigenesis, and evolution.* New York: Cambridge University Press.

Rowlands, M. (2010). *The new science of the mind: From extended mind to embodied phenomenology.* Cambridge, MA: The MIT Press.

Santostefano, S. (2010). Developmental psychopathology—self, embodiment, meaning: a holistic-systems perspective. In W. F. Overton (Ed.), & Richard M. Lerner (Editor-in-Chief). *Handbook of life-span development Cognition, biology, and methods across the lifespan* (Vol. 1, pp. 792–836). Hoboken, NJ: Wiley.

Searle, J. (1992). *The rediscovery of the mind.* Cambridge, MA: MIT Press.

Sheets-Johnstone, M. (1990). *The roots of thinking.* Philadelphia, PA: Temple University Press.

Smith, L. B. (2005). Cognition as a dynamic system: Principles from embodiment. *Developmental Review, 25,* 278–298.

Spencer, J. P., Perone, S., & Johnson, J. S. (2009). The dynamic field theory and embodied cognitive dynamics. In J. P. Spencer, M. S.C. Thomas & J. L. McClelland (Eds.), *Toward a unified theory of development: Connectionism and dynamic systems theory re-considered* (pp. 86–118). Oxford: Oxford University Press.

Stewart, J., Gapenne, O., & Di Paolo, E. A. (2010). *Enaction: Toward a new paradigm for cognitive science.* Cambridge, MA: The MIT Press.

Taylor, C. (1964). *The explanation of behavior.* New York: The Humanities Press.

Taylor, C. (1995). *Philosophical arguments.* Cambridge, MA: Harvard University Press.

Thelen, E., & Smith, L. B. (1994). *A dynamic systems approach to the development of cognition and action.* Cambridge, MA: MIT Press.

Toulmin, S. (1953). *The philosophy of science.* New York: Harper and Row.

Toulmin, S. (1961). *Foresight and understanding.* New York: Harper & Row.

Turiel, E. (2010). The development of morality: reasoning, emotions, and resistance. In W. F. Overton (Ed.), & Richard M. Lerner (Editor-in-Chief). *Handbook of life-span development Cognition, biology, and methods across the lifespan* (Vol. 1, pp. 554–583). Hoboken, NJ: Wiley.

Turkheimer, E. (2011). Still missing. *Research in Human Development, 8*(3–4), 227–241.

Valsiner, J. (1998). *The guided mind: A sociogenetic approach to personality.* Cambridge, MA: Harvard University Press.

van der Maas, H. L. J., & Molenaar, P. C. M. (1992). Stagewise cognitive development: an application of catastrophe theory. *Psychological Review, 99,* 395–417.

van Geert, P. (2003). Dynamic systems approaches and modeling of developmental processes. In J. Valsiner & K. J. Connolly (Eds.), *Handbook of developmental psychology* (pp. 640–672). London: Sage.

van Geert, P., & Steenbeck, H. (2005). Explaining after by before: Basic aspects of a dynamic systems approach to the study of development. *Developmental Review, 25,* 408–442.

Varela, F. J., Thompson, E., & Rosch, E. (1991). *The embodied mind: Cognitive science and human experience.* Cambridge, MA: MIT Press.

von Bertalanffy, L. (1968a). *General system theory.* New York: George Braziller.

von Bertalanffy, L. (1968b). *Organismic psychology and systems theory.* Barre, MA. Barre.

von Wright, G. H. (1971). *Explanation and understanding.* Ithaca, NY: Cornell University Press.

Vygotsky, L. S. (1978). *Mind in society: The development of higher psychological processes.* Cambridge, MA: Harvard University Press.

Wahlsten, D. (2012). The hunt for gene effects pertinent to behavioral traits and psychiatric disorders: from mouse to human. *Developmental Psychobiology, 54*(5), 475–492.

Werner, H. (1948). *Comparative psychology of mental development.* New York: International Universities Press. (Originally published 1940).

West-Eberhard, M. J. (2003). *Developmental plasticity and evolution.* New York: Oxford University Press.

Winch, P. (1958). *The idea of a social science and its relation to philosophy.* London: Routledge and Kegan Paul.

Witherington, D. C. (2007). The dynamic systems approach as metatheory for developmental psychology. *Human Development, 50,* 127–153.

Witherington, D. C. (2011). Taking emergence seriously: the centrality of circular causality for dynamic systems approaches to development. *Human Development, 54,* 66–92.

Wittgenstein, L. (1958/1953). *Philosophical investigations* (Anscombe, G. E. M., Trans.), (3rd ed.). Englewood Cliffs, NJ: Prentice Hall.

CHAPTER THREE

Developmental Systems Theory: What Does It Explain, and How Does It Explain It?

Paul E. Griffiths*,1, James Tabery†
*Department of Philosophy and Charles Perkins Centre, University of Sydney, Sydney, NSW, Australia
†Department of Philosophy, University of Utah, Salt Lake City, UT, USA
1Corresponding author: E-mail: paul.griffiths@sydney.edu.au

Contents

We are indebted to Susan Oyama for reading and providing very useful feedback on an earlier version of this article. Griffiths' work on this paper was supported as part of an Australian Research Council Discovery Project DP0878650.

Advances in Child Development and Behavior, Volume 44
ISSN 0065-2407, http://dx.doi.org/10.1016/B978-0-12-397947-6.00003-9

Abstract

We examine developmental systems theory (DST) with two questions in mind: What does DST explain? How does DST explain it? To answer these questions, we start by reviewing major contributions to the origins of DST: the introduction of the idea of a "developmental system", the idea of probabilistic epigenesis, the attention to the role of information in the developmental system, and finally the explicit identification of a DST. We then consider what DST is *not*, contrasting it with two approaches that have been foils for DST: behavioral genetics and nativist cognitive psychology. Third, we distill out two core concepts that have defined DST throughout its history: epigenesis and developmental dynamics. Finally, we turn to how DST explains, arguing that it explains by elucidating mechanisms.

1. ORIGINS OF DEVELOPMENTAL SYSTEMS THEORY

Developmental systems theory (DST) emerged in the 1990s, building on earlier developmental systems perspectives. In this section, we consider several such contributions to the wider developmental systems perspective that eventually turned into DST: Conrad Hal Waddington's introduction of the "developmental system", Gilbert Gottlieb's concept of probabilistic epigenesis, Susan Oyama's attention to the role of information in the developmental system, and finally Donald Ford and Richard Lerner's explicit identification of a "DST".

1.1. Waddington and the "Developmental System"

Conrad Hal Waddington was a true British polymath, spending portions of his career devoted to ammonite paleontology, Whiteheadian process philosophy, embryology, biochemistry, developmental genetics, population genetics, and theoretical biology (Robertson, 1977; Slack, 2002). Waddington used the phrase "developmental system" in a sense that has much in common with current usage in an address to the 1951 Australian and New Zealand Association for the Advancement of Science (Waddington, 1952). Waddington contrasted preformationist theories of development (the characters of adults are present in the fertilized egg) with the theory of epigenesis (the characters emerge from causal interactions between simpler components in the fertilized egg). "There can be no doubt nowadays that this epigenetic point of view is correct. … An animal is, in fact, a developmental system," Waddington continued, "and it is these systems, not the mere adult forms which we conventionally take as typical of the species, which becomes modified during the course of evolution" (Waddington, 1952, p. 155). To help convey the idea, Waddington also provided a "mental picture of the developmental system" (Fig. 3.1). Waddington later termed this image the

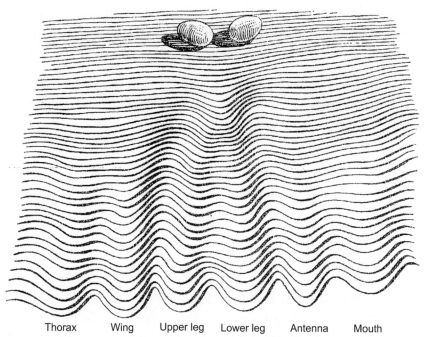

Thorax   Wing   Upper leg   Lower leg   Antenna   Mouth

Figure 3.1 A representation of the developmental system of an animal as a grooved surface over which biased balls are free to roll. *(From Waddington (1952), Fig. 1).*

"epigenetic landscape" (Waddington, 1957), and it is this idea for which he is perhaps best known. But it is worth recognizing that this well-known concept and image was originally conceived of as a *developmental system* by Waddington, and he emphasized this concept of a developmental system late in to his career (Counce & Waddington, 1972, 1973).

Studying the developmental system, Waddington pointed out, was not the same thing as studying heredity: "For the purpose of a study of inheritance, the relation between phenotypes and genotypes can be left comparatively uninvestigated; we need merely to assume that changes in the genotype produce correlated changes in the adult phenotype, but the mechanism of this correlation need not concern us. Yet this question is, from a wider biological point of view, of crucial importance, since it is the kernel of the whole problem of development" (Waddington, 2012 [1942], 10).[1] In contrast to identifying correlations between inputs (genotype) and outputs (phenotype), the "whole problem of development" required

[1]The *International Journal of Epidemiology* conveniently reprinted Waddington's "The Epigenotype" along with commentaries by Gilbert (2012), Haig (2012), and Jablonka and Lamm (2012). Page references are to the 2012 reprint.

elucidating the *mechanisms* that causally linked inputs to outputs. We will return to Waddington's focus on mechanisms later, as the nature of mechanisms and mechanical explanation is now a major topic in the philosophy of science.

Waddington's attention to developmental systems was progressive for its day. Still, his vision was quite gene centric. Indeed, the quote just given concerning the whole problem of development was from an article titled "The Epigenotype" (Waddington, 2012 [1942]). In this article, Waddington emphasized that, "…the genotype is [in] continual and unremitting control of every phase of development. Genes are not interlopers, which intrude from time to time to upset the orderly course of a process which is essentially independent of them; on the contrary, there are no developmental events which they do not regulate and guide" (Waddington, 2012 [1942], p. 12). In the same way that an individual has its genotype, then, Waddington's developmental system had its epigenotype.

1.2. Gottlieb and Probabilistic Epigenesis

Gilbert Gottlieb, recalling his preparations for a 1970 Festschrift devoted to Theodore Schneirla, noted that, "In my literature review, I found two rather different conceptualizations of behavioral embryology, one I called predetermined epigenesis and the other probabilistic epigenesis" (Gottlieb, 2001, p. 42). Despite the passage of decades since Waddington championed epigenesis, Gottlieb saw the persistence of preformationism in biology beneath a superficial layer of epigenetic clothing. Predetermined epigenesists, according to Gottlieb, understood behaviors to arise from invariant schedules of neural growth and maturation; moreover, they believed that the environment played little role in this maturational process. Probabilistic epigenesists such as Schneirla, in contrast, understood behaviors to arise probabilistically; moreover, they believed that the environment was critical in the probabilistic process. Where the predetermined epigenesist saw invariance along a set course, the probabilistic epigenesist saw inherent uncertainty due to factors such as neurochemical stimulation or the musculoskeletal effects of use during development. And, where the predetermined epigenesist saw a unidirectional relationship between structure and function, wherein the former dictated the latter, the probabilistic epigenesist saw a bidirectional structure–function relationship wherein the former not only directed but also received direction from the latter (Gottlieb, 1970). Gottlieb's attention to complex interactions and bidirectional relationships apparently developed as an undergraduate when he read John Dewey and Arthur Bentley's

Knowing and the Known (1949), which envisioned interaction superseding unidirectional self-action and then transaction superseding interaction. "I got really excited about transactionalism," Gottlieb recalled just a month before his death, "the idea that you didn't just have interactions going in one direction, but you had them going in both directions so you needed a new word. You needed transaction, so you can go across, and that just excited me" (Miller, 2006, p. 3).

Importantly, Gottlieb recognized early in his career that the assumptions dividing the predetermined and probabilistic camps were empirically testable (Gottlieb, 2001). This recognition set Gottlieb on a course of innovative experimental research that lasted for decades. In an early experiment, Gottlieb found that ducklings hatched in incubators with no exposure to maternal calls could still identify their maternal call after hatching and could distinguish that maternal call from the maternal call of a chicken (Gottlieb, 1965). Robert Lickliter and Christopher Harshaw have noted, "Had Gottlieb taken the path favored by most nativists, that of proclaiming the behavior in question to be 'instinctive' or the product of some 'innate module' and then moving on to other topics, developmental science would have been deprived of one of its most interesting series of discoveries" (Lickliter & Harshaw, 2010, p. 503).[2] Gottlieb did not take that path. Instead, he devised a method of devocalizing the avian embryos in a way that did not interfere with the otherwise healthy development of the birds (Gottlieb & Vandenbergh, 1968). The completely devocalized ducklings that were not exposed to maternal calls could not distinguish their maternal call from that of a chicken, while ducklings that were devocalized only after being able to hear their own vocalizations could distinguish the duck call from the chicken call (Gottlieb, 1971). This was a win for probabilistic epigenesis, and Gottlieb noted as much, writing, "The present results indicate that the epigenesis of species-specific auditory perception is a probabilistic phenomenon, the threshold, timing, and ultimate perfection of such perception being regulated jointly by organismic and sensory stimulative factors" (Gottlieb, 1971, p. 156).

The experiments described above, although genuinely innovative, ignored the role of genes in the phenomenon, thus leaving genetic activity out of the bidirectional relationship. Gottlieb sought to correct for this by collaborating with neurologists who had expertise with protein synthesis in the nervous system. Gottlieb prepared three groups of duck embryos: the

---

[2]See also Batson and Logan (2007) for another tribute to Gottlieb's innovative experimental legacy.

first group was exposed for several days to species-specific vocalizations, the second group was exposed for several days to extravisual stimulation via a lighted incubator chamber, and the third group was incubated in acoustic isolation and in the dark. Gottlieb delivered the three groups of duck embryos to his collaborators, and they found enhanced protein synthesis in the auditory nuclei and the optic lobes in the first two groups. Gottlieb saw in this result yet another win for probabilistic epigenesis, recalling, "This, of course, implied a bidirectional S[tructure]–F[unction] relationship all the way to the genetic level during the embryonic period, and it meant that genetic activity could be influenced by normally occurring exteroceptive sensory stimulation and thus result in an enhancement of neural maturation" (Gottlieb, 2001, p. 46). Unfortunately, one of Gottlieb's collaborators was overburdened by other commitments, and these results never reached a peer-reviewed journal. Still, Gottlieb brought genes into the bidirectional relationship in 1976, contrasting the predetermined "older view" inspired by the central dogma of molecular biology (DNA → RNA → Protein) with the probabilistic "newer view" (Fig. 3.2(a)). Gottlieb eventually developed this

(a)

Older view: Unidirectional structure–function relationship

Genes → Structural maturation → Function.

Newer view: Bidirectional structure–function relationship

Genes ↔ Structural maturation ↔ Function

(b)

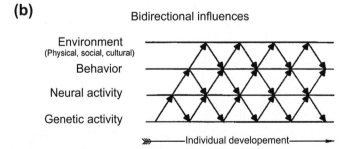

Figure 3.2 (a) The predetermined unidirectional versus the probabilistic bidirectional structure–function relationship *(Reproduced from Gottlieb, 1976, p. 218)*. (b) The bidirectional and coactional relationship. *(From Gottlieb (1992), p. 186).*

early representation of the bidirectionality between genetic activity, structure, and function into his mature vision (Fig. 3.2(b)) of the "completely bidirectional and coactional nature of genetic, neural, behavioral, and environmental influences over the course of individual development" (Gilbert, 2001, p. 50), which pervaded in his later works (Gottlieb, 1992, 1997).

1.3. Oyama and the Ontogeny of Information

DST has been extremely skeptical of the concept of a genetic program and the idea that either the development of behavior or that behavior itself represents "the decoding of the programmed information contained in the DNA code of the fertilized zygote" (Mayr, 1961, p. 1502). One of the founders of developmental psychobiology (a disciplinary precursor to DST), Daniel S. Lehrman, summed up this skepticism by saying that, "although the idea that behavior patterns are 'blueprinted' or 'encoded' in the genome is a perfectly appropriate and instructive way of talking about certain problems of genetics and evolution, it does not in any way deal with the kinds of questions about behavioral development to which it is so often applied" (Lehrman, 1970, p. 35).

The influential work of Susan Oyama, and especially *The Ontogeny of Information* (1985), was a systematic development of this idea. Her work demonstrated how the idea of information was deployed to background the role of nongenetic factors in development and to minimize the impact of accepting that phenotypes develop through the interaction of genes and environment. Oyama pioneered the "parity argument", identifying the criteria used to assign the central causal role in development to genes and showing that these criteria were ignored when they applied to nongenetic factors in development (Oyama, 2000).

The positive aspect of Oyama's program was the idea that developmental information is actually produced during development: information has an ontogeny. Developmental scientists have often compared nativist views of development to the doctrine of preformationism in early modern biology. Instead of a tiny homunculus, the genome contains little "traitunculi" (Schaffner, 1998). Oyama identified the concept of information as the last bastion of preformationism. The causal connections between genes and complex phenotypic traits were indirect and contingent upon many other causal factors. Treating genes as representations of traits, or instructions for making them, reduced the role of these other factors to providing nonspecific support for reading that information. In contrast, Oyama argued that the phenotypic significance of a single developmental factor, genetic or otherwise, was always contextual, conferred upon it by its role in the system as a whole. Locating developmental information in the genome confused a contextual property with an intrinsic property.

The ontogeny of information produced a radical reformulation of the distinction between nature and nurture (Oyama, 2002). Nature and nurture were not interacting causes, as in the conventional idea of gene–environment interaction. They were process and product. Nurture was the interaction between the current state of the organism and its environment: the *organism*–environment interaction of Gottlieb and Lehrman. The nature of the organism at each stage was simply the state of that organism and of its developmental environment, both of which were products of earlier processes of nurture.

1.4. Ford, Lerner, and a "DST"

The phrase DST came from Ford and Lerner (1992), who set a systematic research agenda for developmental psychology that incorporated many of the themes introduced earlier. They defined development itself in a way that placed organism–environment interaction at its core. Development consisted of a series of functional transformations of the organism produced by the interaction of the current state of the person with their current context.

One of the core theses of Ford and Lerner's DST was *developmental contextualism*, which they recognized as closely related to Gottlieb's concept of probabilistic epigenesis (Ford & Lerner, 1992, p. 11). Rather than reduce one level of causal analysis to another, or treat one level as focal and the others as background against which it unfolds, contextualism treated development as a process that proceeded at several levels and treated interaction between levels as the prime focus of research. Another core thesis was *dynamic interactionism*, as opposed to static interactionism. This contrast was closely related to the contrast between organism–environment interaction and gene–environment interaction mentioned earlier. Ford and Lerner stressed that interaction was an ongoing process in which the interactants were themselves transformed, so that what was interacting later in the process depended on the earlier phases of interaction.

Ford and Lerner linked the idea of a developmental *system* to systems theory and cybernetics. The dynamics of the system played an important role in explaining development. This emphasis on a systems level of explanation provides a link back to the ideas of Waddington, with whom we started. Because Waddington's approach was more internalist than contextualist, it may seem odd that he has been regarded so positively by DST. What

Waddington shared with Ford and Lerner, however, was their dynamic interactionism and the realization that this form of explanation depended on a rigorous theory of systems.

Starting in the 1990s, there was an explosion of interest in DST in the philosophy of science, mostly in response to the work of Susan Oyama (Gray, 1992; Griffiths & Gray, 1994; Godfrey-Smith, 2000; Moss, 1992; Robert, 2001). However, this interest was aroused by the implications of DST for causation and explanation in genetics. So while most scientific work in the DST framework has been on behavioral development, and much of it on human development, philosophical discussion of DST has focused on its application to molecular biology, or to developmental biology with its traditional emphasis on embryology (Robert, 2001, 2003, 2004; Stotz, 2006, 2008).

2. WHAT DEVELOPMENTAL SYSTEMS THEORY IS *NOT*

Our review of the contributions of Waddington, Gottlieb, Oyama, and Ford and Lerner provides a sense of what DST is. It is also instructive, however, to focus on what DST is not. And indeed, many of the contributions from the developmentalists discussed earlier took shape in response to other disciplines that they saw as fundamentally flawed. We discuss two such disciplines that were foils for DST: quantitative behavioral genetics and nativist cognitive developmental psychology.

2.1. Quantitative Behavioral Genetics and Its Developmental Critics

A discipline such as quantitative behavioral genetics may be defined either methodologically or sociologically (Hull, 1988, p. 393). Methodologically, quantitative behavioral genetics consists of the application of quantitative genetic methods to behavioral phenotypes. In 1918, Ronald A. Fisher published "The Correlation between Relatives on the Supposition of Mendelian Inheritance" (Fisher, 1918). In the process of demonstrating the compatibility of Mendelian and biometrical models, Fisher also introduced a new statistical concept—variance (Box, 1978, p. 53). In contrast to previous methods of measuring similarity, variance was a measure of difference. The concept was of interest to Fisher because it offered him a means of quantifying genetic and environmental differences and establishing how much each contributed to the total phenotypic variation for a trait in a population. Much of Fisher's career was spent developing statistical methods such as the analysis of variance, tests of statistical significance, and the design of experiments, all with the goal of answering this *how-much question* about the *relative contributions of nature and nurture*. This focus on relative contributions would become a defining feature of quantitative behavioral genetics (Griffiths & Tabery, 2008; Tabery & Griffiths, 2010).

From a more sociological perspective, the emergence of behavioral genetics can be dated to around 1960 (Fuller & Simmel, 1986; Whitney, 1990). This date is chosen to recognize the publication of John Fuller and William Thompson's *Behavior Genetics*. "The time seems ripe", the authors began, "for a modern treatment of the division of knowledge we have called 'behavior genetics'" (Fuller & Thompson, 1960, p. v). A textbook-style treatment, *Behavior Genetics* introduced readers to the basics of genetics and population biology, the methods of analysis of variance and twin studies, and the application of these methods to personality and mental disorders. The pivotal disciplinary event occurred a decade later, in 1970, with the creation of the journal *Behavior Genetics*, along with the founding of the Behavior Genetics Association with Ukranian-American geneticist Theodosius Dobzhansky as its first president (Griffiths & Tabery, 2008; Tabery & Griffiths, 2010).

In later years, advances in molecular biology facilitated the investigation of the role that genes played in the development of a phenotype at the molecular level. When the focus was on humans, however, ethical considerations largely confined behavioral geneticists to traditional quantitative genetic methods. Thus, classical twin and adoption studies continued to be employed to evaluate the relative contributions of different sources of variation, along with gene-hunting studies, that tracked the distribution of genetic markers in families (linkage studies) or populations (association studies) in an attempt to seek out candidate genes associated with behavioral traits (Griffiths & Tabery, 2008; Tabery & Griffiths, 2010).

Also, in later years, quantitative behavioral genetics ventured into the study of development. Developmental behavioral genetics involved the application of the quantitative behavioral genetic methods to developmental data, that is, to repeated observations of the same phenotype at different stages of development—the study of "distributions of individuals developing across time," as Sandra Scarr has characterized the field (Scarr, 1995, p. 158). Scarr argued that developmental behavioral genetics should resemble traditional quantitative behavioral genetics in seeking the causes of phenotypic *differences*, rather than the causes of phenotypes (Scarr, 1992, 1993, 1995). Thus developmental behavioral genetics, according to Scarr, was a discipline that sets out to explain how much of the observed differences between the developmental trajectories of children can be attributed to genetic differences, to differences in environment, to correlations between genes and environment, and so forth (Griffiths & Tabery, 2008; Tabery & Griffiths, 2010).

DST took shape in part as a reaction against the vision of nature and nurture espoused by quantitative behavioral genetics because it was not "truly developmental" (Gottlieb, 2003). Two features of quantitative behavioral genetics have been particularly subject to criticism: (1) its focus on *how-much* questions about relative contribution rather than *how* questions about developmental mechanisms and (2) its dichotomous view of nature and nurture rather than an interactionist view of their dynamic interplay. Waddington nicely captures the difference between quantitative behavioral genetics and the study of development when it comes to how-much questions about relative contributions versus how questions about causal mechanisms. The behavioral geneticists "assume that changes in the genotype produce correlated changes in the adult phenotype, but the mechanism of this correlation need not concern [them]," while, for Waddington, the mechanism is "of crucial importance, since it is the kernel of the whole problem of development" (Waddington, 2012 [1942], p.10). Gottlieb, likewise, argued that behavioral genetics was not truly developmental in the sense that, "[t]he population view of behavioral genetics is not developmental. It is based on the erroneous assumption that the quantitative analysis of the genetic and environmental contributions to individual differences sheds light on the developmental process of individuals" (Gottlieb, 2003, p. 338). Oyama's demand for causal parity relied on identifying the developmental mechanisms involved in ontogeny, which would be missed if the task was statistically partitioning relative contributions to one or the other. In sum, the developmentalists directed attention to *how genes caused traits*, rather than to *how much traits genes caused*.

DSTists were also critical of the reification and dichotomization of nature and nurture inherent in the process of statistically partitioning their relative contributions. While statistically valid, in their view, this procedure utterly misrepresented the actual process of development. Gottlieb's distinction between the unidirectionality of the predetermined epigenesists and the bidirectionality of the probabilistic epigenesists was in this vein. Oyama's theory of the construction and contextualization of information during ontogeny rejects the very idea that nature exists separate from and before nurture. Finally, both Waddington and Ford and Lerner pointed to dynamic interactionism as a process in which nature and nurture do not come together to create ontogeny, but are both partly products of ontogeny.

2.2. Nativist Cognitive Psychology and Its Developmental Critics

DST also defined itself against neonativist cognitive developmental psychology. A good deal of research in contemporary cognitive developmental psychology is devoted to documenting "innate" features of the mind that constrain and/or enable later cognitive development (Carruthers, Laurence, & Stich, 2005, p. 9). Chomsky's "language acquisition device" has served as an exemplar for research on the innate contributions to other psychological domains (Chomsky, 1965). Similar to the language acquisition device, other putatively innate features are thought to embody innate "knowledge" or innate "theories" about specific cognitive domains. For example, the eminent cognitive developmental psychologists Susan Carey and Elizabeth Spelke argue that children possess four domains of innate "core knowledge" that underlies much of their later cognitive development. These domains are "objects, agents, numbers and space" (Carey & Spelke, 1996, p. 517; see also Carey, 2011).

Neonativists support their claims by presenting one or more of three kinds of evidence: (1) the putatively innate features are distinctive of one cognitive domain rather than another, (2) the environment of the child does not contain the right stimuli for the child to learn these features (poverty of the stimulus), or (3) the same features are found in many human cultures. For example, one putatively innate feature of cognition is said to be that living things are subject to a strictly hierarchical classification and that a particular level in that classification—the "generic species"—is associated with "psychological essentialism". People make inferences about members of a generic species that embody the implicit theory that individuals have a hidden "essence" that causes them to have the typical properties of their species (Atran, 1990; Medin & Atran, 2004; Medin & Ortony, 1989; see also Gelman, 2003). To support this claim, studies are presented in which children reason in this way about living things but not about other domains and in which people in different cultures around the world reason in this way about living things.

The DST critique of neonativism is straightforward: "innate" is not an explanatory category. Neonativists simply do not have developmental explanations of the features they document. This is a modern incarnation of the critique of early twentieth century instinct theories by the

Chinese psychologist Zing-yang Kuo (Guō Rènyuǎn), whose work was later to influence Gottlieb (2001). Calling something "innate" or "instinctive" merely raises the question, "How are our instincts acquired?" (Kuo, 1922) For this reason, when Gottlieb identified an apparently "innate" trait in ducks' recognition of maternal vocalizations despite being raised in isolation, he asked *how* this trait was acquired, and undertook experiments to link it back to the embryonic ducks' own vocalizations (see discussion above). A related critique accuses nativists of conflating evolutionary with developmental explanations (Lickliter & Berry, 1990). The evidence for nativism is often evidence that the trait in question is the result of evolutionary design, and one of the many meanings of "innate" is that a trait is the result of evolution (Mameli & Bateson, 2006). So the label "innate" suggests a developmental explanation when the evidence for that label only supports an evolutionary explanation.

Neonativists think that this critique underestimates their understanding of development. They recognize that all traits, including those that they label "innate", result from an interaction of genes and the environment and that psychological development results from the interaction of these innate features with the local developmental environment (Carey & Gelman, 1991; Marcus, 2004). They define their own views against forms of environmentalism that do not recognize any specific biological contribution to psychological development. However, neonativist explanations of the development of "innate" traits engage in just the kind of backgrounding of nongenetic factors that DST sets itself against. Their development is maturational; their epigenesis is predetermined—with environmental factors only permitting the expression of genetic potential.

A striking feature of neonativism is that the innate contribution to psychological development is studied at the level of behavior with little, if any, attention to developmental processes at lower levels of biological organization (Marcus, 2004 is an important exception). This has led to the suggestion that two fundamentally separate issues are conflated in the dispute between DST and neonativism (Perovic & Radenovic, 2011). The first issue is the pattern of gene–environment interaction in the development of the features documented by neonativist research. Neo-nativists claim that the pattern is maturational, with genes as specific causes and environmental factors as merely "permissive" causes (Holtzer, 1968). The second issue is the neonativist claim that later stages of cognitive development are primarily explained by these early-developing features. Most DSTists deny both these claims, but the critiques mentioned earlier only speak directly to the first claim. Conversely, most neonativists make both claims, but neonativist research is primarily about the second claim. The potential therefore exists for a position that not only claims that some later stage in cognitive development is primarily explained by early-developing features that strongly constrain and/or enable later development but also claims that those early developing features emerge in a fully epigenetic manner, with both genetic and environmental factors playing an instructive role.

3. CORE CONCEPTS

The discussions about what DST is and is not allow us to distill out two core concepts that are integral to the DST perspective: epigenesis and developmental dynamics.

3.1. Epigenesis

The term "epigenetics" is derived from the process of epigenesis. It is a continuation of the concept that development unfolds and is not preformed (or ordained), epigenetics is the latest expression of epigenesis.

(Hall, 2011, p. 12).

The idea that development is a process of epigenesis is at the heart of DST. The term "epigenesis" was introduced by the seventeenth century anatomist William Harvey and is derived from the Greek for "upon" and "origin". For the next century and a half, epigenesis referred to the view that the contents of the ovum are relatively simple and that the operation of natural laws on these simple ingredients leads to increased complexity. The alternative view, preformation, saw the egg as a divinely designed Newtonian mechanism that could unfold and reorganize itself to produce an animal. In the eighteenth and nineteenth centuries, epigenesis was strongly associated with the idea of spontaneous generation. If the egg contains simple physical ingredients that develop under the influence of physical laws, why cannot other pieces of matter, under the influence of physical laws, produce life? The idea of spontaneous generation was strongly rejected by nineteenth century biology, partly because of the cell theory ("cells only come from cells") and partly because of the germ theory of disease. Preformationism was recast as the more general doctrine of predeterminism, the idea that development consists of an orderly progression of qualitative change to a predetermined end point. What predeterminism has in common with preformationism is the view that the environment of the egg and physical laws are nonspecific or permissive factors, while all the specific or instructive factors are inside the egg, the nucleus, or the genome (Gottlieb, 2001). Ernst Mayr's view that development is guided by a genetic program is one such predeterminist view (Mayr, 1961).

The term "epigenetics" was coined by Waddington through the fusion of "epigenesis" and "genetics", to refer to the processes by which genotype gave rise to phenotype and to the study of those processes (Waddington, 2012 [1942]). Waddington emphasized that epigenetics was a search for causal mechanisms and suggested that existing knowledge from experimental embryology supported a view of how genes were connected to phenotypes broadly in line with the older idea of epigenesis. The interaction of many genes produced an emergent level of organization that he termed the "epigenotype" (Fig. 3.1) and development was explained by the dynamics of the developmental system at this level. Thus, Waddington's epigenesis was a systems view of development and was also strongly gene centered.

In 1958, the biologist David L. Nanney introduced a narrower understanding of "epigenetics" and gave the word the sense in which it is primarily used in molecular biology today (Haig, 2004). Epigenetics was the study of mechanisms that determine which genome sequences will be expressed in the cell, mechanisms that control cell differentiation and give the cell an identity that is often passed on through mitosis. In the year that Francis Crick first stated his "sequence hypothesis" that the order of nucleotides in DNA determines the order of amino acids in a protein (Crick, 1958), Nanney wrote that, "On the one hand, the maintenance of a 'library of specificities,' both expressed and unexpressed, is accomplished by a template replicating mechanism. On the other hand, auxiliary mechanisms with different principles of operation are involved in determining which specificities are to be expressed in any particular cell. ...they will be referred to as 'genetic systems' and 'epigenetic systems'" (Nanney, 1958, p. 712). This is the picture of gene expression found in contemporary molecular biology.

Epigenetics in both Waddington's and Nanney's senses created opportunities for the environment to play an instructive role in development, even if those opportunities were often ignored. Waddington's epigenotype was a global expression of the genetic causal factors that influence development. The effect of changing any one gene depended on how it interacted with the rest of the system. The epigenotype as a whole interacted with the environment to determine the phenotype. DST expanded the epigenotype to include nongenetic factors that influence development. The expanded epigenotype, or developmental system, was a global expression of the causal factors that influenced development. It still did not determine a unique phenotype, both because development was a probabilistic process and because development was plastic by design. So the environment figured in two ways in DST: (1) first by supplying the background to normal development (the 'ontogenetic niche', see below) and thus partly constituting the developmental system and (2) second by supplying the variable parameters of that system and so determining which particular course development took. It is sometimes overlooked that in Waddington's original picture genes also played two roles. Genes collectively determined the shape of the developmental landscape and gene *mutations* also threw development down one valley ("creode") or another.

Epigenetics in Nanney's sense also created a space for the environment: "As the past 70 years made abundantly clear, genes do not control development. Genes themselves are controlled in many ways, some by modifications of DNA sequences, some through regulation by the products of other genes and/or by [the intra- or extra-cellular] context, and others by external and/or environmental factors" (Hall, 2011, p. 9). The regulated expression of the coding regions of the genome depended on the mechanisms that differentially activated and selected the information in coding sequences depending on context. Biological information was distributed between the coding regions in the genome and regulatory mechanisms, and the specificity manifested in gene products was the result of a process of "molecular epigenesis" (Griffiths & Stotz, 2013; Stotz, 2006). It was often assumed that the additional information provided by epigenetic regulation traced back to some feature or other of the whole genome sequence. The alternative view was that the environment, acting through epigenetic regulatory mechanisms, played an instructive role in regulating gene expression.

Much of the recent interest in "epigenetic inheritance" has been motivated by the desire to document the instructive role of environmental factors in this manner (Jablonka & Lamb, 1995, 2005; Jablonka & Raz, 2009). The phrase "epigenetic inheritance" has two senses, corresponding to the two senses of "epigenetic" outlined earlier. Epigenetic inheritance in the narrow sense is the inheritance of genome expression patterns across generations (e.g., through meiosis) in the absence of a continuing stimulus (Holliday, 1987). Epigenetic inheritance in the broad sense is the inheritance of phenotypic features via any causal mechanism other than the inheritance of nuclear DNA. To avoid confusion, we refer to this broader sense as "exogenetic inheritance". The developmental psychobiologists Meredith West and Andrew King introduced the term "ontogenetic niche" to refer collectively to the products of exogenetic heredity (West & King, 1987).

The ontogenetic niche, or developmental niche, contained the nongenetic developmental factors required for normal development. That is to say it contained the factors omitted from Waddington's original "epigenotype". The full developmental system consisted of the organism (initially the fertilized egg) and the developmental niche. Sixty years of work in developmental psychobiology documents the instructive role of the developmental niche in behavioral development (Michel & Moore, 1995). And, recent work has shown how these environmental factors operate via epigenetic mechanisms of genome regulation (Meaney, 2001, 2004).

The developmental niche concept reflects the central feature of DST that Ford and Lerner called "developmental contextualism" (Ford & Lerner, 1992). Organisms reproduce themselves by reproducing the context of development. The stability of inheritance is not only explained by the insensitivity of development to context, as Waddington's concept of canalization suggested, but also by the active reproduction of context by the parental generation. However, organisms are plastic, as well as stable. One advantage of DST over traditional interactionism is that it can recognize that the evolutionary functions of heredity systems combine the stable transmission of phenotypes across the generations and plastic responses to fluctuating environments (Bateson & Gluckman, 2011). The very same mechanisms can serve both functions, depending on the particular settings of their parameters. "Parental effects" in evolutionary ecology, for example, are mechanisms for producing both correlations and anticorrelations between parent and offspring phenotypes to match fluctuating environments (Maestripieri & Mateo, 2009; Mousseau & Fox, 1998). They are mechanisms for the intergenerational transmission of phenotypes, but they exist in the service of plasticity, not fixity (parental effects have also attracted attention in recent medical research (Gluckman & Hanson, 2005a, 2005b; Gluckman et al., 2009)).

Two themes have emerged in our discussion of epigenesis: (1) the importance of the systems level in the analysis of development and (2) developmental contextualism. Some "developmentalist" visions in biology emphasize the first of these themes more than the second. For example, the successful new discipline of evolutionary developmental biology ("evo-devo") retains a vision of development much closer to that of Waddington than that of Gottlieb. The history of life on Earth is the history of changes in gene regulatory networks via changes in the DNA sequence itself (Arthur, 1997; Carroll, Grenier, & Weatherbee, 2001). Scott Gilbert's recent call for an "ecological developmental biology" represents an attempt to introduce something like developmental contextualism into evolutionary developmental biology (Gilbert, 2001; Gilbert & Epel, 2009).

We described earlier how Gottlieb's concept of probabilistic epigenesis added the environmental context to Waddington's vision of a developmental system. But Gottlieb also added the concept of coaction. Higher levels of developmental organization are not simply the expression of lower levels, but act on those lower levels. The organism itself is an agent in development and not just a product of development. It is to this theme that we now turn.

3.2. Developmental Dynamics

The idea that development is a dynamic process is central to DST (Kuo, 1967, pp. 55–58; Gottlieb, 1970; Lerner, 1978). In a seminal paper, Daniel Lehrman wrote that "The interaction out of which the organism develops is *not* one, as is so often said, between heredity and environment. It is between *organism* and environment! And the organism is different at each stage of its development" (Lehrman, 1953, p. 345, emphasis in original). In a dynamic approach, development at each stage builds on the results of development at an earlier stage. The components produced by interaction at one stage of development are the components that do the interacting at a later stage. Lehrman emphasizes change in the organism, but the environment can also change as a result of development. For example, in Celia Moore's well-known work on sexual development in male rat pups, male sexual development depends on differential licking of the genital area of male and female pups by the mother. But her response to male pups depends on differences in their urine, which are the result of earlier processes of sexual differentiation (Moore, 1984, 1992). The presence of this environmental influence is a feed forward from earlier development in the pup itself.

Dynamic interaction can be seen in the figure by Gottlieb depicting bidirectional influences in development (Fig. 3.2(b)). Causal influences from any one level of analysis feed forward to the other levels of analysis. The influence of one level on a second level can help to produce the later influence of the second level on the first. In his own work, Gottlieb documented the reciprocal influence of structure (e.g., morphology) on function (e.g., behavior), especially the influence of prenatal activity in birds on the development of neural structure.

Ford and Lerner contrast dynamic interaction with a more conventional conception of interaction associated with analysis of variance techniques, such as those used in behavioral genetics (Ford & Lerner, 1992).

In what we might call "static interactionism", the values of two variables measured before development, such as shared genes and shared environment, are shown to interact with one another (there are ambiguities that we will not pursue here, see Tabery, 2007). In contrast, dynamic interaction must be studied as a temporally extended process. In the next section, we will discuss how the fact that dynamic interaction is studied over time makes it easier for a dynamic interactionist approach to recognize bidirectional influences.

If interaction is a dynamic process, then the temporal dynamics of the interaction may play an independent role in explaining the outcome. The introduction of dynamical systems theory (here written as DyST to avoid confusion) adds an additional dimension to developmental explanation. We note that there is nothing about the basic idea of dynamical interaction that requires the use of DyST. The kinds of systems pictured in Fig. 3.2(b) can be analyzed as sequential mechanisms in the sense discussed in the next section. Whether they need to be analyzed as dynamic mechanisms depends on the specific causal structure. However, some developmental psychologists have made extensive use of DyST to explain developmental outcomes by mapping the dynamics of the system as it evolves over time in multiple actual or simulated "runs" and establishing that the outcome is an attractor for the system.

This kind of explanation was used by Esther Thelen in her studies of the emergence of coordinated stepping movements in human infants. Because this behavior emerges *before* the child begins to walk, it satisfies one of the traditional criteria for innate behavior—"prefunctionality" (Mameli & Bateson, 2006). Thelen argues, however, that this behavior emerges not because it is planned or programmed somewhere in the genes, but from the previous dynamical state of the system as one of its parameters varies. This can be demonstrated in "microgenesis" experiments that bring about the emergence of a new behavior before its normal period of "maturation" by manipulating some parameter of the system. Coordinated stepping in infants results from the interaction of motor patterns present from earliest infancy but suppressed during an intermediate period by the weight of their limbs. When the available muscular force catches up with the weight of the limbs, the old pattern results in what appears as the "spontaneous" emergence of effective stepping behavior. However, this and other aspects of walking can be brought out earlier in development by removing simple physical constraints (Thelen & Ulrich, 1991).

Thelen summed up her DyST approach to child development as follows:

...behavior and cognition, and their changes during ontogeny, are not represented anywhere in the system beforehand either as dedicated structures, or symbols in the brain, or as codes in the genes. Rather, thought and behavior are 'softly assembled' as dynamical patterns of activity that arise as a function of the intended task at hand and an individual's 'intrinsic dynamics' [by which is meant] the preferred states of the system given its current architecture and previous history of activity...

(Thelen, 1995, p. 76; see also Thelen & Smith, 1994).

DyST explanations have often been criticized for being merely descriptive and not truly explanatory. The fact that a state is an "attractor" for the system is determined solely by observing its occurrence in multiple runs of the system but the existence of this attractor is treated as an explanation for the system being in that state (for an extended discussion see chapter six of Clark, 1997). In the next section, we will see how dynamical explanations have been incorporated into recent accounts of mechanistic explanation.

4. MECHANISMS—A PHILOSOPHY FOR DEVELOPMENTAL SYSTEMS THEORY

Waddington identified the mechanism linking genotype to phenotype as "the whole problem of development" (Waddington, 2012 [1942], p. 10). Likewise, Nanney distinguished the genetic system from the epigenetic system based upon what the mechanism of each did. Why? What is it about a mechanism and mechanical explanation that scientists within the developmental perspective find so attractive?[3] In this section, we introduce the philosophy of mechanism, an influential school of thought in the philosophy of science, and apply it to DST in order to characterize *how* DST explains. In particular, we apply the philosophy of mechanism to (1) Gottlieb's concept of bidirectionality, (2) Ford and Lerner's concept of dynamic interactionism, and (3) DST's vision of truly developmental explanations.

4.1. The Philosophy of Mechanism

Throughout the twentieth century, the philosophy of science was dominated by a theory of scientific explanation that took explanations to consist of derivations from laws of nature. On this "deductive-nomological" account,

---

[3]For other examples of this focus on mechanism and mechanical explanation, see Feldman & Lewontin, 1975, pp. 1167–1168; Moore, 2008, p. 382.

a phenomenon was explained by deducing it from a set of premises one or more of which was a law of nature (Hempel & Oppenheim, 1948). This theory of scientific explanation seemed satisfactory in the physical sciences where laws of nature were readily available, but the theory did not apply so neatly to the biological sciences where laws of nature remained elusive. In the face of this dilemma, philosophers of science turned to the actual practice of biological science to assess how explanations worked there and what they consistently found were appeals to mechanisms (Bechtel & Abrahamsen, 2005; Bechtel & Richardson, 1993; Glennan, 1996, 2002; Machamer, Darden, & Craver, 2000).

The resulting philosophy of mechanism offers a new theory of scientific explanation: scientists explain a phenomenon by identifying and manipulating the variables in the mechanisms responsible for that phenomenon, thereby determining how those variables are situated in and make a difference in the mechanism; the explanation ultimately amounts to the elucidation of how those variables act and interact in the mechanism to produce the phenomenon under investigation. The philosophy of mechanisms is meant to capture how scientists answer questions such as the following: How do plants convert solar energy into chemical energy (Tabery, 2004)? How do rats form spatial memories of their environments (Craver, 2007)? And how does the cell produce proteins (Darden, 2006)? Such questions are answered by elucidating the mechanisms responsible for photosynthesis, spatial map formation, and protein synthesis, respectively. This theory of scientific explanation as mechanism-elucidation is very much in line with DST.

4.2. Bidirectionality

Gottlieb, throughout his career, advocated a bidirectional probabilistic epigenesis. Figure 3.2(b) shows how Gottlieb visualized this bidirectional relationship: multiple levels (genetic, neural, behavioral, environmental), with arrows pointing up and down, all moving through time to collectively constitute individual development. The arrows pointing upward and representing genetic influence on the higher levels are uncontroversial, but how should we make sense of the arrows from higher levels exerting downward influence? Such top-down causation is often dismissed as spooky because it appears to involve mysterious forces exerted by wholes upon their parts. Philosophers of mechanism Carl Craver and William Bechtel, however, disagree (Craver & Bechtel, 2007). According to Craver and Bechtel, the right way to understand top–down causation is by uniting intralevel causal relations with interlevel *constitutive* relations. "Level", for Craver and Bechtel, refers to a level of mechanism, so the idea is that causal relations exist within mechanisms at a given level, but mechanisms at a given level are constitutively related to mechanisms at other levels. Thus, for Craver and Bechtel, top-down causation arises via mechanistically mediated effects: "As the mechanism as a whole is put into new conditions, it is organized such that its components change with those conditions" (Craver & Bechtel, 2007, p. 561).

We can apply Craver and Bechtel's account of top-down causation to Gottlieb's emphasis on bidirectionality by drawing on Gottlieb's own research. Consider his description of the duck embryos exposed to extravisual stimulation or species-specific vocalizations, which then generated enhanced protein synthesis at the genetic level (discussed earlier, Gottlieb, 2001). The developing duck embryos consisted of many levels of mechanism: mechanisms of protein synthesis at the genetic level, mechanisms of visual and auditory perception at the neural level, mechanisms of vocalization at the behavioral level, and mechanisms of light and sound emission at the environmental level. On Craver and Bechtel's account, there was top-down causation in this case via the mechanistically mediated effects: As the mechanism as a whole (i.e., the duck embryo) was put into new conditions (i.e., the environment with extra visual/auditory stimulation), it was organized such that its components (i.e., genes) changed (i.e., gene expression) with those conditions. The mysterious idea of downward *causation* is replaced by the very unmysterious idea that large causal mechanisms are *constituted* of parts that are themselves smaller causal mechanisms. The operation of the larger mechanism produces changes in its parts.

4.3. Dynamic Interactionism

Gottlieb's concept of bidirectionality is related to Ford and Lerner's concept of dynamic interaction; both emphasize the reciprocal relationship between the various levels of mechanism of an organism. Ford and Lerner added to this an emphasis on the dynamic way parts interact in a system. Ford and Lerner contrasted sequential, linear causality with the idea of a causal field, wherein a change in any variable comes about as a consequence of the operation of the entire field of variables (Ford & Lerner, 1992, p. 57).

The philosophy of mechanism also has resources for making sense of this appeal to dynamical explanations. Bechtel, along with Adele Abrahamsen, has advanced the philosophy of mechanism by developing it so as to capture *dynamical mechanistic explanations* (Bechtel & Abrahamsen, 2010,

2011). According to Bechtel and Abrahamsen, scientists begin to explain a phenomenon by decomposing the system in order to identify the parts, operations, and organization that generate the phenomenon. Take circadian rhythm—the ability of organisms to keep track of day/night cycles—which is found in organisms across the plant and animal kingdoms. Starting in the 1970s, researchers honed in on the suprachiasmatic nucleus (SCN) as playing a critical role circadian rhythm; lesions that removed the SCN left an organism arrhythmic, and transplant experiments that moved the SCN from a donor hamster with an abbreviated rhythm to a recipient hamster with a normal rhythm led to an abbreviated rhythm in the recipient hamster. Bechtel and Abrahamsen point out, however, that scientists do not settle for decomposition; the next step is recomposition, where the task is to put the parts back together in order to produce the phenomenon to be explained. They typically do this with computational modeling that elucidates the spatial organization of parts and the temporal organization of operations. Starting in the 1990s, scientists constructed a series of computer models that identified how the circadian rhythm arose from the synchronization of individual neurons; they also explained phenomena associated with desynchronization—jet lag (Bechtel & Abrahamsen, 2010). So, for Bechtel and Abrahamsen, dynamic mechanistic explanations involve both a reductive decomposition and an integrative recomposition, and it is in the integrative recompositional phase that the dynamic interactions become apparent.

Ford and Lerner's flagship example of a dynamically interacting system is a perfect example of Bechtel and Abrahamsen's dynamical mechanistic explanation. The example is a beating heart. Each cardiac muscle, Ford and Lerner explain, is an autonomous oscillator displaying rhythmic contraction patterns, but the heart as a whole beats as a result of the synchronization of the individual cardiac cells. They also note that this synchronization can break down (fibrillation), and the organism will suffer a fate worse than jet lag if this desynchronization is not corrected—death. Ford and Lerner conclude, "As with the network of power generators, the synchronization results from the influence of the entire organization of heart muscles on component muscle cells, not from the singular influence of individual cells on one another" (Ford & Lerner, 1992, p. 58). This explanation of a beating heart involves both a decompositional identification of individual cardiac cells and a recompositional analysis of how these parts dynamically interact to synchronize such that the heart as a whole can beat and pump blood.

4.4. Truly Developmental Explanations

Gottlieb criticized quantitative behavioral genetics for its inability to provide "truly developmental" explanations. To make sense of Gottlieb's complaint, we must first give an account of quantitative behavioral genetic explanations and then assess what those explanations lack. Quantitative behavioral geneticists, recall, seek to partition the relative contributions of nature and nurture. They do this by studying a trait in a population and determining how much actual differences in genes and how much actual differences in environment contribute to actual differences in the trait. In humans, they often study twins, adoptees, and blood relatives because then the actual differences (or lack thereof) in genes and environment can be specified. In some cases, a single actual difference in either genes or environment fully accounts for the actual difference in the trait; Waters (2007) refers to such a factor as "*the* actual difference makers". Often, however, there are multiple actual differences in genes and the environment that account for the actual difference in the trait; Waters labels each "*an* actual difference maker". When studying actual difference makers, the goal is linking up actual difference(s) in input with actual difference(s) in output; the mechanism linking them, however, is often left uninvestigated. Waters' case study involves the work of Thomas Hunt Morgan and his research on the genetics of *Drosophila*, such as on the purple eye gene. "The explanatory reasoning here does not depend on identifying the material make-up, mode of action, or general function of the underlying purple gene," Waters acknowledges (Waters, 2007, p. 558). Notice how this admission matches on to Waddington's distinction between the study of inheritance, which "merely assume[s] that changes in genotype produce correlated changes in the adult phenotype," and the study of development, which investigates "the mechanism of this correlation." Quantitative behavioral genetics falls into the former category; it seeks out actual difference makers and attempts to quantify the relative contribution made by that/those actual difference makers for a given trait in a population. Thus, a quantitative behavioral genetic explanation amounts to answering a how-much question about actual difference makers.

Actual difference makers, however, are not the only causally relevant difference makers between a given input and a given output in a population. There are also what we can call "potential difference makers". These are causes that, *had they varied,* would have led to variation in the output, but since *they did not vary* they contributed no variation. So they are difference makers, but they are only potential difference makers because there is not an actual difference that can generate an actual difference in the population. In the mechanism between input and output, there are many potential difference makers that do not actually vary. Nevertheless, identifying these potential difference makers and elucidating how they are situated in the mechanism is crucially important for understanding how the mechanism works and which possible interventions on the mechanism would generate different outputs. Take, for example, exposure to species-specific vocalizations during embryonic development. If all the duck embryos in a clutch are exposed to the same maternal vocalizations and respond with the same vocalizations of their own, then none of the actual differences in the duck hatchlings can be attributed to the vocalizations because there was no actual difference in the vocalizations. So vocalizations would not count as an *actual* difference maker in such a scenario. What Gottlieb discovered, however, was that vocalizations nevertheless were *potential* difference makers. His experiments that isolated and devocalized the duck embryos in turn led to differences in the duck hatchlings (i.e., an inability to recognize species-specific vocalizations). Such an experiment, for Gottlieb, was one step in generating a "truly developmental" explanation of avian vocalization. Thus, a *truly developmental explanation* of a phenomenon should be understood as (1) the identification of both the actual *and* the potential difference makers in the mechanism(s) responsible for the phenomenon, (2) the distribution of the actual *and* potential difference makers in a population, and (3) the elucidation of how those difference makers make or would make their difference in the mechanism(s). Gottlieb's criticism of quantitative behavioral genetics was that its methodologically enforced restriction to actual difference makers prevented it from generating truly developmental explanations.

5. CONCLUSIONS

DST has a rich history, and today, different researchers draw on different aspects of this theoretical tradition. But the tradition has some strong unifying themes. We have identified two core concepts of DST, epigenesis and developmental dynamics. First, development is a truly epigenetic process. The outcomes of development are explained at the systems level, and developmental is influenced by the context in which it unfolds, leading to an extensive conception of that system. Second, development is a dynamic process: the interactants at one stage are the products of earlier stages of development.

We have interpreted the explanatory structure of developmental systems using some ideas from recent philosophy of science. The "truly developmental" explanations at which DST aims are mechanistic explanations, and often dynamical mechanistic explanations, of the developmental potential of the system. They are mechanistic because they explain developmental phenomena by displaying how the components of the developmental system are arranged so as to produce those phenomena. They are dynamic mechanistic explanations when the phenomena to be explained are not the immediate consequence of the arrangements of the components but emerge from the dynamic operation of the mechanism. Finally, truly developmental explanations do not merely explain why one individual differs from another. They explain the potential of the developmental system to produce these and other outcomes.

REFERENCES

Arthur, W. (1997). *The origin of animal body plans: A study in evolutionary developmental biology.* Cambridge: Cambridge University Press.
Atran, S. (1990). *Cognitive foundations of natural History: Towards an anthropology of science.* Cambridge and New York: Cambridge University Press.
Bateson, P., & Gluckman, P. D. (2011). *Plasticity, robustness, development and evolution.* Cambridge: Cambridge University Press.
Batson, P., & Logan, C. (2007). Invited address: Gilbert Gottlieb (1929–2006). *Developmental Psychobiology, 49,* 446–449.
Bechtel, W., & Abrahamsen, A. (2005). Explanation: a mechanist alternative. *Studies in the History and Philosophy of Biological and Biomedical Sciences, 36,* 421–441.
Bechtel, W., & Abrahamsen, A. (2010). Dynamic mechanistic explanation: computational modeling of circadian rhythms as an exemplar for cognitive science. *Studies in the History and Philosophy of Science, 41,* 321–333.
Bechtel, W., & Abrahamsen, A. (2011). Complex biological mechanisms: cyclic, oscillatory, and autonomous. In C. A. Hooker (Ed.), *Philosophy of complex systems: Handbook of the philosophy of science.* (Vol. 10). New York: Elsevier.
Bechtel, W., & Richardson, R. C. (1993). *Discovering complexity: Decomposition and localization as strategies in scientific research.* Princeton: Princeton University Press.
Box, J. F. (1978). *R. A. Fisher: The life of a scientist.* New York: Wiley.
Carey, S. (2011). *The origin of concepts.* New York: Oxford University Press.
Carey, S., & Gelman, R. (Eds.), (1991). *The epigenesis of mind: Essays on biology and cognition.* Hillsdale, NJ: Lawrence Erlbaum.
Carey, S., & Spelke, E. (1996). Science and core knowledge. *Philosophy of Science, 63,* 515–533.
Carroll, S. B., Grenier, J. K., & Weatherbee, S. D. (2001). *From DNA to diversity: molecular genetics and the evolution of animal design.* Oxford: Blackwells.
Carruthers, P., Laurence, S., & Stich, S. (Eds.), (2005). 2009). *The innate mind.* (Vols. 1–3). Oxford: Oxford University Press.
Chomsky, N. (1965). *Aspects of the theory of syntax.* Cambridge, MA: MIT Press.
Clark, A. (1997). *Being there: Putting brain, body and world together again.* Cambridge, M.A: MIT Press.
Counce, S. J., & Waddington, C. H. (1972). *Developmental systems: insects.* (Vol. 1). London: Academic Press.

Counce, S. J., & Waddington, C. H. (1973). *Developmental Systems: Insects*. (Vol. 2). London: Academic Press.

Craver, C. F. (2007). *Explaining the brain: Mechanisms and the mosaic unity of neuroscience*. Cambridge: Cambridge University Press.

Craver, C. F., & Bechtel, W. (2007). Top-Down Causation without Top-Down Causes. *Biology and Philosophy, 22*, 547–563.

Crick, F. H.C. (1958). On protein synthesis. *Symposia of the Society for Experimental Biology, 12*, 138–163.

Darden, L. (2006). *Reasoning in biological discoveries: Essays on mechanisms, interfield relations, and anomaly resolution*. Cambridge: Cambridge University Press.

Dewey, J., & Bentley, A. F. (1949). *Knowing and the known*. Boston: Beacon Press.

Feldman, M.W., & Lewontin, R. C. (1975).The heritability hang–up. *Science, 190*, 1163–1168.

Fisher, R. A. (1918). The correlation between relatives on the supposition of Mendelian inheritance. *Transactions of the Royal Society of Edinburgh, 52*, 399–433.

Ford, D. H., & Lerner, R. M. (1992). *Developmental systems theory: An integrative approach*. Newbury Park, CA: Sage.

Fuller, J. L., & Simmel, E. C. (1986). Trends in behavior genetics: 1960–1985. In J. L. Fuller & E. C. Simmel (Eds.), *Perspectives in behavior genetics* (pp. 1–27). Hillsdale, NJ: Lawrence Erlbaum Associates, Publishers.

Fuller, J. L., & Thompson, W. R. (1960). *Behavior genetics*. New York: Wiley.

Gelman, S. (2003). *The essential child: Origins of essentialism in everyday thought*. New York: Oxford University Press.

Gilbert, S. F. (2001). Ecological developmental biology: developmental biology meets the real world. *Developmental Biology, 233*, 1–22.

Gilbert, S. F. (2012). Commentary: "The Epigenotype" by C. H. Waddington. *International Journal of Epidemiology, 41*, 20–23.

Gilbert, S. F., & Epel, D. (2009). *Ecological developmental biology: Integrating epigenetics, medicine and evolution*. Sunderland, MA: Sinauer Associates.

Glennan, S. (1996). Mechanisms and the nature of causation. *Erkenntnis, 44*, 49–71.

Glennan, S. (2002). Rethinking mechanistic explanation. *Philosophy of Science, 69*, S342–S353.

Gluckman, P. D., & Hanson, M.A. (2005a). *The fetal matrix: Evolution, development and disease*. Cambridge, UK: Cambridge University Press.

Gluckman, P. D., & Hanson, M. A. (Eds.), (2005b). *Developmental origin of health and disease*. Cambridge: Cambridge University Press.

Gluckman, P. D., Hanson, M. A., Bateson, P., Beedle, A. S., Law, C. M., Bhutta, Z. A., et al. (2009). Towards a new developmental synthesis: adaptive developmental plasticity and human disease. *Lancet, 373*(9675), 1654–1657.

Godfrey-Smith, P. (2000). Explanatory symmetries, preformation, and developmental systems theory. *Philosophy of Science, 67*, S322–S331. Supplement. Proceedings of the 1998 Biennial Meetings of the Philosophy of Science Association. Part II: Symposia Papers.

Gottlieb, G. (1965). Imprinting in relation to parental and species identification by avian neonates. *Journal of Comparative and Physiological Psychology, 59*, 345–356.

Gottlieb, G. (1970). Conceptions of prenatal behavior. In L. R. Aronson, E. Tobach, D. S. Lerman & J. S. Rosenblatt (Eds.), *Development and evolution of behavior: Essays in memory of t. C. Schneirla* (pp. 111–137). San Francisco: W. H. Freeman.

Gottlieb, G. (1971). *Development of species identification in birds: An inquiry into the prenatal determinants of perception*. Chicago: University of Chicago Press.

Gottlieb, G. (1976). Conceptions of prenatal development: behavioral embryology. *Psychological Review, 83*, 215–234.

Gottlieb, G. (1992). *Individual development and evolution: The genesis of novel behavior*. New York: Oxford University Press.

Gottlieb, G. (1997). *Synthesizing nature-nurture: Prenatal roots of instinctive behavior*. Mahwah, NJ: Erlbaum.

Gottlieb, G. (2001). A developmental psychobiological systems view: early formulation and current status. In S. Oyama, P. E. Griffiths & R. D. Gray (Eds.), *Cycles of contingency: Developmental systems and evolution* (pp. 41–54). Cambridge, Mass: The MIT Press.

Gottlieb, G. (2003). On making behavior genetics truly developmental. *Human Development, 46*, 337–355.

Gottlieb, G., & Vandenbergh, J. G. (1968). Ontogeny of vocalization in duck and chick embryos. *Journal of Experimental Zoology, 168*, 307–325.

Gray, R. D. (1992). Death of the gene: developmental systems strike back. In P. E. Griffiths (Ed.), *Trees of life: Essays in the philosophy of biology*. Dordrecht: Kluwer.

Griffiths, P. E., & Gray, R. D. (1994). Developmental systems and evolutionary explanation. *Journal of Philosophy, XCI*(6), 277–304.

Griffiths, P. E., & Stotz, K. (2013). *Genetics and philosophy: An introduction*. New York: Cambridge University Press.

Griffiths, P. E., & Tabery, J. (2008). Behavioral genetics and development. *New Ideas in Psychology, 26*, 332–352.

Haig, D. (2004). The (dual) origin of epigenetics. *Cold Spring Harbor Symposia on Quantitative Biology, 69*, 67–70.

Haig, D. (2012). Commentary: the epidemiology of epigenetics. *International Journal of Epidemiology, 41*, 13–16.

Hall, B. K. (2011). A brief history of the term and concept of epigenetics. In B. Hallgrimsson & B. K. Hall (Eds.), *Epigenetics: Linking genotype and phenotype in development and evolution*. Berkeley: University of California Press.

Hempel, C. G., & Oppenheim, P. (1948). Studies in the logic of explanation. *Philosophy of Science, 15*, 135–175.

Holliday, R. (1987). The inheritance of epigenetic defects. *Science, 238*(4824), 163–170.

Holtzer, H. (1968). Induction of chondrogenesis: a concept in quest of a mechanism. In R. Fleischmajer & R. E. Billingham (Eds.), *Epithelial–mesenchymal interactions* (pp. 152–164). Baltimore: The Williams and Wilkins Company.

Hull, D. L. (1988). *Science as a process: An evolutionary account of the social and conceptual development of science*. Chicago: University of Chicago Press.

Jablonka, E., & Lamb, M. J. (1995). *Epigenetic inheritance and evolution: The Lamarckian dimension*. Oxford, New York, Tokyo: Oxford University Press.

Jablonka, E., & Lamb, M. J. (2005). *Evolution in four dimensions: Genetic, epigenetic, behavioral, and symbolic variation in the history of life*. Cambridge, Mass: MIT Press.

Jablonka, E., & Lamm, E. (2012). Commentary: the epigenotype—a dynamic network view of development. *International Journal of Epidemiology, 41*, 16–20.

Jablonka, E., & Raz, G. (2009). Transgenerational epigenetic inheritance: prevalence, mechanisms, and implications for the study of heredity and evolution. *Quarterly Review of Biology, 84*(2), 131–176.

Kuo, Z. Y. (1922). How are our instincts acquired? *Psychological Review, 29*, 344–365.

Kuo, Z. Y. (1967). *The dynamics of behavioral development*. New York: Random House.

Lehrman, D. S. (1953). Critique of Konrad Lorenz's theory of instinctive behavior. *Quarterly Review of Biology, 28*(4), 337–363.

Lehrman, D. S. (1970). Semantic & conceptual issues in the nature-nurture problem. In D. S. Lehrman (Ed.), *Development & evolution of behaviour*. San Francisco: W. H. Freeman and Co.

Lerner, R. M. (1978). Nature, nurture and dynamic interactionism. *Human Development, 21*, 1–20.

Lickliter, R., & Berry, T. (1990). The phylogeny fallacy. *Developmental Review, 10*, 348–364.

Lickliter, R., & Harshaw, C. (2010). Canalization and malleability reconsidered: the developmental basis of phenotypic stability and variability. In Kathryn E. Hood, Carolyn Tucker Halpern, Gary Greenberg & Richard M. Lerner (Eds.), *Handbook of developmental science, behavior, and genetics* (pp. 491–525). Malden, MA: Blackwell Publishing.

Machamer, P. K., Darden, L., & Craver, C. F. (2000). Thinking about mechanisms. *Philosophy of Science*, *67*, 1–25.

Maestripieri, D., & Mateo, J. M. (Eds.), (2009). *Maternal effects in mammals*. Chicago: The University of Chicago Press.

Mameli, M., & Bateson, P. (2006). Innateness and the sciences. *Biology and Philosophy*, *21*(2), 155–188.

Marcus, G. F. (2004). *The birth of the mind: How a tiny number of genes creates the complexities of human thought*. New York: Basic Books.

Mayr, E. (1961). Cause and effect in biology. *Science*, *134*(3489), 1501–1506.

Meaney, M. J. (2001). Maternal care, gene expression, and the transmission of individual differences in stress reactivity across generations. *Annual Review Neuroscience*, *24*, 1161–1192.

Meaney, M. J. (2004). The nature of nurture: maternal effects and chromatin remodelling. In J. T. Cacioppo & G. G. Berntson (Eds.), *Essays in social neuroscience*. Cambridge, MA: MIT Press.

Medin, D., & Atran, S. (2004). The native mind: biological categorization and reasoning in development and across cultures. *Psychological Review*, *111*, 960–983.

Medin, D., & Ortony, A. (1989). Psychological essentialism. In S. Vosniadou & A. Ortony (Eds.), *Similarity and analogical reasoning*. Cambridge: Cambridge University Press.

Michel, G. F., & Moore, C. L. (1995). *Developmental psychobiology: An interdisciplinary science*. Cambridge, MA: MIT Press.

Miller, D. (2006 10 June). *Gilbert Gottlieb: developmental psychologist and theorist*. Interview of Gottlieb by Miller [video File]. Retrieved from: Transcript retrieved from http://icube.uconn.edu/GGvideo.movhttp://web.uconn.edu/millerd/GGTranscript.pdf.

Moore, C. L. (1984). Maternal contributions to the development of masculine sexual behavior in laboratory rats. *Developmental Psychobiology*, *17*, 346–356.

Moore, C. L. (1992). The role of maternal stimulation in the development of sexual behavior and its neural basis. *Annals of the New York Academy of Sciences*, *662*, 160–177.

Moore, D. S. (2008). Individuals and populations: How biology's theory and data have interfered with the integration of development and evolution. *New Ideas in Psychology*, *26*, 370–386.

Moss, L. (1992). A kernel of truth? on the reality of the genetic program. *Philosophy of Science Association Proceedings*, *1*, 335–348.

Mousseau, T. A., & Fox, C. W. (Eds.), (1998). *Maternal effects as adaptations*. New York, Oxford: Oxford University Press.

Nanney, D. L. (1958). Epigenetic control systems. *Proceedings of the National Academy of Sciences*, *44*(7), 712–717.

Oyama, S. (1985). *The ontogeny of information: Developmental systems and evolution*. Cambridge: Cambridge University Press.

Oyama, S. (2000). Causal democracy and causal contributions in developmental systems theory. *Philosophy of Science*, *67*(s1), S332.

Oyama, S. (2002). The nurturing of natures. In A. Grunwald, M. Gutmann & E. M. Neumann-Held (Eds.), *On human nature: Anthropological, biological and philosophical foundations*. Berlin: Springer.

Perovic, S., & Radenovic, L. (2011). Fine-tuning nativism: the 'nurtured nature' and innate cognitive structures. *Phenomenology and the Cognitive Sciences*, *10*(3), 399–417.

Robert, J. S. (2001). Interpreting the homeobox: metaphors of gene activation in development and evolution. *Evolution and Development*, *3*, 287–295.

Robert, J. S. (2003). Developmental systems theory. In B. K. Hall & W. M. Olsen (Eds.), *Keywords and concepts in evolutionary developmental biology*. Cambridge, MA and London: Harvard University Press.

Robert, J. S. (2004). *Embryology, epigenesis and evolution: Taking development seriously*. Cambridge, New York: Cambridge University Press.

Robertson, A. (1977). Conrad Hal Waddington, 8 November 1905–26 September 1975. *Biographical Memoirs of Fellows of the Royal Society*, *23*, 575–622.

Scarr, S. (1992). Developmental theories for the 1990s: development and individual differences. *Child Development*, *63*, 1–19.

Scarr, S. (1993). Biological and cultural diversity: the legacy of Darwin for development. *Child Development*, *64*, 1333–1353.

Scarr, S. (1995). Commentary. *Human Development*, *38*, 154–158.

Schaffner, K. F. (1998). Genes, behavior and developmental emergentism: one process, indivisible? *Philosophy of Science*, *65*(2), 209–252.

Slack, J. M.W. (2002). Conrad Hal Waddington: the last renaissance biologist? *Nature Reviews Genetics*, *3*, 889–895.

Stotz, K. (2006). Molecular epigenesis: distributed specificity as a break in the central dogma. *History and Philosophy of the Life Sciences*, *28*(4), 527–544.

Stotz, K. (2008). The ingredients for a postgenomic synthesis of nature and nurture. *Philosophical Psychology*, *21*(3), 359–381.

Tabery, J. (2004). Synthesizing activities and interactions in the concept of a mechanism. *Philosophy of Science*, *71*, 1–15.

Tabery, J. (2007). Biometric and developmental gene-environment interactions: looking back, moving forward. *Development and Psychopathology*, *19*, 971–976.

Tabery, J., & Griffiths, P. E. (2010). Historical and philosophical perspectives on behavioral genetics and developmental science. In K. E. Hood, C.T. Halpern, G. Greenberg & R. M. Lerner (Eds.), *Handbook of developmental science, behavior and genetics* (pp. 41–60). Chichester: Wiley-Blackwell.

Thelen, E. (1995). Time-scale dynamics and the development of an embodied cognition. In R. F. Port & T. van Gelder (Eds.), *Mind as motion: Explorations in the dynamics of cognition*. Cambridge, M.A: MIT Press.

Thelen, E., & Ulrich, B. D. (1991). Hidden Skills: A Dynamic Systems Analysis of Treadmill Stepping During the First Year (with a commentary by Peter H. Wolff). *Monographs of the Society for Research in Child Development*, *56*(1), 1–104.

Thelen, E., & Smith, L. (1994). *A dynamic systems approach to the development of cognition and action*. Cambridge, M.A: MIT Press.

Waddington, C. H. (1952). *The evolution of developmental systems*. Report of the Twenty-Eighth Meeting of the Australian and New Zealand Association for the Advancement of Science, Brisbane: A. H.Tucker, Government Printer. 155–159.

Waddington, C. H. (1957). *The strategy of the genes: A discussion of some aspects of theoretical biology*. London: George Allen and Unwin.

Waddington, C. H. (2012). [1942]. The epigenotype. *International Journal of Epidemiology*, *41*, 10–13.

Waters, C. K. (2007). Causes that make a difference. *The Journal of Philosophy*, *CIV*, 551–579.

West, M. J., & King, A. P. (1987). Settling nature and nurture into an ontogenetic niche. *Developmental Psychobiology*, *20*, 549–562.

Whitney, G. (1990). A contextual history of behavior genetics. In M. E. Hahn, J. K. Hewitt, N. D. Henderson & R. H. Benno (Eds.), *Developmental behavior Genetics: Neural, biometrical, and evolutionary approaches* (pp. 7–24). New York: Oxford University Press.

CHAPTER FOUR

Emergence, Self-Organization and Developmental Science

Gary Greenberg*[,1], Kristina Schmid Callina[†], Megan Kiely Mueller[†]

*Department of Psychology, Wichita State University, Wichita, KS, USA
[†]Eliot-Pearson Department of Child Development, Tufts University, Medford, MA, USA
[1]Corresponding author: E-mail: Gary.Greenberg@wichita.edu

Contents

Abstract

Our understanding is that psychology is a biopsychosocial science as well as a developmental science. Behavioral origins stem from ontogenetic processes, behavioral as well as biological. Biological factors are simply participating factors in behavioral origins and not causal factors. Psychology is not a biological science; it is a unique psychological science, a natural science consistent and compatible with the principles of the other sciences. Accordingly, we show in this chapter how principles and ideas from other sciences play important roles in psychology. While we focus on the concepts from physics of self-organization and emergence, we also address the cosmological and evolutionary biology idea of increased complexity over time, the organizing principle of integrative levels, and the epigenetic processes that are in part responsible for transgenerational trait transmission. Our discussion stresses the developmental science concepts of embodiment and contextualism and how they structure thinking about psychological processes. We conclude with a description of how these ideas support current postpositivist conceptions of relational processes and models in contemporary developmental science.

1. INTRODUCTION

While many scholars still prefer to identify psychology as a social science, we believe that this descriptor has done our science a disservice, suggesting it is a "soft science". Indeed, a recent *New York Times* article suggested that the social sciences suffer from "physics envy": "... the social sciences need to overcome their inferiority complex, reject hypothetico-deductivism and embrace the fact that they are mature disciplines with no need to emulate other sciences" (Clarke & Primo, 2012, p. 8). After all, there is simply science and nonscience. We come from traditions that understand psychology to be a natural science, consistent and compatible with the principles of all the other natural sciences.

From this perspective, behavior is as natural a phenomenon as rolling balls down inclined planes was for Galileo. Since adopting the methodology of science in the late nineteenth century, psychology has been influenced by ideas from the other sciences. At first, this influence manifested itself in the use of the atomistic metaphor of structuralism, the identification with positivism and the reliance on Western science's materialism and a clockwork universe (Boring, 1950). Later, functionalism drew on ideas from Darwinian evolution (Benjamin, 2007; White, 1968). The twentieth century is understood by many to be the century of behaviorism: materialistic, positivistic, and reductionistic (e.g., Baum, 2005; Benjamin, 2007). Accordingly, B. F. Skinner has been identified as the most eminent psychologist of the period (Haggbloom et al., 2002).

Just as this nineteenth and twentieth century approach in physics and the other sciences has failed to live up to its initial promise and has given way to a more holistic, field-oriented and contextual paradigm (Davies & Gribbin, 1992; Goodwin, 1994; Kauffman, 2000; Sheldrake, 1995), some in psychology too have begun to give up their adherence to an old-fashioned science in favor of this newly emerging scientific perspective (e.g., Chorover, 1990; Lerner, 2006). Thelen and Smith (1994) go so far as to suggest that we "... are in the midst of a paradigm shift" (p. 339). Overton and Lerner (2012) have made a similar claim. Indeed, Overton and Müller (2012) point out that psychology has now taken a more postpositivist and relational stance—becoming a process-oriented rather than a substance-focused discipline—characterized by a relational developmental systems approach linking individuals to context in a nonrecursive mutually influential manner across history. That the Relational Developmental Systems model we favor in this chapter represents a paradigm shift is made abundantly clear by Overton's (2012) comprehensive review of the evolution of paradigms in science, and especially in psychology.

As Lewis and Granic (1999) point out, conceptualizations from the other sciences have often been imported into psychology. Indeed, "Modern psychology, from its inception, has been informed by ideas imported from related fields of inquiry" (Moore, 2008, p. 327). Structuralism, for example, had seeds of the atomic theory of matter and biology, and functionalism (and Freud's, 1954, psychoanalysis) drew from Darwinian evolution. It is now beyond question that principles from other sciences have relevance for understanding many of the phenomena of psychology. Early discussions of this integration include how thermodynamics and evolution affect behavior (Swenson, 1998) and how physics can help us understand aspects of locomotor behavior (Kelso, 1995; Vogel, 1998). Thelen and Smith (1994) show how "… new forms of behavior can arise during development in a self-organizing manner, consistent with the universal laws of physics… " (p. 129).

This newer perspective in science, and its extremely broad application, can be summarized as follows:

> … *since the 1960s, an increasing amount of experimental data … imposes a new attitude concerning the description of nature. Such ordinary systems as a layer of fluid or a mixture of chemical products can generate, under appropriate conditions, a multitude of self-organisation phenomena on a macroscopic scale – a scale orders of magnitude larger than the range of fundamental interactions – in the form of spatial patterns or temporal rhythms…. [Such states of matter] provide the natural archetypes for understanding a large body of phenomena in branches which traditionally were outside the realm of physics, such as turbulence, the circulation of the atmosphere and the oceans, plate tectonics, glaciations, and other forces that shape our natural environment; or, even, the emergence of self-replicating systems capable of storing and generating information, embryonic development, the electrical activity of the brain, or the behavior of populations in an ecosystem or in economic development.*
>
> ***(Nicolis, 1989, p. 316).***

While psychology comes late to a full-blown adoption of the principles associated with self-organization phenomena, there are seeds of these ideas in the writings of Z.Y. Kuo on epigenesis (1967), J. R. Kantor on interbehaviorism (1924, 1926, and 1959), T. C. Schneirla on integrated behavioral levels (Aronson, Tobach, Rosenblatt, & Lehrman, 1972; Schneirla, 1949), and Gilbert Gottlieb on probabilistic epigenesis (1984, 1992, and 1997). Biologists such as Brian Goodwin (1994) and Stuart Kauffman (1993, 1995) have elucidated the linkages between developmental psychobiology and newly emerging concepts of complex adaptive systems and self-organization (Prigogine & Stengers, 1984). This way of thinking is now even playing a role in medicine (Ahn, Tewari, Poon, & Phillips, 2006). In psychology, this integrative approach is the hallmark of developmental psychobiology (e.g., Michel and Moore, 1995), developmental systems theory (e.g., Ford & Lerner, 1992; Oyama, Griffiths, & Gray, 2001), and probabilistic epigenesis (Gottlieb, 1984, 1992, 1997). In this context, the assessment by Lewis and Granic (1999) is telling: "It is particularly fortunate that psychologists are exploring ideas of self-organization while they are still congealing in biology and the other sciences. This fulfills a long-standing wish for currency and parity with the sciences at large" (p. 369). Although these ideas originated in the physical sciences, Gell-Mann (1995) recognized their wider application: "Even more exciting is the possibility of useful contributions to the life sciences, the social and behavioral sciences, and even matters of policy for human society" (p. 322). Haken (1993, 1994) speculates that all the following can be seen to emerge from the principles of complex adaptive systems and self-organization: language, national character, ritual, form of government, public opinion, corporate identity, and social climate. As we discuss later, many other aspects of behavior also emerge from these principles.

There are three crucial linkages among the diverse components of complex adaptive systems: (1) the important organizing principle of integrative levels (e.g., Novikof, 1945), (2) the tendency toward increased complexity with evolutionary advance (Saunders & Ho, 1981), and (3) the contextual nature of behavioral events (Elder, 1974, 1980; Lerner, 1991). The synthesis of these three ideas leads to a developmental perspective in which behavior is seen to be the result of the fusion of biological and psychosocial factors, driven by probabilistic epigenetic events rather than by preprogrammed genetic or other biochemical ones (Gottlieb, 1992, 1997; Kuo, 1967). The physics model of this perspective is referred to as "nonlinear dynamic systems theory". Used in psychology, it provides a theoretically consistent language with which to describe and analyze behavioral development (Michel & Moore, 1995; Novak, 1996).

For example, concepts of nonlinear systems from physics bear directly on our contemporary understanding of psychological development (e.g., Overton & Müller, 2012). These concepts pertain to the ideas of *system* (that the parts of organisms function interdependently and that behavior is a *process,* and not a substance or "thing" of the system), of *hierarchy or directionality* (as a fundamental principle of science that the universe exists as a family of hierarchies in which natural phenomena exist in levels of increasing organization and complexity), of *emergence and self organization,* and of *epigenesis,* described later.

One of the authors has earlier identified how the ideas of emergence and self-organization in physics and of evolution and microbiology in biology are helpful in understanding the aspects of behavioral development (Greenberg, 2011; Greenberg, Partridge, & Ablah, 2006; Partridge & Greenberg, 2010). This chapter discusses these ideas fully, especially as they relate to developmental science in general; psychology is, after all, a developmental science (Greenberg, Partridge, Mosack, & Lambdin, 2006). The significance of the idea of emergence for psychology is that "Emergentism is a form of materialism which holds that some complex natural phenomena cannot be studied using reductionist methods" (Sawyer, 2002, p. 2). For us, in this chapter, the implication is that psychological phenomena cannot be reduced to biological factors such as genes and brains (e.g., Greenberg, 2011).

Our intention in this chapter is to demonstrate, from both phylogenetic and ontogenetic perspectives, how concepts such as epigenesis and embodiment are useful in understanding the nature and course of individual development. We begin by reviewing factors that contribute to the ontological structure of psychology. We then discuss emergence, evolution, and epigenesis as means for understanding the nature of structural transformation in human development. Finally, we describe how these ideas support current postpositivist conceptions of relational processes and models in contemporary developmental science (e.g., Overton, 2010).

2. ONTOLOGICAL STRUCTURE OF PSYCHOLOGY

Our understanding of psychology is that it is a biopsychosocial science, a natural science consistent and compatible with the principles of the other natural sciences. Psychology is a unique science with its own principles that are still under development, given that it is only about 133 years since the adoption of the scientific method by Wilhelm Wundt. Psychology's principles are not those of physics, biology, or physiology. Bunge's (2003, p. 141) comment about this point summarizes our understanding very well:

> *What about psychology: is it reducible to biology? Assume, for the sake of the argument, that all mental processes are brain processes…. Does this entail that psychology is a branch of biology and, in particular, of neuroscience…? Not quite, and this for the following reasons. First because brain processes are influenced by social stimuli… Now, such psychosocial processes are studied by social psychology [and its categories] are not reducible to biology. A second reason is that psychology employs concepts of its own, such as those of emotion, consciousness, and personality… that go beyond biology.*

To be sure, the ideas and principles of all sciences, are germane to psychology, but in the hierarchical arrangement of the sciences, psychology has emerged from those lower than it, as illustrated in Fig. 4.1. Gell-Mann (1995) refers to this hierarchical nature of the sciences by raising the question of which science is the fundamental one, "… science A is more fundamental than science B when… the laws of science A encompass in principle the phenomena and laws of science B… " (p. 109). Novak (1996) calls this hierarchy the "continuum of scientific disciplines" (p. 4). Psychology, then, is more complex than sciences lower

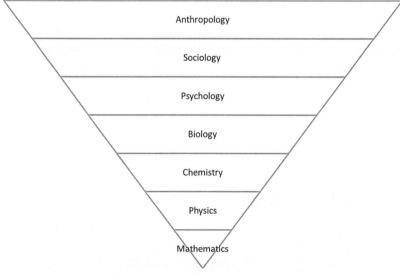

Figure 4.1 Hierarchy of the sciences.

than it in this hierarchy; but, while more complex, it is not more difficult to comprehend.

Thus, while a student may struggle more in her physics class, that subject matter has identified variables on both sides of its equations. Psychology is still attempting to identify the myriad of variables that influence behavior, many of which are, to use Gilbert Gottlieb's (1997) term, nonobvious: "Who would have dreamed that squirrel monkeys' innate fear of snakes could derive from their experience with live insects (Matasaka, 1994)? Or that chicks' perception of mealworms as food depends on their having seen their own toes move (Wallman, 1979)?" (p. 144). This complexity is in large part due to the various factors that are responsible for psychological phenomena that can be summarized as five crucial sets of factors: phylogenetic, ontogenetic, experiential, cultural, and individual.

2.1. The Phylogenetic Set

The phylogenetic set reflects an organism's evolutionary status, i.e., what it is as a species, and recognizes the significance of evolution for psychological phenomena. Phylogenetic factors are embodied in Kuo's (1967) "principle of behavioral potentials," that asserts that each species has the potential to behave in species-typical ways (Haraway & Maples, 1998). There is, of course, no guarantee that these potentials will be actualized in a particular manner, another example of the probabilistic nature of behavior development (Gottlieb, 1992). Thus, Kuo asks, "Is a cat a rat killer or a rat lover?" His research shows that "it depends… " Kittens reared with rats and out of sight of cats that kill and eat rats never eat rats, even when hungry. For the cat that has never seen a rat killed or eaten, it becomes an object of "love" (Kuo's term). With respect to humans, Montagu (1952/1962) has said, "The wonderful thing about a baby… is its promise" (p. 17)—we are born *Homo sapiens*, but we have to *become* human beings (Bronfenbrenner, 2005).

2.2. The Ontogenetic Set

This set of factors refers to the development of an organism, from its conception to its embryonic state, to its state as an adult and to its eventual death. Once again, we underscore the probabilistic nature of this ontogeny. *Nothing in development*—embryological or behavioral—is guaranteed by biology, by genes; nothing is preformed or preordained (Gottlieb, 1992; Nieuwkoop, Johnen, & Albers, 1985). The developmental stage of an organism profoundly impacts its behavior and the way in which it reacts to stimuli. For example, a baby in its crawling period can only get under the kitchen sink, but when she begins to walk care must be taken to fasten the kitchen drawers.

2.3. The Experiential Set

We adopt Schneirla's (1957) definition of experience as "all stimulative effects upon the organism through its life history" (1957, p. 86). As Overton (2006) has pointed out, the organism is not merely a passive recipient of experiences, but rather a source in that all actions initiated by the organism are also experiences; experience, then, is both what happens to the organism and what it does. Kantor (1959) referred to one's experiential history as the "reactional biography" (RB). RB begins at conception and continues to be built up until the organism's death. Every stimulus and each act affects the organism and changes it, although some stimulation and some acts have much more profound and obvious effects than others. Learning, for example, is an important process in behavioral change, but it is nothing more than a special set of experiences.

2.4. The Cultural Set

Organisms function in environments and the organism–environment relation forms a functional whole. Consequently, environments are necessary features of the organism's biological and behavioral development. This relation is most obvious in humans, who have developed cultural systems (e.g., religion, dietary practices, societal institutions) that impact behavioral development in multiple ways. But all living organisms, although perhaps at less-complex levels, function within environments of their own making. Different species may inhabit different environments, eat different foods, and so on. This important point was stressed by the ethologist Jacob von Uexküll (1957), who termed the behavioral environment of an animal its *Umwelt*, its sensory–perceptual world. A recent discussion of this concept is provided by Michel (2010). Chimpanzees, for example, display different behavioral adaptations related to their unique environments (Matsuzawa, 1998). Chimpanzees in two communities separated by only 10 km can display markedly different behaviors. These differences include nest building, ant dipping, use of leaves for water drinking, food choices, and many others. This phenomenon is so pervasive in chimpanzees that such differences are found even in groups that are in closer proximity, suggesting that this effect is not just the result of ecological factors (Luncz, Mundry, & Boesch, 2012).

A classical example of cultural traditions in nonhuman animals is that of the development of potato washing by Japanese macaques. For over 50 years, Japanese primatologists (Itani, 1961; Kawamura, 1959; Nishida, 1986) have been studying the social behavior and emergent traditions of Japanese macaque monkeys. Provisioned with novel foods—potatoes and rice—the monkeys soon began to toss handfuls of rice gathered from the sandy beach into the water where the rice would float and the sand would sink. The monkeys thus discovered a way to wash sand from their food. These practices spread throughout the colony and are now part of the animals' normal behavioral repertoire. The practice is handed down from generation to generation—a primitive form of cultural transmission. Once they began spending more time near and in the water, young macaques began playing in it. This play led to the development of new behavioral skills, such as swimming. The animals also incorporated new foods into their diets, fish for example, and may now be capable of swimming to distant islands—a type of Darwin's finches scenario.

2.5. The Individual Set

This set reflects the uniqueness of each individual organism and how that uniqueness relates to its development (Molenaar, 2010; Nesselroade & Molenaar, 2010). One animal may be more or less sensitive to sounds, may have a developmental abnormality that limits its interactions with its world, or may be larger or smaller than its conspecifics. This set of factors recognizes the contribution of the individual's unique genotype and how the organism's biology, in dynamic interplay with contextual influences, may render it a different behaving creature than all others. Of course, while an organism's genome plays a role in its development, the idea of genetic determinism is no longer feasible, a point we will revisit later (Charney, 2012; Greenberg, 2011; Wahlsten, 2012).

These five organizational sets provide the ontological structure of psychology. We are comforted by the use of a similar analysis by Overton (2006; Overton & Müller, 2012), one of the world's leading developmental scientists (to whom this edition of *Advances in Child Development and Behavior* is dedicated). Overton uses different labels, but is substantially in agreement that several sets of factors, at different levels of analysis and influence, play a dynamic role in human development. We are especially in agreement with respect to the significance of his discussion of the physical ideas of "fluid dynamic holism and associated concepts such as *self-organization*, system, and the synthesis of wholes" (Overton, 2006, p. 19) as they apply to understanding development. The relationship among these factors is one of fusion (Tobach & Greenberg, 1984), no one set being of more importance in an organism's behavior development than another. Exerting their influence along the developmental trajectory, the source of many aspects of behavior cannot be identified; rather they reflect the influence of emergence.

The significance of relationism for developmental science is that development is a multicausal complex *open* system with many elements embedded within the system. Nevertheless, development proceeds in a coordinated fashion, there being no

> *executive agent or ... programme that produces the organized pattern. Rather, the coherence is generated solely in the relationships between the organic components and the constraints and opportunities of the environment. This self-organization means that no single element has causal priority.*
> **(Smith & Thelen, 2003, p. 344).**

We turn next to a discussion of the concepts of emergence, evolution, and epigenesis in terms of structural transformations in the process of human development.

3. STRUCTURAL TRANSFORMATION 1: EMERGENCE IN PSYCHOLOGY

It is always helpful to "begin at the beginning", advice that comes from the unusual source of the Red King in *Alice and Wonderland*. In the present context, the beginning is the Big Bang (Singh, 2004) and the first elements to be formed are hydrogen and helium. Everything in the universe has come from that "simple" beginning, thus, a natural consequence of the Big Bang is that given enough time these elements have become life, and subsequently, sentient beings, a result of the natural law of increasing complexity after the Big Bang and the physical principles of self-organization and emergence. As summarized by Reid (2007), "Life emerged in a universe whose cosmology is subject to physical laws. But life, without disobeying such principles, is not predicted or contained by them" (p. 92). Across history, the continuation of the cosmic and biological evolution of the universe has resulted in increasingly more complex forms of cosmological structure and animal life. An early discussion of this process by Oparin (1961) is surprisingly contemporary in its understanding of these processes and reflects the themes of this chapter: "The facts at our disposal indicate that the origin of life was a gradual process in which organic substances became more and more complicated and formed complete systems which were in a state of complete interaction with the medium surrounding them... .Following the path of the emergence of life in this way... .there arose new biological laws which had not existed before... " (pp. 36–37).

While the idea of self-organization is relatively new to physics (Prigogine & Stengers, 1984), that of emergence, that new properties come into being when old parts are arranged in new ways, has a long history in philosophy, the physical and biological sciences, and more to the point of this chapter, in psychology (Bedau & Humphreys, 2008). Sawyer (2002) notes that Wilhelm Wundt and Henry James recognized the significance of emergence in the new science they were developing. Some trace the origin of the idea of emergence to John Stuart Mill (e.g., McLaughlin, 2008), the father of British emergentism, a nineteenth century philosophical tradition. Applied to developmental science, behavior is seen to be the result of the fusion of these five sets of factors. Therefore, human behavior is a product of probabilistic epigenetic events, not a result of preprogrammed genetics (Gottlieb, 1992, 1997; Kuo, 1967). Nonlinear dynamic systems theory provides a theoretically consistent language with which to describe and analyze the development of behavior, discussed at length by Michel and Moore (1995).

Nonlinear dynamics contains a lexicon of concepts pertaining to change processes that do not exist in any other known theoretical system. Dynamical models allow us to compare and contrast seemingly unrelated phenomena that often share common dynamical structures. Nonlinear dynamics and complex systems analysis are continuing to help revolutionize our understanding in many of the life sciences. This situation was summarized by Kauffman (1993), a leading figure in the widespread application of these ideas as follows:

Eighteenth-century science, following the Newtonian revolution, has been characterized as developing the sciences of organized simplicity, nineteenth-century science, via statistical mechanics, as focusing on disorganized complexity, and twentieth and twenty-first-century as confronting organized complexity (p. 173).

As Prigogine (1994) has noted, nonequilibrium physics is itself an emergent science, and has replaced certitudes (i.e., determinism) in the laws of nature with possibilities (i.e., probabilities).

As already noted, a crucial idea for understanding complex systems is the view that the universe is ordered hierarchically. This concept is illustrated in Fig. 4.2 and summarized by Aronson (1984) as follows:

[The levels concept]... is a view of the universe as a family of hierarchies in which natural phenomena exist in levels of increasing organization and complexity. Associated with this concept is the important corollary that these successions of levels are the products of evolution (1984, p. 66).

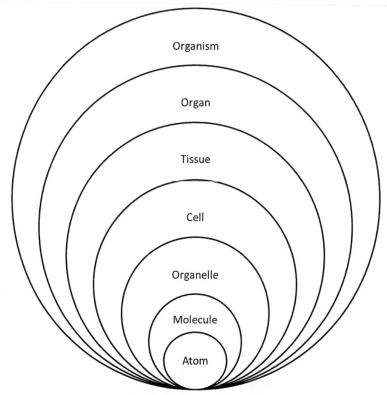

Figure 4.2 Increasing complexity in nature.

As Feibleman (1954) says, referring to the hierarchy of the sciences, the levels concept places "rules" of explanation on studies of the more complex systems (e.g., biology, psychology, and sociology).

A now classical application in behavior development of dynamic systems, emergence, and self-organization is that of Thelen and Smith (1994). In acknowledging the origin of these principles in mathematics, physics, and biology, these authors substantiate the unity of the sciences and the idea that the unique principles of psychology are, as well, natural phenomena. In describing development as orderly, incremental, and progressive, Thelen and Smith point out that one may be left with the impression that such processes are guided from within the organism, or if not a result of genetic and biological determinism, surely that of the *interaction* of "nature and nurture." However, such a formulation grants nature and nurture factors individual and independent significance in influencing behavior (Pronko, 1988). Along with Thelen and Smith, others have recognized that the dynamic interplay between these factors is a *fusion* (Tobach & Greenberg, 1984)—one cannot therefore say how much of behavior is determined by phylogeny, how much by ontogeny, how much by nature, and how much by nurture, in much the same way that we cannot determine how much of the area of a rectangle is a function of its width or its height.

We agree with Pronko's comment that "We must not neglect genetic and other biologic factors, but, instead of treating them as causal, we regard them as *aspects* of an integrated field event or events" (1988, p. 78). This idea is, of course, the essence of the relational development systems approach to development (Overton, 2010, 2011). The prescription for development outlined by Thelen and Smith (1994) represents, "… a radical departure from current … theory. Although behavior and development appear structured, there is no structure. Although behavior and development appear rule-driven, there are no rules. There is complexity. There is a multiple, parallel, and continuously dynamic interplay of perception and action, and a system that, by its thermodynamic nature, seeks stable solutions" (p. xix).

4. STRUCTURAL TRANSFORMATION 2: EVOLUTION

It is instructive to begin our discussion of evolution with a quotation from Ernst Mayr (1979), one of twentieth century's leading evolutionary biologists:

The most consequential change in man's view of the world, of living nature and of himself came with the introduction over a period of some 100 years beginning only in the 18th century, of the idea of change itself, of change over long periods of time: in a word, of evolution. Man's world view today is dominated by the knowledge that the universe, the stars, the earth and all living things have evolved through a long history that was not foreordained or programmed, a history of continual, gradual change shaped by more or less directional processes consistent with the laws of physics "(1979, p. 47)."

While it was appropriate in the late twentieth century to say that "nothing in biology makes sense except in the light of evolution" (Dobzhansky, 1973), it is now the case that this idea applies as well to psychology and to the origins of behavior. This view is especially true now that evolutionary theory has once again embraced development (see our discussion of this later in the chapter).

As with all major scientific theories, Darwinism is an exemplar of simplicity. It can be summarized in five fundamental principles (Mayr, 1991): (1) evolution as such, (2) common descent, (3) multiplication of species,

(4) gradualism, and (5) natural selection. Unfortunately, as Jablonka and Lamb (2005, p. 9) point out, Mayr's summary leaves us with "… the impression that there is a tidy, well-established theory of evolution – Darwin's theory of natural selection – which all biologists accept and use in the same way. The reality is very different, of course."

Darwin knew that characteristics that permit the organisms possessing them to survive were likely to be passed on through the successful reproduction of those organisms. He knew nothing, however, about the mechanism of such inheritance. Our current understanding of genetics, which began with the work of the monk Gregor Mendel, provides this mechanism. The canonical theory of evolution, referred to as "The Modern Synthesis", combines Darwin's ideas of natural selection and Mendel's ideas of genetics (e.g., Mayr & Provine, 1980). Interestingly, Tauber (2011) recently suggested that in light of newer epigenetic thinking, the so-called Modern Synthesis, a melding of genetics and developmental biology and evolution, is likely to come to be referred to as the "Old Synthesis" i.e., a "Refreshed Synthesis". Thus, while Modern Synthesis dictates that characteristics are transmitted across generations only genetically, there is now substantial empirical support for the idea that organisms are affected genetically by experiences of their ancestors, i.e., Lamarckism, or epigenetic inheritance (Masterpasqua, 2009). Such findings challenge the idea that psychological disorders that run in families do so as a result of genetics and point instead to environmental or experiential etiologies (e.g., Greenberg, 2011). Jablonka and Lamb's (2005) summary of epigenetic inheritance is especially clear:

> But information transfer also occurs at higher levels of organization [higher than at cellular inheritance systems]. There is a good example of this in Mongolian gerbils, where the mother's uterine environment may have strong heritable effects on her offspring's development. A female embryo that develops in a uterus in which most embryos are made is inevitably exposed to a high level of the male hormone testosterone. This high level of the hormone is information for the embryo, and it affects her subsequent development. As she grows up, she develops some special characteristics, such as late sexual maturity and aggressive territorial behavior and, most remarkably, when she reproduces, her litter has more males than females. Since most of her embryos are male, her female offspring develop, just as she did, in a testosterone-rich environment, so they grow up to have the same behavioral and physiological traits as their mother. They, too, will produce more male-biased litters, and so the cycle continues. In this way the developmental legacy of the mother is transferred to her daughters – there is nongenetic inheritance of the

mother's phenotype. Consequently, two female lineages that are genetically identical can be very different behaviorally and have different sex ratios, *simply because they transmit different nongenetic information (pp. 145–146, emphasis added).*

Theories in science are of necessity dynamic. All the facts are never fully collected; this is why science is characterized as a self-correcting discipline and why "truth" in science is written with a lower case "t". New discoveries and new facts rarely result in the discarding of a strong theory, rather the course taken is to tweak the theory to accommodate the new findings (see Kuhn's 1962 book, *The Structure of Scientific Revolutions*, on this point; we appear to be on the verge of an important paradigm shift in our understanding of evolution). So it is with Darwinism, as Jablonka and Lamb (2005) point out: "Not only has opinion about the theory of natural selection as a whole changed over the years, there have also and inevitably been changes in the details" (p. 40). We address three important modifications to Darwin's original proposal.

1. While Darwin provided the fundamental law of his theory, that of Natural Selection, Saunders and Ho (1976) suggested that an increase in complexity over geological time (i.e., with evolution), can be understood to be a second law of Darwinism. One of us has discussed complexity theory and its corresponding idea of emergence in other publications (Greenberg, Partridge, & Ablah, 2006; Partridge & Greenberg, 2010). It is sufficient here to state that increases in complexity and the epigenetic emergence of novelty are the rule in evolution. This assertion should come as no surprise, for many—including Gould (1997; see also Krasny, 1997), Maynard Smith (1970), Carroll (2001) and others—have pointed out that when you begin with a single cell, with simplicity, there is only one direction to go in and that is up toward greater complexity. For instance, "… it is in some sense true that evolution has led from the simple to the complex: prokaryotes precede eukaryotes, single-celled precede many-celled organisms, taxes and kineses precede complex instinctive or learnt acts… .And if the first organisms were simple, evolutionary change could only be in the direction of complexity" (Maynard Smith, 1970, p. 271).

2. A cornerstone of Darwinian theory is that evolutionary change is slow and gradual, taking millions of years. The absence of a corresponding fossil record is one source of challenge to this idea. However, Eldredge and Gould (1972) provided an explanation for these gaps in the fossil record and at the same time demonstrated the dynamism of Darwinian

theory, that it can be tweaked; their idea is referred to as *punctuated equilibrium*. The proposal, now widely accepted as another modification of Darwinian theory (Gould & Eldredge, 1993), is that species remain unchanged for long periods (i.e., in equilibrium) and that these long periods of no change are punctuated by episodes of relatively rapid (e.g., in geologic time, tens or hundreds of thousands of years) change. Thus, there is no gradual fossil record to be discovered. This is saltatory, rather than gradual, evolution and is an example of how the principle of emergence plays a role in our contemporary understanding of evolution (Reid, 2007).

3. The third major modification to Darwinian theory is that of the reintroduction of a form of Lamarckian inheritance, what Ho and Saunders refer to as the Epigenetic Theory of Evolution (Ho, 2010). While we have already discussed epigenesis in detail, it is important to emphasize the role it plays in the contemporary understanding of evolution. In this approach, natural selection is seen to play little or no role, "based on evidence suggesting on the one hand that most genetic changes are irrelevant to the evolution of organisms... " (Ho, 2010, p. 72). As Ho explains, her theory renders evolution compatible with nonlinear dynamic systems, and to the point of this chapter, with the ideas of emergence and self-organization. Consistent with this conception, the following are now known: (1) genes are not directly responsible for phenotypic expression, but rather, the environmental context of development plays a crucial role in this process; (2) it is not only that genes work from the inside out, but rather that behavior too can influence the expression of genes (referred to as "downward causation" by Campbell, 1990); (3) not all genes of a genome get expressed; (4) natural selection is but one of the several mechanisms responsible for evolutionary change; and (5) the path from genes to physical or behavioral traits is enormously complex and indirect.

Pushing natural selection from its pedestal is not as heretical as it may seem. No less an ardent Darwinian as Stephen Jay Gould (2002, pp. 137–141) discusses its (diminished?) role in modern evolutionary biology. In criticizing those who cling to old-fashioned Darwinian fundamentalism, Gould (1997) shows that, as with all sound theories, Darwinism is dynamic, even now allowing Lamarckian ideas to be entertained (and of course Darwin, 1859; was an adherent of Lamarck's ideas). In the words of Jablonka and Lamb (2005, p. 102), "... Darwinian evolution can include some Lamarckian processes, because the heritable variation on which selection acts is not entirely blind to function; some of it is induced or 'acquired' in response to the conditions of life". Ho (2010, p. 62) has noted, however, that there remain some objections to the resurrection of Lamarckism: "Epigenetic inheritance, instances of Lamarckian evolution, has now been widely documented in numerous studies, and continues to get many biologists hot under the collar."

5. STRUCTURAL TRANSFORMATION 3: EPIGENESIS

Among the significant ideas of contemporary developmental science are those of epigenetics and epigenesis. These ideas, first introduced into modern biology by Waddington in 1957 (Haig, 2004; Waddington, 1957), supplement Darwinian evolution by showing that there are routes to inheritance other than by DNA and genetics. Genetics and developmental biology in the early part of the twentieth century were separate disciplines. Mayr (1982) noted that early progress in genetics required the banishment of development from discussions of evolution. In this context, we note that Maynard Smith and Holliday (1979) declared that, with respect to evolution, development has been safely ignored. Waddington had the foresight to join the two and in so doing coined the term epigenetics (Holliday, 2006). It took until modern times for the biological sciences to follow in Waddington's footsteps and understand the critical role that development plays in all aspects of biology, including evolution (Robert, 2004, 2008). Nevertheless, despite the renewed interest by some in studying development in an evolutionary framework (e.g., evolutionary-developmental biology, evo-devo), "The field is still dominated by the idea that genes control development" (Ho, 2010, p. 70). Of course, the present authors, along with others in this volume disagree and believe instead that development itself is an epigenetic phenomenon from which novel processes emerge. A cogent discussion of this point is provided by Moore (2003).

While epigenetics and epigenesis have similar roots, they refer to different processes. Not all authors recognize these differences and use the terms interchangeably, as even Waddington recognized as early as 1956 (Haig, 2004). While there are, as Holliday points out, different meanings of epigenetics, we are most comfortable understanding this term as referring to the developmental process in which the entire series of interactions among cells and their products results in morphogenesis and differentiation. With respect to epigenesis, we note that Overton (2011) and Overton and Müller (2012) follow in a long line of developmentally oriented scientists (e.g., Kuo,

1967; Moltz, 1965) in embracing the concept of epigenesis. Their discussion of this concept reflects the breadth of their interests and their widespread impact on this discipline. Overton (2010, p. 7) states that epigenesis:

> *is the principle that the role played by any part of a relational developmental system – gene, cell, organ, organism, physical environment, culture – is a function of all of the interpenetrating and coacting parts of the system. It is through complex relational bidirectional and multidirectional reciprocal interpenetrating actions among the coacting parts that the system moves to levels of increasingly organized complexity.... .Epigenesis also points to a closely related feature of transformational developmental change:* emergence. *Transformational change results in the* emergence of system novelty.

To summarize then, in our usage, epigenetics refers to transgenerational Lamarckian-type evolutionary processes, while epigenesis refers to transformational (qualitative) changes in ontogeny, as Overton points out.

In one of his earliest theoretical papers (actually written in 1965, although not published until 1970), Gottlieb distinguished between predetermined and probabilistic epigenesis, the former term lying firmly on the unbending nature side of the nature–nurture issue. This approach to behavior understands it to be the outcome of biology (or nature), with very little influence at all of experience (or nurture). Gottlieb's own work (1973), as well as that of others (e.g., Kuo, 1967; Smotherman & Robinson, 1988), has shown this conception to be simply wrong.

The alternative position explicated and favored by Gottlieb is that of probabilistic epigenesis, in which "behavioral development of individuals within a given species does not follow an invariant or inevitable course, and, more specifically, that the sequence and outcome of individual behavioral development is probable... rather than certain" (Gottlieb, 2001, p. 43). Indeed, this view of epigenesis is seen now to hold not just for behavior, but "is recognized in many quarters as a defining feature of development" (Gottlieb, 2003, p. 341).

Epigenesis has been described in a variety of ways, but none has been so well put as that by Moltz (1965, p. 44):

> *An epigenetic approach holds that all response systems are synthesized during ontogeny and that this synthesis involves the integrative influence of both intraorganic processes and extrinsic stimulative conditions. It considers gene effects to be contingent on environmental conditions and regards the genotype as capable of entering into different classes of relationships*

depending on the prevailing environmental context. In the epigeneticist's view, the environment is not benignly supportive, but actively implicated in determining the very structure and organization of each response system.

We cannot overestimate the enormous influence of Gottlieb in contemporary discussions of epigenesis, more specifically of probabilistic epigenesis (Wahlsten, 2010). While he was an animal psychologist and a developmental psychobiologist, his influence as a critic of genetic determinism is widespread in the biological and behavioral sciences. Predetermined epigenesis ignores the fact that development is contextual; *all events,* from the level of the cell to those of the external environment, continuously affect development. Gottlieb illustrated this integration in a now famous diagram (1992, p. 186) of development as four interacting levels across time (i.e., genetic activity, neural activity, behavior, and the environment). Development therefore involves bidirectional processes, with events at all levels playing crucial, fused, roles in developmental outcomes.

The evidence supporting the concept of probabilistic epigenesis then, constitutes a potent argument against behavior genetics and the idea that genes alone determine all characteristics, structural and behavioral (Greenberg, 2011). Predetermined epigenesis suggests a direct one-to-one linkage between biology, genes, for example, and structure or behavior. In this formulation, biology is destiny. Having a gene for breast cancer or phenylketonuria (PKU), for example, results in the development of these diseases in the lifetime of the individual, *no matter what*! Probabilistic epigenesis, on the other hand, speaks of a probabilistic linkage between biology and both physiological structure and behavior. One may thus have a gene potentially linked to breast cancer or PKU, but as a result of diet, or other lifestyle factors, this gene may or may not be involved in such development or even express itself. Meaney (2010) provides an overview of data documenting this point.

The significance of epigenetics for developmental science and for our discussion was summed up concisely by Ho (2010, p. 61): "In the new holistic perspective, epigenetics mediates between the psychosocial and biological realms, and holds the key to how we can shape our own development." Returning to an earlier discussion in this chapter, it is of interest here to note that the demarcations between the sciences is not as clear cut as one might expect; transitions from one level to another are not smooth. Indeed, we never really know where one level, or one science, ends and another begins. Thus, the relational holistic position we favor in this chapter takes a dramatically different perspective on the relation between biology and psychological development. From this perspective, development is an active system of processes superordinate to biology and evolution. Thus, it is not that genes and brains explain development, but rather that the developmental system explains the functioning of the gene, the brain, and even evolution at the level of individual ontogeny. The developmental system integrates biological functions into coordinated patterns that support behavior. It is, then, the process of development that shapes biological organization and provides a temporal context for biology–behavior–ecology interrelationships (Lerner & Bush-Rossnagel, 1981; Overton & Müller, 2012).

It is appropriate to conclude this discussion of epigenetics and epigenesis with a quote from Bunge (2003), a statement that summarizes our understanding of reductionistic analysis and the role of genes and biology in behavioral origins:

At first sight, the discovery that genetic material is composed of DNA molecules proves that genetics has been reduced to chemistry....However, chemistry only accounts for DNA chemistry: it tells us nothing about the biological functions of DNA – for instance that it controls morphogenesis and protein synthesis. In other words, DNA does not perform any such functions when outside a cell, anymore than a stray screw holds anything together. (Besides, DNA does nothing by itself: it is at the mercy of the enzymes and RNAs that determine which genes are to be expressed or silenced. In other words, the genetic code is not the prime motor it was once believed to be. This is what epigenesis is all about.) (p. 138, emphasis added).

6. RELATIONAL PROCESSES AND MODELS IN DEVELOPMENTAL SCIENCE

Contemporary developmental science has successfully provided a dialectical synthesis of earlier organismic and mechanistic theories by positing that behavioral development is a function of an organism interacting with an active sociohistorical ecology (Greenberg, 2011). A family of theories has arisen that includes perspectives such as the developmental bioecological model (Bronfenbrenner, 1979), developmental contextualism (Lerner, 1998), the life-span perspective (Baltes, 1987), the person-oriented approach (Magnusson, 1996), and transactional models (Sameroff, 1975). The success of these theoretical formulations is indicated by the change in the scope and content of developmental science.

Concurrent with advances in developmental science, science in general has been revolutionized by developments in the study of nonlinear dynamic systems, as we have discussed earlier in this chapter. Under the general rubric of nonlinear dynamics are several subfields: chaos theory, the study of complex behavior resulting from simple and deterministic processes; fractal geometry, the study of geometrical forms invariant across scale; and complex systems theory, the study of stable organized behavior resulting from complex and stochastic processes. It is the last of the three theories that seems to hold the most relevance for current formulations of developmental science. Today's developmental science is what Lerner (2006) has described as involving "the ascendency of a developmental systems frame" (p. 5). It is relational and integrative, and the approach we will discuss represents a successful attempt at unifying the field.

Traditionally, questions that have framed the study of human development involve Cartesian split conceptions about the bases of developmental processes (see Overton, 2010; for a review). These questions ask whether behaviors, attitudes, emotions, and other aspects of psychological functioning are the result of nature (genes) versus nurture (environment), continuity versus discontinuity, stability versus instability, etc. (Lerner, 2002). According to Lerner, "This split, reductionist ontology about development meant that the epistemological route to learning about the basis of development was to identify the essential (nature or nurture) explanatory variable(s)" (Lerner, 2012, p. 5). Such split conceptions of approaching developmental science do not account for the embedded integrative nature of the developmental system (e.g., Overton, 2010).

However, beginning in the last third of the twentieth century, the field of developmental science saw a paradigmatic shift in thinking that has led to a more integrative approach to understanding human development. As is the case with paradigm shifts (Kuhn, 1962), acceptance is neither immediate nor universal. While Lerner and others (e.g., Lerner, Easterbrooks, & Mistry, 2012; Overton & Lerner, 2012) see the newer ideas about human development discussed in this chapter akin to a Kuhnian-like paradigm shift, not all agree. This difference of opinion is due to two facts: (1) because these ideas are still new, many "…. have difficulty intuitively understanding the properties of complex nonlinear systems" (Singer, 2009, p. 327) and (2) because traditional psychology focuses on structures rather than processes, the clearest example of which is in the formation of cognitive models of memory located in specific parts of the brain (e.g., Friedenberg, 2009).

Nevertheless, in Overton and Müller's (2012) view, contemporary perspectives in developmental science have moved beyond the traditional linear approach to exploring development to models that view human development as dynamic and integrative. As Mascolo and Fischer (2010) note, developmental scientists are transitioning away from thinking about psychological structures and functions as involving individual actors in isolation, independent and discrete processes, and linear movement through global stages. Instead, human development is seen as a series of dynamic relations that occur within dynamically integrated structures within developmental systems. As Mascolo and Fischer (2010) note, researchers now recognize the embededness of multiple levels of the ecology and the "profound lack of independence of the systems that make up human action as well as the systems and contexts within which human action is embedded" (p. 150).

Given the emphasis on conceptualizing human development within an embedded integrated system, there has also been a shift in how to conceptualize development in psychology. Within the embedded system, what develops and how? Instead of viewing the units of development as individual actors and independent processes, the "what" of psychological development is viewed instead as emergent systems that become integrated across the life span (Mascolo & Fischer, 2010). Therefore, development does not involve just the person in isolation, but involves integrated skills that are developed in a dynamic relationship within various contextual systems. These integrative structures are composed of appraisal-affect-action processes that promote skill building in different domains (Mascolo & Fischer, 2010).

As we previously noted, traditional theories of psychological structures describe development in terms of a linear series of steps. The Piagetian theory of cognitive development, for example, requires the individual to develop "singular, broad-based, homogeneous competencies" (Mascolo & Fischer, 2010, p. 161), within each of the four stages: sensorimotor, preoperational, concrete operational, and formal operational. Erickson (1959) and Freud (1954) held analogous theories regarding global stages of identity and personality development, respectively. However, Mascolo and Fischer (2010) note that evidence from several decades of research on the integration of cognition, emotion, and action indicates that changes in psychological structures are in fact the result of an "emergent developmental web" (p. 163). Rather than global stages that develop along a linear trajectory, human development involves integrative psychological structures that develop dynamically within contexts that support their construction.

7. NEURAL DEVELOPMENT

While we acknowledge the role of biological factors in behavioral origins, we are by no means biological reductionists. That being said, we understand the important role played by neural factors in behavior (Greenberg, Partridge, Weiss & Haraway, 1998). It is important to point out that the core ideas we discussed here also apply to neural development. Friedenberg (2009) has invoked various ideas of emergence and self-organization to attempt an explanation of how these ideas relate to brain functioning. Different aspects of neural activity perhaps underlie coding in the system: single and multiple neural firing rates and other temporal features, neural synchrony, and neural coupling represent just a sampling of the possible ways in which the nervous system represents information. The brain's complexity cannot be understood by reductionistic analysis. Thus, while the 600 electrical and 5000 chemical synapses in the nervous system of the invertebrate worm *Caenorhabditis elegans* (containing a mere 302 neurons) have been mapped, "this knowledge itself failed to provide realistic ideas about the function and dynamics of this minimal nervous system" (Koch & Laurent, 1999, p. 97). Even in the simplest nervous systems, complexity makes for the emergence of neural processes.

As Benno (1990) makes abundantly clear, the development of the brain is itself regulated by a complex set of epigenetic processes. The internal dynamics accounting for increased neural organization, however, are linked to the larger behavioral context. During the development of the embryo, animal brains grow an enormously complex web of neurons. Each of these neurons establishes connections with many other neurons by means of hundreds and thousands of dendrites. The arrangement of circuits among this tangle of neurons and dendrites, that is, the fine tuning of a brain, is the result of experiential input. Neural development involves several sets of processes, including (1) determination, the transition from totipotentiality of a nerve cell to a state of more limited fate; (2) proliferation, the production of new cells; (3) migration, the movement of neurons from where they are produced to where they eventually settle in the brain; and (4) selective cell death, more neurons are produced than the newly formed brain eventually contains. As Benno (1990) discusses, these processes are governed by probabilistic epigenetic factors. Much can go wrong in neural development, a result of environmental input. It is of interest to note that even at the time of this writing, in 2012, the mechanisms responsible for these neural changes are not yet fully understood.

Neural development is rendered even more complex by postnatal epigenetic effects. We now know that genomes remain active throughout life and are affected by many environmental stimuli (Charney, 2012). Social deprivation, contextual assets or deficits associated with socioeconomic status, or malnutrition all have dramatic effects on development in general and on brain development in particular. Indeed, we have known for some time that pervasive and sustained negative social conditions have a profound effect not only on brain development but also on physical and behavioral development in general, a phenomenon labeled psychosocial dwarfism (Reinhart & Drash, 1969).

Of course, consistent with an embodiment model of development, biological factors are not to be ignored; they are, however, participating and not causal factors in behavioral origins (Pronko, 1988). On this point Bunge (2003, p. 52) has said, "However, far from being fully autonomous, the brain is intimately connected with the endocrine and immune systems, as well as with such support systems as the cardiovascular, digestive, and musculo-skeletal systems." With respect to embodiment, Overton (2008) states, "embodiment [is not] about a set of genes causing behavior, or a split-off brain causing or being the mind .... Embodiment is a concept of synthesis, a bridge that joins broad areas of inquiry into a unified whole (e.g., the biological, the phenomenological, the sociocultural and environmental) as relative standpoints that together constitute the whole" (p. 3, emphasis added). He further adds that the embodiment model, i.e., a person-centered perspective, "rescues psychology generally, and developmental psychology specifically, from becoming a mere adjunct to biology" (p. 7). We will leave the discussion of the mind to another time (Greenberg & Partridge, 2010; for an interesting discussion of mind). It is suffice to say for the time being that we agree with Overton's (2008) point that "mind and body are coconstituted, and as such, form indissociable complements" (p. 3).

Thus, we appreciate the crucial role played by neural factors in the development of complex cognitive processes. In addition, we recognize the evolutionary point that brains have increased not only in size, but also in cognitive computational power (Rumbaugh & Pate, 1984a,b). Greenberg, Partridge, Weiss and Haraway (1998) provide an introduction to the role of brain evolution and complexity in the appearance of language, complex learning, and cultural complexity, especially in humans.

8. CONCLUDING COMMENTS

Our goal of writing this chapter was to present psychology as a natural science, a unique science, grounded of course in biology, but not a biological science. As a natural science, the principles of psychology are consistent and compatible with those of other sciences, although not reducible to them. Specifically, behavior cannot be reduced or understood as the product of genes and brains. Rather, psychology is a developmental science (Greenberg, Partridge, Mosack & Lambdin 2006) that is embodied. Accordingly, we discussed how concepts of development from biology and other sciences play the crucial role in behavioral origins.

A key idea in this developmental approach is that of emergence, as we discussed at length. Coming from physics, and before that from philosophy, emergence suggests that new properties simply appear when older parts are organized in new ways, i.e., they emerge. No reductionist analysis will find the reality of the new whole in the old parts. The principal thesis of emergentist theorists has been that even with a full knowledge of all the lower order parts and their potential relationships, the laws of the higher order wholes cannot be deduced. Thus, even armed with the full sequence of the human genome and a full understanding of the multiplicity of regulatory networks involved in protein synthesis, a full understanding of even morphological phenotypes, let alone behavioral phenotypes, could not be ascertained. Similarly, even if all the neural circuits of the human brain could be sketched in a grand schematic and all the probabilistic rules governing the synaptic flow of information could be cataloged, neuroscience would be no better prepared to predict behavior. This point has always been the defining claim of emergent holism, and for the better part of scientific history, it has been the Achilles heel of the position of genetic or biological reductionism.

While we present our formulations to apply across the animal spectrum, i.e., to psychology in general (Greenberg & Haraway, 2002; Greenberg, Partridge, Weiss, & Pisula, 2004), our focus here is on humans. Consistent with our definition of psychology as a biopsychosocial science, development is an embodied phenomenon (Overton, 2006, 2007). Among the implications of embodiment is that at any point in time, development cannot be understood in reference to a single variable or a single dimension, either internal or external. As we have pointed out, the study of cognitive development, brain development, personality development, or any psychological phenomenon must recognize the fusion of these processes with other internal and external dimensions of change in which they are fused. Of course, embodiment with all levels of the ecology necessarily includes the individual's embeddedness with temporality (history) that involves at the very least ontogenetic and phylogenetic time (Elder, 1998). Furthermore, embodiment is not static; that is, due to epigenesis, developmental scientists should expect qualitative discontinuities in the nature of the embodied developing individual across time and place, both ontogenetically and phylogenetically.

An important goal we had in mind in writing this chapter was to make suggestions regarding the future direction to be taken by others. Developmental scientists are advised to consider phylogenetic and ontogenetic dimensions in both time and place (Elder, 1998) in trying to understand the trajectory of an individual's life course.

ACKNOWLEDGMENTS

This paper was supported in part by a grant from the John Templeton Foundation. The authors would like to thank Richard M. Lerner for his contribution to the writing of this chapter

REFERENCES

Ahn, C. A., Tewari, M., Poon, C.-S., & Phillips, R. S. (2006). The limits of reductionism in medicine: could systems biology offer an alternative? *PloS Medicine, 3*(6), 001–005.

Aronson, L. R. (1984). Levels of integration and organization: a revaluation of the evolutionary scale. In G. Greenberg & E. Tobach (Eds.), *Behavioral evolution and integrated levels* (pp. 57–81). Hillsdale, NJ: Erlbaum.

Aronson, L. R., Tobach, E., Rosenblatt, J. R., & Lehrman, D. H. (1972). *Selected writings of T. C. Schneirla.* San Francisco: Freeman.

Baltes, P. B. (1987). Theoretical propositions of life-span developmental psychology: on the dynamics between growth and decline. *Developmental Psychology, 23*, 611–626.

Baum, W. M. (2005). *Understanding behaviorism: Behavior, culture, and evolution.* Malden, MA: Blackwell.

Bedau, M. A., & Humphreys, P. (Eds.), (2008). *Emergence: Contemporary readings in philosophy and science.* Cambridge, MA: MIT Press.

Benjamin, L. T., Jr. (2007). *A brief history of modern psychology.* Malden, MA: Blackwell.

Benno, R. H. (1990). Development of the nervous system: genetics, epigenetics, and phylogenetics. In M. E. Hahn, J. K. Hewitt, N. D. Henderson & R. H. Benno (Eds.), *Developmental behavior genetics: Neural, biometrical, and evolutionary approaches* (pp. 113–143). New York: Oxford University Press.

Boring, E. G. (1950). *A history of experimental psychology* (2nd ed.). New York: Appleton-Century-Crofts.

Bronfenbrenner, U. (1979). *The ecology of human development: Experiments by nature and design.* Cambridge, MA: Harvard University Press.

Bronfenbrenner, U. (2005). *Making human beings human.* Thousand Oaks, CA: Sage.

Bunge, M. A. (2003). *Emergence and convergence: Qualitative novelty and the unity of knowledge.* Toronto: University of Toronto Press.

Campbell, D. T. (1990). Levels of organization, downward causation, and the selection-theory approach to evolutionary epistemology. In E. Tobach & G. Greenberg (Eds.), *Scientific methodology in the study of the mind: Evolutionary epistemology* (pp. 1–17). Hillsdale, NJ: Erlbaum.

Carroll, S. B. (2001). Chance and necessity: the evolution of morphological complexity and diversity. *Nature, 409*, 1102–1106.

Charney, E. (2012). Behavior genetics and post genomics. *Behavioral and Brain Sciences, 35*(6), 1–80.

Chorover, S. L. (1990). Paradigms lost and regained: changing beliefs, values, and practices in neuropsychology. In G. Greenberg & E. Tobach (Eds.), *Theories of the evolution of knowing* (pp. 87–106). Hillsdale, NJ: Erlbaum.

Clarke, K. A., & Primo, D. M. (April 1, 2012). Overcoming 'physics envy'. *New York Times Sunday Review*, 8.

Darwin, C. (1859). *The origin of species by means of natural selection or the preservation of favoured races in the struggle for life.* London, England: J. Murray.

Davies, P., & Gribbin, J. (1992). *The matter myth*. New York: Simon & Schuster.

Dobzhansky, T. (1973). Nothing in biology makes sense except in the light of evolution. *The American Biology Teacher, 35*(3), 125–129.

Elder, G. H., Jr. (1974). *Children of the great depression*. Chicago: University of Chicago Press.

Elder, G. H., Jr. (1980). Adolescence in historical perspective. In J. Adelson (Ed.), *Handbooks of adolescent psychology* (pp. 3–46). New York: Wiley.

Elder, G. H., Jr. (1998). The life course and human development. In W. Damon (Series Ed.), & R. M. Lerner (Vol. Ed.), *Handbook of child psychology. Theoretical models of human development* (Vol. 1, 5th ed., pp. 939–991). Hoboken, NJ: John Wiley & Sons, Inc.

Eldredge, N., & Gould, S. J. (1972). Punctuated equilibria: an alternative to phyletic gradualism. In T. J.M. Schopf (Ed.), *Models in paleobiology* (pp. 82–115). San Francisco: Freeman.

Erickson, E. H. (1959). Identity and the life cycle. *Psychological Issues, 1*, 50–100.

Feibleman, J. K. (1954). Theory of integrative levels. *British Journal for the Philosophy of Science, 5*, 59–66.

Ford, D. H., & Lerner, R. M. (1992). *Developmental systems theory: An integrative approach*. Newbury Park, CA: Sage.

Freud, S. (1954). *Collected works, standard edition*. London: Hogarth Press.

Friedenberg, J. (2009). *Dynamical psychology: Complexity, self-organization and mind*. Litchfield Park, AZ: ISCE Publishing.

Gell-Mann, M. (1995). Plectics. In J. Brockman (Ed.), *The third culture: Beyond the scientific revolution* (pp. 316–332). New York: Simon & Schuster.

Goodwin, B. (1994). *How the leopard got its spots: The evolution of complexity*. New York: Scribner's.

Gottlieb, G. (1973). Introduction to behavioral embryology. In G. Gottlieb (Ed.), *Studies on the development of behavior and the nervous system Behavioral embryology* (Vol. 1, pp. 3–45). New York: Academic Press.

Gottlieb, G. (1984). Evolutionary trends and evolutionary origins: relevance to theory in comparative psychology. *Psychological Review, 91*, 448–456.

Gottlieb, G. (1992). *Individual development and evolution: The genesis of novel behavior*. New York: Oxford University Press.

Gottlieb, G. (1997). *Synthesizing nature-nurture: Prenatal roots of instinctive behavior*. Mawah, NJ: Earlbaum.

Gottlieb, G. (2001). A developmental psychobiological systems view: early formulation and current status. In S. Oyama, P. E. Griffiths & R. D. Gray (Eds.), *Cycles of contingency: Developmental systems and evolution* (pp. 41–54). Cambridge, MA: MIT Press.

Gottlieb, G. (2003). On making behavior genetics truly developmental. *Human Development, 46*, 337–355.

Gould, S. J. (1997). Darwinian fundamentalism. *New York Review of Books, 44*(10), 34–37.

Gould, S. J. (2002). *The structure of evolutionary theory*. Cambridge, MA: Harvard University Press.

Gould, S. J., & Eldredge, N. (1993). Punctuated equilibrium comes of age. *Nature, 366*, 223–227.

Greenberg, G. (2011). The failure of biogenetic analysis in psychology: why psychology is not a biological science. *Research in Human Development, 8*, 173–191.

Greenberg, G., & Haraway, M. H. (2002). *Principles of comparative psychology*. Boston, MA: Allyn & Bacon.

Greenberg, G., & Partridge, T. (2010). Biology, evolution, and psychological development. In W. F. Overton (Ed.), *Handbook of life-span development Cognition, biology, and methods across the lifespan* (Vol. 1, pp. 115–148).

Greenberg, G., Partridge, T., & Ablah, E. (2006). The significance of the concept of "emergence" for comparative psychology. In D. Washburn (Ed.), *Primate perspectives on behavior and cognition* (pp. 81–97). Washington, DC: APA Books.

Greenberg, G., Partridge, T., Mosack, V., & Lambin, C. (2006). Psychology is a developmental science. *International Journal of Comparative Psychology, 19*, 185–205.

Greenberg, G., Partridge, T., Weiss, E., & Harawway, M. M. (1998). Integrative levels, the brain, and the emergence of complex behavior. *Review of General Psychology, 3*, 168–187.

Greenberg, G., Partridge, T., Weiss, E., & Pisula, W. (2004). Comparative psychology: a new perspective for the 21st century. Up the spiral staircase. *Developmental Psychobiology, 44*, 1–15.

Haggbloom, S. J., Warnick, R., Warnick, J. E., Jones, V. K., Yarbrough, G. L., Russell, T. M., et al. (2002). The 100 most eminent psychologists of the 20th century. *Review of General Psychology, 6*, 139–152.

Haig, D. (2004). The (dual) origin of epigenetics. *Cold Spring Harbor Symposium on Quantitative Biology, LXIX*, 1–4.

Haken, H. (1993). Synergetics as a strategy to cope with complex systems. In H. Haken & A. Mikhailov (Eds.), *Interdisciplinary approaches to nonlinear complex systems* (pp. 5–11). Berlin, Germany: Springer-Verlag.

Haken, H. (1994). Can synergetics serve as a bridge between the natural and social sciences? In R. K. Mishra, D. Maaß & E. Zwierlein (Eds.), *On self-organization: An interdisciplinary search for a unifying principle* (pp. 51–65). Berlin, Germany: Springer-Verlag.

Haraway, M. H., & Maples, E. G., Jr. (1998). Species-typical behavior. In G. Greenberg & M. H. Haraway (Eds.), *Comparative psychology: A handbook* (pp. 191–197). New York: Garland.

Ho, M. W. (2010). Development and evolution revisited. In K. E. Hood, C. T. Tucker, G. Greenberg & R. M. Lerner (Eds.), *Handbook of developmental science, behavior, and genetics* (pp. 61–109). Malden, MA: Wiley-Blackwell.

Holliday, R. (2006). Epigenetics: a historical review. *Epigenetics, 1*, 76–80.

Itani, J. (1961). The society of Japanese monkeys. *Japanese Quarterly, 8*(4), 421–430.

Jablonka, E., & Lamb, M. J. (2005). *Evolution in four dimensions: Genetic, epigenetic, behavioral, and symbolic variation in the history of life.* Cambridge, MA: MIT Press.

Kantor, J. R. (1924). *Principles of psychology.* (Vol. 1). Bloomington, IN: Principia Press.

Kantor, J. R. (1926). *Principles of psychology.* (Vol. 2). Bloomington, IN: Principia Press.

Kantor, J. R. (1959). *Interbehavioral psychology* (2nd ed.). Chicago: Principia Press.

Kauffman, S. (1993). *The origins of order: Self organization and selection in evolution.* New York, NY: Oxford University Press.

Kauffman, S. (1995). *At home in the universe: The search for the laws of self-organization and complexity.* New York, NY: Oxford University Press.

Kauffman, S. (2000). *Investigations.* Oxford, UK: Oxford University Press.

Kawamura, S. (1959). The process of sub-culture propagation among Japanese macaques. *Primates, 2*, 43–60.

Kelso, J. A. (1995). *Dynamic patterns: The self-organization of brain and behavior.* Cambridge, MA: MIT Press.

Koch, C., & Laurent, G. (1999). Complexity and the nervous system. *Science, 284*, 96–98.

Krasny, M. (1997). Stephen Jay Gould. *Mother Jones, 22*(1), 60–63.

Kuhn, T. S. (1962). *The structure of scientific revolutions.* Chicago: University of Chicago Press.

Kuo, Z. Y. (1967). *The dynamics of behavior development: An epigenetic view.* New York: Random House.

Lerner, R. M. (1991). Changing organism context relations as the basic process of development: a developmental contextual perspective. *Developmental Psychology, 27*, 27–32.

Lerner, R. M. (1998). Theories of human development: contemporary perspectives. In W. Damon (Series Ed.), & R. M. Lerner (Vol. Ed.), *Handbook of child psychology. Theoretical models of human development* (Vol. 1, 5th ed., pp. 1–24). New York, NY: Wiley.

Lerner, R. M. (2002). *Concepts and theories of human development* (3rd ed.). Mahwah, NJ: Erlbaum.

Lerner, R. M. (2006). Developmental science, developmental systems, and contemporary theories of human development. In R. M. Lerner (Ed.), & Richard M. Lerner, & William Damon (Editor-in-Chief). *Theoretical models of human development. Handbook of child psychology* (Vol. 1, 6th ed., pp. 1–17). New York: Wiley.

Lerner, R. M. (2012). Essay review: developmental science: past, present, and future. *International Journal of Developmental Science, 6*, 29–36.

Lerner, R. M., & Bush-Rossnagel, N. A. (1981). Individuals as producers of their development: conceptual and empirical issues. In R. M. Lerner & N. A. Bush-Rossnagel (Eds.), *Individuals as producers of their development: A life-span perspective* (pp. 1–36). New York: Academic Press.

Lerner, R. M., Easterbrooks, A. M., & Mistry, J. (2012). In I. B. Weiner (Editor-in-chief.), *Handbook of psychology. Developmental psychology* (Vol. 6, 2nd ed). Hoboken, NJ: Wiley.

Lewis, M. D., & Granic, I. (1999). Who put the self in self-organization? A clarification of terms and concepts for developmental psychopathology. *Development and Psychopathology, 11*, 365–374.

Luncz, L. V., Mundry, R., & Boesch, C. (2012). Evidence for cultural differences between neighboring chimpanzee communities. *Current Biology, 22*, 922–926.

Magnusson, D. (1996). *The life-span development of individuals: Behavioral, neurobiological, and psychosocial perspectives. A synthesis.* Cambridge, England: Cambridge University Press.

Mascolo, M. F., & Fischer, K. W. (2010). The dynamic development of thinking, feeling, and acting over the life span. In W. F. Overton (Ed.), & R. M. Lerner (Editor-in-Chief). *Handbook of life-span development Biology, cognition and methods across the life-span* (Vol. 1, pp. 149–194). Hoboken, NJ: Wiley.

Masterpasqua, F. (2009). Psychology and epigenetics. *Review of General Psychology, 13*, 194–201.

Matasaka, N. (1994). Effects of experience with live insects on the development of fear of snakes in squirrel monkeys, *Saimiri sciureus. Animal Behaviour, 46*, 741–746.

Matsuzawa, T. (1998). Chimpanzee behavior: a comparative cognitive perspective. In G. Greenberg & M. M. Haraway (Eds.), *Comparative psychology: A handbook* (pp. 360–375). New York, NY: Garland.

Maynard Smith, J. (1970). Time in the evolutionary process. *Studium Generale, 23*, 266–272.

Maynard Smith, J., & Holliday, R. (1979). Preface. In J. Maynard-Smith & R. Holliday (Eds.), *Evolution of adaptation by natural selection* (pp. v–vii). London: The Royal Society.

Mayr, E. (1979). Evolution. *Scientific American, 239*(3), 46–55.

Mayr, E. (1982). *The growth of biological thought: Diversity, evolution, and inheritance.* Cambridge, MA: Harvard University Press.

Mayr, E. (1991). *One long argument.* Cambridge, MA: Harvard University Press.

Mayr, E., & Provine, W. (1980). *The evolutionary synthesis.* Cambridge, MA: Harvard University Press.

McLaughlin, B. P. (2008). Emergence and supervenience. In M. A. Bedau & P. Humphreys (Eds.), *Emergence: Contemporary readings in philosophy and science* (pp. 81–97). Cambridge: MA MIT Press.

Meaney, M. J. (2010). Epigenetics and the biological definition of gene x environment interactions. *Child Development, 81*(1), 41–79.

Michel, G. F., & Moore, C. L. (1995). *Developmental psychobiology: An interdisciplinary science.* Cambridge, MA: MIT Press.

Michel, G. F. (2010). The roles of environment, experience, and learning in behavioral development. In K. E. Hood, C. T. Halpern, G. Greenberg & R. M. Lerner (Eds.), *Handbook of developmental science, behavior, and genetics* (pp. 123–165). Malden, MA: Wiley-Blackwell.

Molenaar, P. C. M. (2010). On the limits of standard quantitative genetic modeling of inter-individual variation: extensions, ergodic conditions and a new genetic factor model of intra-individual variation. In K. T. Hood, C. T. Halpern, G. Greenberg & R. M. Lerner (Eds.), *Handbook of developmental science, behavior, and genetics* (pp. 626–648). Malden, MA: Blackwell.

Moltz, H. (1965). Contemporary instinct theory and the fixed action pattern. *Psychological Review, 72*, 27–47.

Montagu, A. (1962). Our changing conception of human nature. In *The humanization of man* (pp. 15–34). New York: Grove Press. (Reprinted from Impact [UNESCO], 1952, 3, 219–232).

Moore, C. (2003). Evolution, development, and the individual acquisition of traits: what we've learned since Baldwin. In B. H. Weber & D. J. Depew (Eds.), *Evolution and learning: The Baldwin effect reconsidered* (pp. 115–139). Cambridge, MA: MIT Press.

Moore, D. S. (2008). Integrating development and evolution in psychology: looking back, moving forward. *New Ideas in Psychology, 26*, 327–331.

Nesselroade, J. R., & Molenaar, P. C. M. (2010). Emphasizing intraindividual variability in the study of development over the lifespan. In W. F. Overton (Ed.), & R. M. Lerner (Editor-in-Chief). *Biology, cognition and methods across the life-span. Handbook of life-span development*. (Vol. 1). Hoboken, NJ: Wiley.

Nicolis, G. (1989). Physics of far-from-equilibrium systems and self-organisation. In P. Davies (Ed.), *The new physics* (pp. 316–347). Cambridge: Cambridge University Press.

Nieuwkoop, P. D., Johnen, A. G., & Albers, B. (1985). *The epigenetic nature of early chordate development*. Cambridge, MA: Cambridge University Press.

Nishida, T. (1986). Learning and cultural transmission in non human primates. *Folia Primatologica, 12*, 273–283.

Novak, G. (1996). *Developmental psychology: Dynamical systems and behavior analysis*. Reno, NV: Context Press.

Novikof, A. (1945). The concept of integrative levels in biology. *Science, 101*, 209–215.

Oparin, A. I. (1961). *Life: Its nature, origin and development*. New York: Academic Press.

Overton, W. F. (2006). Developmental psychology: philosophy, concepts, and methodology. In R. M. Lerner (Ed.), & William Damon (Editor-in-Chief). *Theoretical models of human development. Handbook of child psychology* (Vol. 1, 6th ed., pp. 18–88). New York: Wiley.

Overton, W. F. (2007). Embodiment from a relational perspective. In W. F. Overton, U. Mueller & J. L. Newman (Eds.), *Developmental perspective on embodiment and consciousness* (pp. 1–18). Hillsdale, NJ: Erbaum Associates.

Overton, W. F. (2008). Embodiment from a relational perspective. In W. F. Overton, U. Müller & J. L. Newman (Eds.), *Developmental perspective on embodiment and consciousness* (pp. 1–18). Hillsdale, NJ: Erlbaum.

Overton, W. F. (2010). Life span development: concepts and issues. In W. F. Overton (Ed.), & R. M. Lerner (Editor-in-Chief). *Cognition, biology, and methods Handbook of life-span development* (Vol. 1, pp. 1–29). Hoboken, NJ: Wiley.

Overton, W. F. (2011). Relational developmental systems and quantitative behavior genetics: alternative or parallel methodologies? *Research in Human Development, 8*, 258–263.

Overton, W. F. (2012). Evolving scientific paradigms: retrospective and prospective. In L. L'Abate (Ed.), *Paradigms in theory construction* (pp. 31–68). New York: Springer.

Overton, W. F., & Lerner, R. M. (2012). Relational developmental systems: a paradigm for developmental science in the post genomic era. *Behavioral and Brain Sciences, 35*(6).

Overton, W. F., & Müller, U. (2012). Meta-theories, theories, and concepts in the study of development. In R. M. Lerner, M. A. Esterbrooks, & J. Misatry (Eds). Comprehensive handbook of psychology: Developmental psychology (Vol. 6., 2nd ed). In R. M. Lerner, M. A. Esterbrooks & J. Misatry (Eds.), & I. B. Weiner (Editor-in-Chief). *Meta-theories, theories, and concepts in the study of development*. New York: Wiley.

Oyama, S., Griffiths, P. E., & Gray, R. D. (Eds.), (2001). *Cycles of contingency: Developmental systems and evolution*. Cambridge, MA: MIT Press.

Partridge, T., & Greenberg, G. (2010). Contemporary ideas of physics and biology in Gilbert Gottlieb's epigenesis. In K. T. Hood, C. T. Halpern, G. Greenberg & R. M. Lerner (Eds.), *Handbook of developmental science, behavior, and genetics* (pp. 166–202). Malden, MA: Blackwell.

Prigogine, I. (1994). Mind and matter: beyond the Cartesian dualism. In K. H. Pribram (Ed.), *Origins: Brain and self organization* (pp. 3–15). Hillsdale, NJ: Erlbaum.

Prigogine, I., & Stengers, I. (1984). *Order out of chaos: Man's new dialogue with nature*. New York: Bantam Books.

Pronko, N. H. (1988). Heredity versus environment. In N. H. Pronko (Ed.), *From AI to Zeitgeist: A philosophical guide for the skeptical psychologist* (pp. 75–80). New York: Greenwood Press.

Reid, R. G.B. (2007). *Biological emergences: Evolution by natural experiment*. Cambridge, MA: MIT Press.

Reinhart, J. B., & Drash, A. L. (1969). Psychosocial dwarfism: environmentally induced recovery. *Psychosomatic Medicine, 31*, 165–172.

Robert, J. S. (2004). *Embryology, epigenesis, and evolution: Taking development seriously*. Cambridge, UK: Cambridge University Press.

Robert, J. S. (2008). Taking old ideas seriously: evolution, development, and human behavior. *New Ideas in Psychology, 26*, 387–404.

Rumbaugh, D. M., & Pate, J. L. (1984a). The evolution of cognition in primates: a comparative perspective. In H. L. Roitblatt, T. G. Bever & H. S. terrace (Eds.), *Animal cognition* (pp. 569–587). Hillsdale, NJ: Erlbaum.

Rumbaugh, D. M., & Pate, J. L. (1984b). Primates' learning by levels. In G. Greenberg & E. Tobach (Eds.), *Behavioral evolution and integrative levels* (pp. 221–240). Hillsdale, NJ: Erlbaum.

Sameroff, A. J. (1975). Transactional models in early social relations. *Human Development, 18*, 65–79.

Saunders, P. T., & Ho, M. W. (1976). On the increase in complexity in evolution. *Journal of Theoretical Biology, 63*, 375–384.

Saunders, P. T., & Ho, M. W. (1981). On the increase in complexity in evolution. II. The relativity of complexity and the principle of minimum increase. *Journal of Theoretical Biology, 90*, 515–530.

Sawyer, R. K. (2002). Emergence in psychology: lessons from the history of non-reductionist science. *Human Development, 45*, 2–28.

Schneirla, T. C. (1949). Levels in the psychological capacities of animals. In R. W. Sellars, V. J. McGill & M. Farber (Eds.), *Philosophy for the future* (pp. 243–286). New York: Macmillan.

Schneirla, T. C. (1957). The concept of development in comparative psychology. In D. B. Harris (Ed.), *The concept of development: An issue in the study of human behavior* (pp. 78–108). Minneapolis, MN: University of Minnesota Press.

Sheldrake, R. (1995). *Seven experiments that could change the world*. New York: Riverhead Books.

Singer, W. (2009). The brain: a complex self-organizing system. *European Review, 17*, 321–329.

Singh, S. (2004). *Big bang*. New York: Harper Collins.

Smith, L. B., & Thelen, E. (2003). Development as a dynamic system. *Trends in Cognitive Science, 7*, 343–348.

Smotherman, W. P., & Robinson, S. R. (Eds.), (1988). *Behavior of the fetus*. Caldwell, NJ: The Telford Press.

Swenson, R. (1998). Thermodynamics, evolution, and behavior. In G. Greenberg & M. M. Haraway (Eds.), *Comparative psychology: A handbook* (pp. 207–218). New York: Garland.

Tauber, A. L. (2011). Epigenetics and the "New Biology": enlisting in the assault on reductionism. In S. B. Gisis & E. Jablonka (Eds.), *Transformations of Lamarckism: From subtle fluids to molecular biology* (pp. 385–387). Cambridge, MA: MIT Press.

Thelen, E., & Smith, L. B. (1994). *A dynamic systems approach to the development of cognition and action*. Cambridge, MA: MIT Press.

Tobach, E., & Greenberg, G. (1984). The significance of T. C. Schneirla's contributions to the concept of levels of integration. In G. Greenberg & E. Tobach (Eds.), *Behavioral evolution and integrative levels* (pp. 1–7). Hillsdale, NJ: Erlbaum.

Vogel, S. (1998). Locomotor behavior and physical reality. In G. Greenberg & M. M Haraway (Eds.), *Comparative psychology: A handbook* (pp. 713–719). New York: Garland.

Von Uexkuell, J. (1957). A stroll through the world of animals and men. In C. H. Schiller (Ed. & Trans.). Instinctive behavior (pp. 5–80). New York: International Universities Press.

Waddington, C. H. (1957). *The strategy of the genes*. London: Allyn and Unwin.

Wahlsten, D. (2010). Probabilistic epigenesis and modern behavioral and neural genetics. In K. E. Hood, C. T. Halpern, G. Greenberg & R. M. Lerner (Eds.), *Handbook of developmental science, behavior and genetics* (pp. 110–122). Malden, MA: Wiley-Blackwell.

Wahlsten, D. (2012). The hunt for gene effects pertinent to behavioral traits and psychiatric disorders: from mouse to human. *Developmental Psychobiology, 54*, 475–492.

Wallman, J. (1979). A minimum visual restriction experiment: preventing chicks from seeing their feet affects later responses to mealworms. *Developmental Psychobiology, 12*, 391–397.

White, S. H. (1968). The learning-maturation controversy: Hall to Hull. *Merrill-Palmer Quarterly, 14*, 187–196.

CHAPTER FIVE

The Evolution of Intelligent Developmental Systems

Ken Richardson

Bellevue, Dunblane, UK
E-mail: k.richardson@mac.com

Contents

Abstract

This chapter aims to understand the relations between the evolution and development of complex cognitive functions by emphasizing the context of complex, changeable environments. What evolves and develops in such contexts cannot be achieved by linear deterministic processes based on stable "codes". Rather, what is needed, even in the molecular ensembles of single-cell organisms, are "intelligent" systems with nonlinear dynamic processing, sensitive to informational structures, not just elements, in environments. This is the view emerging in recent molecular biology. The research is also constructing a new "biologic" of both evolution and development, providing a clearer rationale for transitions into more complex forms, including epigenetic, physiological, nervous, cognitive, and human sociocognitive forms. This chapter explains how these transitions form a nested hierarchical system in which the dynamics within and between levels creates emergent abilities so often underestimated or

Advances in Child Development and Behavior, Volume 44
ISSN 0065-2407, http://dx.doi.org/10.1016/B978-0-12-397947-6.00005-2

even demeaned in previous accounts, especially regarding human cognition. The implications of the view for human development in modern societies are also briefly considered.

1. INTRODUCTION

An enduring problem in the study of both evolution and development has been to explain how we get such complex forms and functions as cognitive systems from the original "bag of molecules" without making improbable assumptions, and while reconciling the processes in one (across generations) with those in the other (within generations). Throughout the twentieth century, nativism, associationism and constructivism vied with one another, each claiming to fill gaps in the others' accounts. Eventually, many associationists, via connectionism, claimed to be constructivists; nativists increasingly claimed some of the constructivists' ground; constructivists interpreted new findings by incorporating some nativist principles; and some theorists attempted to amalgamate all three (for review, see Richardson, 1998).

None of these amalgamations proved to be totally convincing, albeit highly stimulating to an already exciting field with profound scientific, social and policy implications. All rested on some fundamental, but unsubstantiated, principles, as the source of evolving or developing complexity. But, between 1991 and 2010, two new themes have appeared, each claiming to be based on firm foundations and with rival claims about the role of evolution in the nature of cognitive systems and their development.

The first of these—which came to be known as evolutionary psychology (EP)—asserted a fundamentalist Darwinism, together with a radical geneticism, throwing all explanation onto the selection environment for the shaping of cognitive functions. Thus, "Natural selection shapes (mental) mechanisms so that their structure meshes with the evolutionarily stable features of their particular problem domains" (Cosmides & Tooby, 1994, p. 96). This results in "functionally specialized problem-solving strategies… for ancestrally recurrent adaptive problems" (Cosmides, Barrett, & Tooby, 2010, p. 9007). Accordingly, "The mind is organized into modules or mental organs, each with a specialized design… The module's basic logic is specified by our genetic program" (Pinker, 1997, p. 21; see also Pinker, 2010). "Complex adaptations," therefore, "are intricate machines that require complex 'blueprints' at the genetic level" (Tooby & Cosmides, 1995, p. 78). And the role of the blueprint is to "reconstruct in offspring the evolved functional organization which was present in their parents," in environments "that have endured long enough to have been organized by natural selection" (Tooby & Cosmides, 1995, pp. 77–78).

Critics have, of course, reminded EP theorists of many other factors in both evolution and development, and there have been efforts to accommodate these (e.g., Buss & Reeve, 2003). But it has remained highly controversial for framing "the public debate on human life and development" while resting "on shaky empirical evidence, flawed premises and unexamined political presuppositions" (Rose & Rose, 2001, cover). Accordingly, numerous issues and limitations continue to be exposed and discussed, as in a special paper recently in the *American Psychologist* (Confer et al., 2010). So much so that other theorists can argue that "evolutionary psychology currently offers no coherent framework for how to integrate genetic, environmental, and experiential factors into a theory of behavioral or cognitive phenotypes" (Lickliter & Honeycutt, 2003, p. 869).

The other new theme involved a radical break from the linear determinism that framed most theories hitherto, including EP. In order to model and theorize mental, including cognitive, functions, psychology has followed classical science in reducing natural entities to as few variables as possible, sequestered from their natural contexts, while pretending relationships between them to be linear and time-free (stable over time and space). The new approach, however, examined the behavior of great numbers of variables, simultaneously, in their natural contexts, including nonlinear relationships and changes over time. It showed how multivariate, nonlinear systems can have properties quite different from simple linear ones. For example, they can respond in unexpected ways to external or internal perturbations, and self-organize into radically different phases, with the emergence of new, more complex forms and properties not expected from the old.

Such descriptions of the physical world, chimed in perfectly with some views in biology, as in Bertalanffy's (1968) "general theory of systems". It became taken up in life sciences generally (Kauffman, 1995), and human development in particular (e.g., Fogel, King, & Shanker, 2008). It has given rise to exciting new theories of development, and its connection with evolution, coming to be called dynamic systems theory (DST). It suggests more plausibly than EP or antecedent theories how complex cognitive systems emerged from simpler systems, while simultaneously relating evolution and development through environmental realities. It is just such a vision that I aim to support and foster in this chapter. To start, it is worth mentioning briefly the specific problems it is up against as they serve as a useful foil for defining the issues surrounding evolution and development in general.

2. WHAT DEVELOPS?

The part of the environment that does the natural selection, in the EP (and general nativist) view, also determines what develops. It becomes reflected in the brain as "conceptual schemes" that "determine how the person as a whole behaves" (Pinker, 2006, p. 2). Bjorklund and Pelligrini (2001, p. 191) write "What infants come into the world with are processing biases and constraints—products of natural selection—that serve as the foundation for developing a human mind." Geary (2007, p. 3) describes these as "folk knowledge" that "results from the organization of the brain systems that have evolved to process and integrate specific forms of information," such as "different systems for detecting features of typical prey or predatory species". Such forms are prestructured in outline by genes but remain sufficiently plastic to adjust to environmental experience (Bjorklund, Ellis, & Rosenberg, 2007). As mentioned earlier, these functions are envisaged as "mental organs" or "modules," perhaps hundreds or thousands of them, each genetically designed by natural selection.

The problem is that these are purely nominal, and therefore hollow, accounts of what we might expect to develop, with little insight into their constitution and nature of function. Accordingly, they leave much mystery about what develops, as admitted by some nativists themselves (e.g., Fodor, 2000). Above all, they cannot say why what develops is so much more complex in humans. Similar problems apply, of course, to past associationist and constructivist accounts, which equivocate, by and large, about what infants come into the world with and why. In this chapter, I aim to show that what develops has been exceedingly understated in these approaches. Above all, I suggest that it is a poverty of analyses of *environments* that has made them blind to the richness, sophistication and computational complexity of even the simplest cognitive systems.

3. HOW?

The details of how development arises from supposed genetic codes, to mature functions, are given as rather vague outlines in EP stories. Generally, what is envisaged is a kind of automatic assembly line of genetically specified parts culminating in brain structures and cognitive functions, thereby reinstating the form and function that assisted survival in parents. So Pinker (2006, p. 3) claims that "a linear string of DNA can direct the assembly of an intricate three-dimensional organ that lets us think, feel and learn." Hence, the fashion for viewing the task of biology and psychology as one of "reverse engineering at every level" (Dennett, 2011), i.e., that the process of development can be understood by (theoretically) rewinding the assembly line, to reveal each programmed step on the way (and, presumably, what determines it or possibly interferes with it).

The problem is if development proceeds mainly from the information in genetic codes, why have organisms evolved the highly complex regulations of gene transcription and epigenesis now being revealed in molecular biology (see below, sections 9-11)? Failure to face this question is why Ellis and Bjorklund (2005, p. 9) can argue that "although evolutionary psychology does not ignore development, it has failed to treat it seriously in its theorizing."

Again, of course, the "how" of development, and its increasing complexity, has been less than fully coherent in associationist and constructivist accounts. There has been much talk of gene–environment interactions, but less about why they evolved. Moreover, it has rarely been clear how such interactions, rather than vague intermingling of effects, actually become directed along well-adapted channels. Here, I aim to show how a new biologic of evolution and development is emerging from contemporary molecular biology. Instead of a simple instrumental role in achieving adapted ends specified by genes, "intelligent" development itself has become the main means of adaptation and variation in complex functions, often creating new trajectories of evolution in the process. Ultimately, in human sociocognitive systems, the whole relationship with natural selection is turned on its head, the system adapting the world to itself, rather than vice versa.

4. EVOLUTION OF COMPLEXITY

As already implied, the bland Darwinian logic cannot easily explain how complex functions like cognition evolved from simpler adaptation (except by the tautology that they evolved because they were more successful in hypothetical "arms races"). This gap has become increasingly problematic in evolutionary theory generally. Failure to answer such questions has produced claims that "there is no attempt in neo-Darwinian theory to explain the ever-increasing complexity of living things" (Bird, 2003, p. 53), or that "natural selection theory by itself cannot account for increases in structural complexity" (Salthe, 2008, p. 363). We need good *reasons* for increasing complexity—in cognitive systems, as all else—because, without them, "the mechanisms behind the complexification and its relation to evolution are not well understood" (Gershenson & Lenaerts, 2008, p. 241). And, of course, it produces the EPs dilemma of having to claim that humans still carry "stone-age" cognitive adaptations that are misfits in the modern technological world (Pinker, 1997).

Increasing awareness of the gap has brought diverse attempts to fill it. One of these has been an *Extended Evolutionary Synthesis* (see, for example, Mazur, 2008) Another has been a new upsurge of creationism on both sides of the Atlantic: the belief that human mental functions are so complex than their evolution and development can only be explained by the intervention of a supernatural "intelligent designer" (for discussion, see Greener, 2007). Here, I want to show how an act of faith in the simple Darwinian logic has distracted us from the really complex products of evolution, namely the increasingly complex intelligent systems, able to abstract dynamic structure in order to render changeable environments more predictable (Richardson, 2010). As we shall see, such systems are fundamentally developmental ones. But understanding this requires a radical new look at the environments of evolution and development.

5. FROM STABLE TO DYNAMIC ENVIRONMENTS

Accounts of evolution and cognitive development, and the connections between them, have been based on purely nominal, unanalyzed, and impressionistic accounts of environments, that do not specify what it is about them that is important. In consequence, they have failed to identify what really matters. As two prominent EPs, Richerson and Boyd (2005, p. 337) had to admit, "it is not entirely clear what selective regimes favor complex cognition." Or, as Flinn, Geary, and Ward (2005, p. 10) put it, "Human cognitive abilities are extraordinary…The conditions favoring the evolution of human cognitive adaptations remain an enigma…it has proven difficult to identify a set of selective pressures that would have been sufficiently unique to the hominin lineage." Accordingly, "Most of the rules governing the evolutionary process toward more complex brains are still unknown" (Sporns & Kötter, 2004, p. 1910). Above all, the standard accounts do not say how or why human and other primate cognitive systems are so much more complex than those of worms or ants, or even noncognitive systems of adaptation.

The problem stems from the assumption that environments are relatively constant and repetitive both across and within generations. The problem is inherent in the simple logic of natural selection: if the environment experienced by offspring is not closely similar to that of parents, there could be no selection across generations because the targets are constantly moving (the definition of the "best" phenotype—and, therefore, the "best" genes—is constantly changing). This is what is implicit in the stipulations among EPs (mentioned in the quotations earlier) about "ancestrally *recurrent* adaptive problems", "*evolutionarily stable* features of … problem domains", environments "that have endured long enough" and so on (my emphases). This view of stable environments is the basis of genetic determinism of development: since the environment remains more or less the same, development is simply a matter of repeating in the current generation—through the genetic program—the designs selected in the past.

However, true such stability may be for many aspects of environments (as with seed sizes to the form of finch's beaks, or temperature to coat thickness), it is not general. Other important aspects of environments are constantly and rapidly changing, often by the activities of organisms themselves. Darwin himself pointed out that the "simpler ways of life" must have soon become filled in the course of evolution, forcing living things in to more complex, changeable, ones. Sometimes, these changes merely take the form of occasional perturbations (such as temperature changes) around a stable mean that can threaten to throw development and adaption off track. More often (as we shall see later) they involve changes in underlying structure, as well as surface appearance, within the lifetimes of most organisms, requiring means of tracking much faster than that of genetic selection and fixed adaptation. As recent DST accounts have insisted, changing environments produce a complexity of experience over time and space in a different league from Darwinian stable conditions. Let us consider these briefly in informational terms.

5.1. The Information in Environmental Change

All living things need predictability in their environments in order to survive. The perfect form of predictability is constancy of environment: but it has been shown how such environments foster only dull, settled states that bear little resemblance to the living and do not evolve (Langton, 1989). Life, that is, originated, not with the fortuitous conjunction of water and a few chemical compounds (whatever Martian explorers may think), but with the need to track environmental change. Classic Darwinian selection relies on a set of direct correlations—the simplest form of information—between a mutated gene, an advantageous change in the phenotype, and a change in the environment. Through such correlations, the DNA code is basically predicting that such and such a phenotype will be adaptive in the next generation, so long as the environment stays much the same.

But where aspects of environment change more rapidly, *within* generations, information-for-prediction is not so straightforward. The change has to be predicted, not in stable genetic codes, but by some contemporaneous means. Again, direct, bivariate, correlations may be useful: one cue now (such as increasing light) might predict forthcoming availability of nutrients in seconds, minutes or days. In psychology, of course, such simple associations provided the basis of old-fashioned "learning theory". But in more complex changing environments, pairwise associations will tend to be fewer and weaker. Instead, predictability has to be found in whole patterns of associations and their changes over time. Thus, a kingfisher darting for a fish will need to predict the latter's position according to its own distance from it now, as conditioned by its speed of flight, the flow of the stream, the prey's depth in the water (and associated light refraction), the fish's swimming speed, and so on. That is, predictability can only be found in the higher order correlations (two-way, three-way and higher interactions, or multivariate dependencies). When the dimensions are changing in time and space (are spatiotemporal), and are nonlinear, as in this case and most cases, information of great depth, but predictability, is possible. Changing environments, that is, can harbor predictability in their underlying structure.

Such information-in-structure is indeed tractable by intelligent organic systems. It is measurable in terms of "mutual information" derived from Shannon entropy, which generalizes naturally to higher order dependencies among variables (Prokopenko, Boschetti, & Ryan, 2009; Slonim, Atwal, Tkačik, & Bialek, 2005). Indeed, wherever researchers have looked, they have found such deeper, dynamic structure in environments, that can render complex environments predictable (e.g., Becks *et al.*, 2005; for review, see Richardson, 2010). These sources of information correspond with intuitive notions of underlying structure in the natural world, but systems more "intelligent" than stable genetic codes are needed to utilize it.

The recent remarkable discoveries in molecular biology are revealing that it is just such intelligent systems that have evolved in changeable environments. They demonstrate a striking creativity of responses, even in single cells, and suggest the crucial biological (and conceptual) basis for the evolution of even more complex intelligent systems, including cognitive systems: a new biologic of evolution and development that also explains the directiveness in gene–gene and gene–environment interactions. The discoveries also suggest a different relationship of the organism with the world, and the primary role of development in it, than the purely passive one described by EPs. The rest of this chapter is about the details of such intelligent systems and their evolution into more complex forms.

6. EARLY PERSPECTIVES: CANALIZATION AND DEVELOPMENTAL PLASTICITY

Of course, awareness of different kinds of gene–environment interaction in development has been around for a long time. Various alterations of development, responding to environmental changes during the lifetime of a single organism, have been known and studied for many decades, and are generally referred to as *developmental plasticity* or *epigenesis*. Others have been discovered more recently. It is important to discuss these briefly in order to appreciate their connections with environmental change, and why other intelligent systems evolved from them.

Even where the environment remains fairly stable across generations, a major form of intragenerational change consists of the constant perturbation any developing organism is subjected to. These perturbations can arise from the outside world, as physical bumps and shocks, droughts, nutrient shortages, temperature changes, and so on, or as genetic mutations, or internal environmental challenges. C.H. Waddington in the 1940s claimed that for many characteristics at least, development of the normal phenotype is so strongly buffered against genetic variation or environmental perturbation as to be maintained on its "chreod" or fated pathway (Waddington, 1957). Examples are found in the development of all basic aspects of the body like the eye, limbs, internal organs, and so on.

In other cases, aspects of the environment may change drastically between parental and offspring generations, evoking a prominent plasticity of development of offspring. For example, if a predator appears in the environment of offspring, defensive forms or appendages may develop that were not present in immediate parents (Stearns, 1989). Sometimes gross changes in morphology are involved, as in caste formation in bees and ants, totally altering behavior and physiology as well as anatomy. Metamorphosis in frogs and other amphibians involves the remodeling of almost every organ in the body. In some cases of developmental plasticity, the differences between the morphs have been so stark as to suggest different species (Gilbert, 2001). There are also spectacular examples of *lifelong* developmental plasticity, as with the changing colors of the chameleon, seasonal or other cyclical changes, or the total sex change (including gonadal, behavioral, and coloration) undertaken by females of certain coral-reef fishes, as reported by Gottlieb (1991).

A more recent example of epigenesis concerns alterations to parental DNA (such as addition of a methyl group to "silence" a gene) that can then be passed across generations to influence the phenotypic outcomes in offspring (Feil & Fraga, 2012; Meaney, 2010). These changes may be caused by preconceptual or in utero exposure to maternal stress, malnourishment, environmental toxins, and so on, as if a "warning" signal to offspring to adjust development accordingly (Bollati & Baccarelli, 2010). That is, effects of experiences of mothers, even before conception, can be epigenetically inherited by offspring to influence development, behavior and physical health in later life.

In each of these cases, responses superficially appear to be due to singular external cues (such as environmental stressors) in a simple, homeostatic cue–response function. But the systems are far more complex than that. Waddington (1957) himself suggested that canalization must be based on more complex gene interactions. Recent analyses of gene expression data for canalized traits confirm this, suggesting the formation of gene regulatory networks through which groups of genes can be utilized to regulate expression of one another in order to reduce developmental variation (Manu et al., 2009; Wilkins, 2008). Likewise, with developmental plasticity, it turns out that the structured responses of cells and tissues are themselves coordinated with the structural (spatial and temporal) information in the outside world as a way of predicting the future. In tadpoles, for example, developmental plasticity, such as timing of metamorphosis, depends on the integration of information from multiple environmental sources (Crespi & Denver, 2005; Rose, 2005). That is, responses are based on the deeper structural context of numerous factors at different hierarchical levels. Simple switch-like homeostatic systems do not need such complexity, so what is it about?

The answer, as already mentioned, is that environments rarely consist of independent cues or triggers. Even for the single cell, the environment is continuously changing in its deeper form and structure. The various epigenetic devices described so far are needed to survive changeable environments because such environments are far more complex and computationally demanding than their superficial appearance suggests. Cases of developmental plasticity suggest more intelligent systems, sensitive to the structural context of evolution and development. More recent molecular biology has been elaborating a biologic, quite different from that of direct natural selection and gene-determined development, through which far more complex forms and functions, including cognitive ones, may have evolved.

7. LIFE WITHOUT GENES

Some of these logic have been compelled by recent studies into the origins of life. Although the idea of genes as the "seeds" of life persists, the crucial question has been how could some original DNA have gathered about itself all the hugely complex supply chains of ingredients, enzymes and transport processes needed for the purpose of their existence, i.e., faithful gene transcription?

It now seems highly likely that they did not, and that living forms existed long before genes had been invented. They survived for 2 billion years or more, the only form of life, as self-maintaining, autocatalytic, molecular ensembles, obtaining materials adventitiously from the environment and dispersing waste products as open systems. They also included polymer strings, other than DNA/RNA, perfectly capable of self-replication (Kauffman, 1995). Just such an "autocatalytic" reaction was discovered in 1996 (Lee, Granja, Martinez, Severin, & Ghadiri, 1996), and we now know of a variety of self-replicating polymers (Martin & Russell, 2002). Instead of genes, these systems had what Shenhav, Solomon, Lancet, and Kafri (2005) call "compositional information," in which the interactional structure among components determines system responses rather than the sequential information we are more used to in genes. Shenhav and Lancet's group has shown in the laboratory how such ensembles are capable of propagating their constituents and self-reproducing without RNA or DNA "codes". Such systems could also evolve by "mutation" (e.g., accidental protein changes) and reproduce and develop by occasional fragmentation and further replication of ingredients. Either way, the incorporation of DNA/RNA polymers that could act as templates for production of components "in-house," as it were, came later. This already suggests putting genes into perspective as resources for development, subordinate to control processes centered elsewhere, rather than functioning, themselves as programs for development.

8. ABSTRACTING STRUCTURE FROM THE ENVIRONMENT: SIGNALING SYSTEMS

Even organisms like bacteria and amoeba are constantly encountering novel physical obstacles, and other organisms, as well as changes in light, temperature, nutrient availability, and so on. Unlike the repetitive environment assumed by EPs, these combinations are always unique, and have to be uniquely negotiated. For example, in orienting its locomotor (chemotaxic) processes toward food gradients, the bacterium *E. coli* cannot respond on a simple cue–response basis: a single "hit" of a nutrient molecule on a surface receptor gives no information about direction. Rather the cell must assimilate, in its signaling networks, the spatiotemporal statistical structure of multiples of those signals in order to gauge direction of origin. That assimilation, in turn, involves numerous internal signaling proteins in spatiotemporal interactions. The direction of motor organs—the turning of flagella—then involves the recruitment of dozens of genes through further signaling pathways and precisely regulated transcription (Soujik & Armitage, 2010).

Similar considerations apply to any developing cell in a multicellular organism. Each is faced with a constantly changing surround of other cells, and storms of signals from them, as well as from changes further afield. To the migrating and differentiating cells in developing embryos, they include guidance signals as to where to go, and when, and what to differentiate into (Berzat & Hall, 2010). But the very dynamic context, in which hundreds or thousands of external signals are hitting the cell surface concurrently, ensures that this cannot be a direct cue–response process through fixed molecular pathways. Rather cells must differentiate and respond according to the precise spatiotemporal structure of those signals.

For example, initiation of limb development in a vertebrate requires that "signaling gradients...act in concert to establish a basic pattern of tissue layers by coordinating cell proliferation and cell fate determination" (Yang, 2009, p. 3). The structure of signals outside the cell becomes further integrated with those inside through intracellular signal transduction pathways. The pathways are interlinked in elaborate networks, so that the reactions in one are contextualized by the reactions in many others (Peisajovich, Garbarino, Wei, & Lim, 2010). In this, cross-talk enzymes and other cell components have activities conditioned by others at potentially many statistical depths to reflect the deeper structure in experience (Wagner & Wright, 2007). Such structural assimilation has been assisted by a proliferation, in the course of evolution, of signaling proteins with increased numbers of "interaction domains," facilitating recombination in an increasing variety of ways (Peisajovich et al., 2010).

In all cases, the point is to assimilate the deeper correlations experienced in complex environments (Marijuán, Navarro, & del Moral, 2010). Thus, Tagkopoulos, Liu, and Tavazoie (2008) showed how, in coming to anticipate future events from recent experience, the bacterium *E. coli* abstracts "temporally structured correlations on multiple time scales," in "the highly structured (nonrandom) habitats of free-living organisms" (p. 1313). Other research has described the mechanisms through which molecular networks can be sensitive to such correlations (see Gough, 2012). Helikar, Kochi, Konvalina, and Rogers (2011) refer to changeable configurations of signaling networks as cellular "memory modules" that assist future structure abstraction, while Taylor et al. (2009) describe how network properties "depend strongly on cell history".

9. AND MAKING A STRUCTURED RESPONSE: WHERE GENES COME IN

Genes are recruited in the synthesis of cell components for developmental purposes only in response to these highly structured signals. They do not open the scene as "recipes" or "controllers" of development; rather the initiation and regulatory control resides in the cell as a whole, including the signaling networks just mentioned. Accordingly, the evolution of organisms in increasingly complex, changeable environments has been accompanied by increasing numbers of regulatory regions on genes, involving increased varieties of transcription factors, activators, repressors, enhancers and other factors, operating in different combinations in the initiation of gene transcription. In parallel with the expansion of signaling networks, these have vastly expanded the "transcriptome," or ways of varying gene expression (del Moral, González, Navarro, & Marijuán, 2011).

However, even this high degree of regulatory control is only one aspect of gene utilization. The highly fluid, dynamic process that winds/unwinds the protein packaging (histones) around DNA regulates access to genes before transcription can even begin (Deal & Henikoff, 2010). This also happens in a spatiotemporally organized manner according to the structure of signaling patterns (Braun & Madhani, 2012). The immediate products of gene transcription (forms of RNA) can then be further rearranged, by a process known as "exon-shuffling," to allow a greater variety of proteins to be produced from the same gene, with potentially widely different functions (Le Hir, Nott, & Moore, 2003). Newly discovered forms of RNA perform additional regulatory functions, such as modulation of promoters, gene silencing, or coactivators of transcription (Pauli et al., 2011).

All these processes have become more important in more complex organisms in more changeable environments, for which the use of gene-specified, predetermined pathways would be quite unadaptable. The point is that the correct developmental response, such as the production of just the right type of cells, in just the right place at the right time, is far from a programmed assembly line: rather it reflects a highly dynamic process dependent upon the structure of an internal network of influences guided by the ever-changing structure of the outside environment.

10. DYNAMIC SYSTEMS IN DEVELOPMENT

The primary purpose of such massive regulation is, of course, to extend the range of developmental possibilities in changing environments. This could not be achieved through the linear deterministic processes that genetic programs might suggest. In reality, the multitudes of environmental changes, and signaling and gene regulatory pathways, keep cells in "far from equilibrium" states where nonlinear dynamics create "criticality" in molecular networks, sometimes described as "on the edge of chaos". This is a state of maximum information availability, in which the full range of novel response possibilities can be explored and novel decisions rapidly formulated, in the face of changing environments (Kaneko, 2006). Simulations demonstrate that it is only near this critical boundary that such networks can maximize information processing and perform complex computations (Beggs, 2008; Bertschinger & Natschläger, 2004). This creativity could not be obtained by linear deterministic operations based on additive units in fixed algorithms (Freeman, 2000). As the statistical structures of the environment are experienced, their relational parameters are assimilated as "basins of attraction" among the internal molecular networks. These provide activity tendencies to guide and optimize future interactions and their products.

What has evolved, then, is a whole dynamic system with changing biases that arise out of the interaction of a multitude of variables: some from evolutionary history; some from the current environment; some from whatever genetic resources are available; and some from the developing basins of attraction. Even the response of a bacterium to environmental changes requires "global reorganization" of signaling and transcription networks on many hierarchical levels (Buescher et al., 2012). Instead of independent cue–response processes, what are actually observed are harmonious, coordinated responses across multitudes of nonlinear relationships. This is why Kholodenko, Hancock, and Kolch (2010) refer to the "signaling ballet"; Wiedemann (2010) talks about "the symphony of transcription"; Coen (1999) likens cellular self-organization to an orchestra without a conductor; while Bromley et al. (2008) talk about "orchestrating the orchestrators".

11. INTELLIGENT DEVELOPMENTAL SYSTEMS

Recently, investigators have been astounded at the abilities afforded by these molecular intelligent systems. Reporting research by Japanese laboratories on a slime molds, Ball (2008, p. 385) informed the journal *Nature* that, "Learning and memory—abilities associated with a brain or, at the very least, neuronal activity—have been observed in protoplasmic slime, a unicellular organism." Similarly, del Moral et al. (2011) talk about "prokaryotic intelligence," cells with "cognitive resources," "bioinformation intelligence," "cell intelligence," "cell knowledge," and so on (see also Marijuán et al., 2010). Ramanathan and Broach (2007, p. 1801) tell us how current research indicates "a level of memory and information processing that has not been normally associated with single cells, suggesting that such organisms do in fact have the capacity to "think" "a level of memory and information processing that has not been normally associated with single cells, suggesting that such organisms do in fact have the capacity to."

Whether cells really "think" or not, the point is that intelligent sensitivity to the structure of environmental change turns development from the passive assembly process envisaged by gene determinists into a positive and creative process, optimizing the outcome in various ways. Being able to intelligently "rewire" the signaling and transcription networks creates alternative ways of surmounting obstacles. Thus, a deficiency in the provision of a metabolite, either from the environment or through a genetic mutation, can be overcome by recruitment of an alternative biochemical pathway (Luscombe, Babu, Yu, Snyder, & Teichmann, 2004). Ostensibly the same "environment," such as a hormonal signal, can initiate a variety of responses like growth, cell division, differentiation, migration, and so on, depending on deeper context. This is how the same phenotype can be obtained from a variety of genotypes, and vice versa; and a population of animals of identical genotypes reared in similar environments can exhibit the full range of behavioral variation (Visscher, Smith, Hall, & Williams, 2001). That is, rather than a detached cue–response relationship with the environment, there is an ongoing, constructive "conversation".

There are also longer term evolutionary implications. Large changes in morphology and function can be achieved by small alterations in the spatiotemporal timing of signalling and/or transcription networks, so altering evolutionary trajectories without the long process of gene selection over countless generations (López-Maury, Marguerat, & Bähler, 2008). Such changes in one species can also alter the conditions of survival of others, creating self-organized ecological networks, with additional feedback loops, transcending direct genetic selection (Odling-Smee, Laland, & Feldman, 2003).

Although always constrained to some extent by the genetic resources available, we don't have to think of the "evolvability" of species, nor what develops, as hinging purely on what genes are available. This explains why there is little relationship between species complexity and numbers of genes. While humans have around 25,000 genes, the simple nematode worms have nearly 20,000; fruit flies around 14,000, and carrots have around 45,000! (Levine & Tjian, 2003). Only around 1% of the human genome is different from that of the chimpanzees, yet there are vast expression differences, especially in the brain (Minugh-Purvis & McNamara, 2002). Hence, the study of evolution and development (evodevo) has now become eco-evodevo. It is a perspective that offers what so many other accounts have failed to offer, namely seamless conceptual bridges from cellular intelligence to the development of higher forms. Let us now explore those bridges.

12. FROM EPIGENETICS TO NERVOUS SYSTEMS

From around 2.2 billion years ago, some of the single cells that had existed hitherto started to associate and cohere as multicellular life forms. A crucial obligation of cells and tissues in multicellular organisms is to coordinate their activities by constant intercommunication. One cell separated from the rest in a culture dish soon dies without communication from others. So, very early in evolution, a wide range of chemical messenger systems, including pheromones, prostaglandins, interferons, and endocrine systems, became a crucial aspect of the physiology of organisms (Hartenstein, 2006). Such systems were able to create internal environments adjusting the parts of the organism to the whole, and the whole to the changing environment outside. More than 50 human hormones have been identified, all acting by binding to receptor molecules on target cell surfaces, which then articulate with the internal signalling systems mentioned above.

It used to be thought that physiology performs an essentially equilibrium or homeostatic function, each aspect independently maintaining, as far as possible, some constancy of part of the internal milieu—blood sugar, temperature, liquid balance, or whatever—in the face of disturbances from inside or outside. However, we now know that its components interact with nonlinear dynamics, to respond rapidly and constructively to the structure of the outside world (Goldberger et al., 2002). Such dynamics, furnishing rapid and precise responses, are crucial in development (Krain and Denver, 2004).

However, physiological systems soon became embedded within the first nervous systems, which probably appeared as loose nerve nets around 600 million years ago. Through rich beds of chemoreceptors, even in primitive brains, physiology interfaces with signals from all sensory systems (vision, hearing, smell), themselves responding to rapid changes in the outside world. But what pressures favored such evolution? It is important to ask this question because there still seems to be uncertainty about "what the brain is really for" and, therefore, what really develops within it. While research over the last few decades has produced mountains of particular "findings," Edelman (2006, p. 8) suggests that "relatively little progress has been made to integrate the results of this work into a global synthetic view of how the brain works." And while nervous systems and brains are seen as special seats of intelligence, investigators still complain about "the absence of a consistent central theory in the neurosciences" (del Moral et al., 2011, p. 10).

As with epigenetic cellular intelligence, the answer seems to lie in the structure of the environment. Animals were forced into more complex environments very early in evolution, and this required new forms of coordination between inside and outside through more acute and responsive sensory and motor systems. As animals themselves evolved motion behavior, the world became teeming with objects in a new league of changeability. However, the new sensory systems would only seem to present what William James once famously described as a "blooming, buzzing confusion"—or what Warren & Shaw (1985, p. 6) more recently described as "an onslaught of spatiotemporal change"—without further intelligent action.

Some consideration of sensory reality, indeed, tells us why even sensory perception (inferring and conveying to the brain what is present in the outside world) is not easy. Objects in the environment are not experienced as stable, easily reproducible forms. For one thing, the image of any single object is never static. It is under constant spatiotemporal transformation as an animal moves around it, it rotates, passes behind other objects, or moves to different places at different distances (with changes in apparent size). And a single scene may contain dozens of moving objects in changing poses.

Through all these rapidly changing visual experience, the brain must distinguish between real outside changes and those due to personal eye and body movements. Moreover, objects and background layouts are three-dimensional. But these become collapsed as 2D images on the retinal "sheet," and the same 2D image can be created by a number of different objects/features. Such lack of uniqueness in the image would be very confusing to other centers of the brain if conveyed there directly. The 2D image is also grossly distorted because the back of the eye is hemispheric (imagine watching a movie projected onto a bowl-shaped screen). Finally, the receptor surface is not a continuous "sheet" (like a camera film), at all, but an aggregate in the form of millions of light-sensitive cells. So the visual field is detected in the form of moving clouds of light spots. For all these reasons, percepts cannot be exactly determined by the sensory input, but require some kind of inferential construction as well. Irving Rock called it perceptual "problem-solving" or "intelligent perception". As with the other intelligent systems we have discussed so far, this means "going beyond the information given".

A still popular idea is that the sensory apparatus does this by first registering simple features, such as light spots, lines, edges, or direction of motion, from the visual scene. These are then sent to higher perceptual and cognitive centers to be reassembled into a veridical image of the original object. More recently, however, attention has passed from elements to structures, as the key informational fodder of perception (Albright & Stoner, 2002). This is hardly surprising as the "onslaught of spatiotemporal change" is not at all without structure or useful information. Any light fragments reflecting from a naturally moving object do not change independently as a series of static displays; their spatial and temporal dispositions are highly correlated. The edges and corners of a book, for example, "move together" as an observer views it from different angles. In fact, when the moment-to-moment changes in multiples of sample light points like this are measured, there is evidence of considerable deeper structure (Richardson, 2010). This is the kind of higher order information, rather than the mere surface coincidence of independent light points, that distinguishes a straight line from other structures such as curved lines. Given the predictability afforded by such structure, it is hardly surprising that this is really what the nervous system is "looking" for and evolved for.

Such a structure-searching-and-abstraction function has been amply confirmed by recent studies. Of particular relevance are studies using video samples of visual scenes projected onto animal retinas and recording from multiples of neurons simultaneously. These have elicited evidence of responses in retinal ganglion cells to higher order correlations, rather than single-parameter stimuli. As Lee, Stepleton, Potetz, and Samonds (2010, p. 2) say "The key insight is that correlated structures in visual scenes result in correlated neuronal activities, which shapes the tuning properties of individual neurons and the connections between them."

So it is a dynamical image that is passed from eye to brain: statistical structural parameters, so much richer for interpretation than isolated, static features. Similar findings have emerged with respect to other sense domains. Receptors in the cochlea of the ear transform sound (in reality, already a spatiotemporal acoustic stream) into spatiotemporal response patterns sent along the auditory nerve (Shamma, 2001). Tactile senses, too, are able to abstract the deeper statistical structure from sequences of vibration (Conway & Christianson, 2005). The relational parameters abstracted probably form attractor basins, each defining invariant qualities of familiar objects, such as "bookness," but nested at various hierarchical levels to form attractor landscapes. By attuning neuron connectivity to interaction parameters, rather than independent features, the networks can more easily interpret novel data, yet permit an enormous variety of expressions for forward transmission. A major role of "higher" sensory centers in the brain is then to elaborate that kind of attractor landscape for use in making a seemingly disorderly world, and action upon it, more predictable.

13. THE DYNAMICAL BRAIN

Of course, what is becoming clearer is that this intelligence in nervous systems is an extension of the dynamic logic described earlier for cellular (epigenetic) and intercellular (physiological) intelligence. As in those earlier systems, nervous systems use the emergent structure of activity of the whole network to modulate the activity of specific neurons, within the context of environmental change. But the intelligence is vastly scaled up through the sheer numbers of cells and their plastic synapses. Apart from much deeper information abstraction (as multivariate, multilevel correlations), modifiable connections permit continuing predictability in changing environments, often from fleeting or fragmentary sensory data. This level of activity constitutes a lifelong developmental process, incorporating those already evolved at the epigenetic and physiological levels, and nested within it.

As in those earlier systems, too, the very multivariate and nonlinear nature of interactions suggests that predicting identities and properties from skimpy, fleeting sense data can probably only be achieved through nonlinear dynamical processing. Using the attractor landscapes induced from experience, dynamic processing fosters robust operation, and fast and reliable decision making, in the face of changing environments (Skarda & Freeman, 1987; Tschacherm & Haken, 2007).

This natural history of brains also explains the dependence of brain development, in all species, not merely on general environmental "stimulation," but on environmental *structure* in experience. Spatiotemporally structured light, rather than just any light stimulation, appears to be required for proper development of visual cortex. Confining visual experience to white noise, that presents all light frequencies without patterned input, retards development of connectivity (Grubb & Thompson, 2004; White, Coppola, & Fitzpatrick, 2004). Random sound or "noise" is not sufficient for normal development of auditory cortex (Wang, 2004). The powers of predictability furnished by the assimilation of such structure explains the explosion in network sizes as living things evolved into more complex, changeable environments.

In passing, it is worth noting the deep interpenetration of brain function and world dynamics implied in these discoveries. O'Regan and Noë (2001) have reviewed various sources of evidence indicating how the deep statistical structure of the outside world becomes incorporated as a deeply coded "world in the head" that transcends limited immediate experience. The blind man probing a manhole cover with the end of a stick has an impression of the object as projected "out there," not in the hand and fingers where the sensations are created. When riding a bicycle, you somehow become conscious of the whole texture and topology of the road surface (even when cycling in the dark), although all sensation is received by your hands at the handlebars and through the seat of your pants. So the intelligent system of the brain becomes "as one" with the living body and a changing outside world.

14. FROM NERVOUS SYSTEMS TO COGNITION

In the standard computational model (SCM), which has dominated recent cognitive theory, cognitive functions are envisaged as input/output systems that transform sensory inputs according to internal rules. Initial transformations are then transmitted onward as symbols to a "higher" set of rules, and so on, until a motor or other response is output. In most theory, the basic rules, upon which others may be built, unfold in development as genetically determined structures, the genes having been established by natural selection. As with the EP model of evolution and development, what the SCM overlooks is that cognitive systems evolved to both assimilate and construct rules in situations in which the "rules"—i.e., environmental structures—are constantly changing, and are being changed by the actions of the developing system itself. We now know that the flow of activations through basins of attraction is far more efficient for reaching inferences from fuzzy inputs, and for making rapid decisions, than linear steps following a fixed algorithm.

Because of this oversight, the SCM is now being eclipsed by the dynamic systems approach emphasizing self-organization of development and function. Even Ulric Neisser (2006, p. 4), one of the founders of the SCM, has made the point that "the dynamic approach puts cognitive science itself on a new trajectory." The function of cognitive systems is to be very good at assimilating environmental structures that change frequently and then use that structure to predict immediate and distant futures from fuzzy sensory experience.

We see this function immediately in cognition. Neural actions, for example, deal only with two-dimensional patterns of light on the retina, whereas cognition induces, and deals with, the four-dimensional spatial and temporal structure emergent from it. Light patterns reflected from aspects of objects on the eye may consist only of moving arrays of light "points". Cognition induces from them complex features, whole object images, abstract concepts, multisensory concepts, event and action concepts, and so on, each reflecting a level of correlational structure in a self-organizing, multilevel, attractor landscape (Richardson, 2010).

It is such derived patterns, rather than singular cues, to which the system responds. In their experiments on the olfactory system, for example, Freeman et al. show that it is not external smells per se that animals respond to, at least directly. Rather they respond to internal activity patterns created by the dynamics within the olfactory bulb. The nonlinear dynamics are essential because "neither linear operations nor point and limit cycle attractors can create novel patterns." Instead, "...the perceptual message that is sent on into the forebrain is a construction, not the residue of a filter or a computational algorithm" (Freeman, 1995, p. 67).

It is this construction, or activity pattern, that is now a *cognitive* agent, rather than a mere neural one. It contains far more information-for-prediction than the raw neural signals, and enters into a new level of regulations with other such agents, creating new properties of life in the process. Of course, these regulations build upon and incorporate those evolved and developed at antecedent levels. But, as Freeman (2000, p. 11) says, this nature of the system has "profound implications for science and human understanding; they imply that our knowledge of the outside world is constructed by neurodynamics, not derived from logical operations on sensory data." It also has important implications for "what develops" in cognitive systems.

Such inferential and creative powers emerged early in evolution as indicated in studies on insects like the honeybee, with a brain of less than a million neurons. Honeybees exhibit complex forms of knowledge creation "that cannot be explained on the sole basis of elementary associations" but reflects "a dynamic and self-organizing process of information storage" (Lehrer, Srinivasan, Zhang, & Horridge, 1988, p. 356). After foraging trips, bees do not find their way home by simply retracing their outward meanderings: they integrate the directions and distances experienced, along with motion cues, to construct the most direct route, as if having internalized the "higher order" relational invariants of the topology of the world (Chen, Zhang, & Srinivasan, 2003; see also Giurfa, 2003; Miller, 2009). That is, the direction home was not "in" original stimuli but has been constructed from relationships between them. Another research points to the previously undiscovered complexity of brain functions in insects generally (Greenspan & van Swinderen, 2004).

In more evolved animals, from birds to primates, the creative nature of cognition is evident in much innovative behavior. For example, what has been called "insight learning" involves embedding one action in one or more others so that the emergent pattern is greater than the sum of its parts. The created pattern suggests solutions to an otherwise impossible problem. Crows can use sticks, or series of sticks, as tools to push food along a tube into a retrievable position (Wimpenny, Weir, Clayton, Rutz, & Kacelnik, 2009). Leising and Blaisdell (2009, p. 80) report how pigeons and rats encode "spatial maps" in learning to locate food sources, and how "simple maps can be integrated into complex maps through higher order associative processes" to create novel searching strategies.

These ideas have been strongly supported by work on artificial neural networks (ANNs). Recent ANNs have gone well beyond simple associations to mimic real brain dynamics more closely. "Self-supervised" networks, that acquire structure, and generate feedback, automatically, are now in common use (see contributions in Spencer, Thomas, & McClelland, 2009). These use the history of network experience to adjust and constrain the variability of experience, much as cortical feedback connections do in the brain.

What is also found is that as the correlational structure builds up in the network, so the nonlinear dynamics of the ANN become increasingly chaotic, and yet increasingly structured by structured inputs. The implication is that the more the external dynamics are incorporated into brain networks, the more likely the internal dynamics will become chaotic, and the greater the speed and precision of processing.

The specific emergent function of cognitive intelligence—to generate much more complex, and more powerful, intelligence from its own activity—corresponds with Piaget's concept of "reflective abstraction," by which knowledge can obtain more abstract levels, transcending immediate experience (Piaget, 1988). A classic example is that of transitive inference. The conception that object A, say, is bigger than object B, may be gained from direct empirical experience. Likewise with the experience of B being bigger than object C. But predicting that object A is *necessarily* bigger than object C requires reflective abstraction, and the emergence of the more implicit structure, beyond what is available in direct experience. The result is a higher conceptual level with new logical powers, yielding predictability about the world deeper than that in immediate experience. Such powers were probably a founding property of cognitive systems: Paz-y-Miño, Bond, Kamil, and Balda (2004), for example, claim to have observed transitive inference in jays. They must have been a tremendous boost to predictability and adaptability and probably explain the accelerated evolution of expanding brain networks.

15. HUMAN COGNITIVE DEVELOPMENT

Indeed it is the sudden leap in brain size, reflecting a three-fold expansion over that of other apes that stands out in the recent evolution of our own species. Nearly all those differences lie in a vast neocortex, in areas most involved in coordination of activities, and integration of what we feel with what we think and do. As we have seen, relatively simple networks of a few hundred thousand neurons can do sophisticated cognition. So what could the 10 billion neurons and over 50 trillion connections in the human brain have possibly evolved for? What could it have been about the environment and lifestyle of evolving humans that fostered such a remarkable change? And what does that tell us about the nature of the functions that have emerged and the nature of their development?

Each stage in the evolution of intelligent systems so far has coincided with a leap in the complexity of environmental change, and this seems to have been the case in the advent of the human lineage. Although still some-what murky, the evidence suggests that our hominid ancestors evolved in periods of great climate change, thinning forests, and less secure food sup-plies. These circumstances necessitated wider ranges, exposure to predators, and so on, and something else far more significant in the long run: degrees of cooperation only vaguely foreshadowed in ancestral species. Ants and bees may cooperate because of environmental change. But those changes are mere fluctuations within stable underlying structures. They interact through relatively simple, and stable, rule systems such as gradient following and pheromone dropping (Guerin & Kunkle, 2004). And, although living rich social lives in some respects, monkeys and apes live in confined niches demanding little real cooperation. They rarely help group members other than close family, and joint action and "teaching" are also rare. There is little evidence among chimpanzees of agreement to share or of reciprocation (Tomasello & Warnekan, 2008), whereas among humans, it is mandatory. Young children are far more successful than chimpanzees and monkeys in problem situations because of greater cooperative cognition, including teaching through verbal instruction, imitation, and "prosociality" (Dean, Kendal, Schapiro, Thierry, & Laland, 2012).

It is important to emphasize the extreme informational demands placed upon the cognitive systems of obligatory cooperators, requiring new lev-els of environmental structure abstraction, over and above those already evolved. The nested structures of individual epigenetics, physiology, brain, and cognition are further nested within the interpersonal structures cre-ated by and with other people around mutual activities. In simply moving a rock together, for example, the natural relations between forces of mass, gravity, shape and friction, perceived by any one participant, all become conditioned by the actions of others so that the perceiving and acting of one partner continuously and reflexively affects those of others. In hunting and defensive actions against predators, it is even more complex because the object is itself active, and reactive, against the joint action. The dynamics of action and reaction are not even remotely experienced by noncooperating animals and cannot be regulated through a narrow range of stereotyped chemical, gestural or other signals.

Fortunately, as with individual experiences of objects, *joint* actions and perceptions will also contain structure at various statistical depths. Just as the activities of individual neurons can be coordinated by the cognitive patterns emerging between them, individual attention and actions can be coordinated by new *epicognitive* patterns emerging between individuals.

These will include conventions like turn-taking; emerging signals and gestures; marking aspects of what to attend to (e.g., pointing); means for planning shared routes to end goals; plans for divisions of labor; signals for monitoring and modulating progress; and, of course, the use of a human language with its unique productivity and speed of transmission, evolved specifically for coordinating joint perception and action. Other patterns become enshrined in shared mechanical devices and artifacts, from simple stone tools and shelters to emergent technologies, literally designed "with others in mind". The dynamic detail and complexity of these contexts is not remotely captured in the EPs' vague allusions to "ancestral environments," or the IQ theorists' reliance on anonymous components of statistical variance.

This additional "layer" of dynamic regulation has several cognitive and developmental implications. One is the way that cooperative action in the world reveals statistically much deeper structures within it than individual actions can reveal. As the ability to abstract structure evolved, and encouraged further evolution of network capacity, so did the ability to *create* structure through levels of imagination barely foreshadowed in other species. It is such imagination that probably created novel conceptions of objects in tool use, tool manufacture, defense, shelter, body clothing, and so on, permitting our human ancestors to break out from a confined niche to inhabit the whole world. Others soon appeared as institutionally stated obligations and conventions, as in marriage rules, kinship identities, asset ownership conventions, and so on, defining rights and responsibilities, and cementing patterns of relationships. Other forms were invented for simply reveling in the joint creation of structure in art, music, dance, social games, and so on. This is what is called *human culture.*

Another implication is the way that these cultural forms can vary radically from group to group or place to place. Lev Vygotsky referred to them as "cultural tools" because they also become psychological tools through which individuals think and act with others. Just as neural connections become shaped by the patterns of activity between them, so patterns of cognition in humans become shaped by the dominant structure of activities in cultures and subcultures (Cole & Cagigas, 2010; Vygotsky, 1981). As Vygotsky (1981, p. 160) put it, "By being included in the process of behavior, the psychological tool alters the entire flow and structure of mental functions. It does this by determining the structure of a new instrumental act just as a technical tool alters the process of natural adaptation by determining the form of labor operation."

Accordingly, recent research has shown how experience with specific cultural tools or procedures results in changes in brain networks that differentially prepare individuals for given cognitive tasks (May, 2011; Woollett & Maguire, 2011). The point is that there is no "pure" or standard form of human cognition, logic or reasoning, or variation in it, distinct from such cultural tools. The whole spectacular success of specifically human evolution has relied on that variability and adaptability. So the alleged measure of some absolute cognitive ability, as in IQ tests, simply reflects degree of immersion in the specific cultural tools represented in the test—a "distance" rather than a "strength" metric (Richardson, 2002).

Yet another implication is that just as the epigenetic regulations described earlier vastly expand the transcriptome, and reflective abstraction expands cognitive powers, so epicognitive regulations between individuals vastly expand the powers of the human "cognomen". Vygotsky did not use the term "reflective abstraction". But he fully understood the implications of cultural tools for the further amplification of human cognitive abilities, over and above the sheer expansion of network capacity. The dynamics between brains interact with those within brains—just as the dynamics of physiology interact with those within cells—emerging as further levels of reflective abstraction. For example, use of language expanded individual memory functions, a process that has been vastly augmented with the invention of writing, print and, more recently computers and the Internet, providing new media of cognitive organization and planning. Likewise, shared concepts expanded knowledge networks, so that predictabilities about the world obtain increasing depths, far more potent than any found in nonhuman animals. The cultural tool we call science is one of the best examples: a theory is a collective model of part of nature emergent from the dialectics of scientific method, taking us far beyond specific empirical experience. All these extend lifelong developmental plasticity beyond single generations into history *across* generations.

This interaction between dynamic levels explains why cognitive development in human offspring is a far more engaging and creative process than that occurs in other species. As Vygotsky explained, psychological tools appear "first among people as an interpsychical category and then within the child as an intrapsychical category." But this is not a process of simply emulating an encapsulated structure: a readymade adaptation directly (if progressively) replicated in the child, like the evolutionary psychologist's "meme". Rather development lies in the "collision" (as Vygotsky put it) between mature cultural forms of cognition with the less-developed forms in the child. The interaction means that the cultural order is, in a sense, reconstructed in each developing child, but often with novel individual forms and variety that can feed forward to broader cultural change.

16. HUMAN DEVELOPMENT IN MODERN SOCIETIES

The emerging dynamic systems view has important implications for understanding and promoting development in modern class-structured societies. An obvious implication is that just as the development of adaptive physiological traits, such as vision or hearing, will only develop through exposure to light or sound of appropriate structure (as mentioned earlier), and not just random "stimulation," so the development of human cognition will only fully develop through immersion in the important cultural tools of a society, in the context of socially shared goals. Generally, research has shown that with the scaffolding provided in social contexts, complex cognitive skills, creative thinking, language, and productive action develop naturally, rapidly and easily (e.g., Rogoff et al., 2007; contributions in Fogel et al., 2008).

It is undoubtedly the case, however, that access to, and engagement with, important cultural tools—the key decision-making processes in economic, political, educational, and other social institutions—is, in class-structured societies, very unequal. For example, families and subcultures vary in their exposure to, and usage of, the tools of literacy, numeracy, and other "ways of thinking" that help prepare children for schooling and important roles in those institutions (Cole & Cagigas, 2010). In an evolved nested system, denial of access to those crucial structures of engagement will have negative effects at all levels of development, including the cognitive, brain, physiological and epigenetic levels.

For example, the various stresses arising from such denial, felt by mothers, can have epigenetic consequences that are inherited by offspring to influence their development, behavior and physical health in later life (Feil & Fraga, 2012). Stress associated with financial and employment insecurity, and simple sense of subordination in a hierarchical system, also impacts on the physiology and general health and vigor of parents and children. The surveys of Wilkinson and Pickett (2009), for example, show how the greater the degree of social inequality in a society, the more disparate the levels of health and well-being. Likewise, lower class parents and children do not, on average, experience the sense of control or power over environments or activities, enjoyed by middle and upper classes. The research of Smith et al. (2008, p. 443) shows how those who lack power tend to feel lack of control over superordinate goals and values, or thinking in an abstract way, and merely "view themselves as the means for other people's goals." The consciousness of hierarchical subordination also depresses self-evaluations of personal cognitive competence, or cognitive self-efficacy beliefs that can be transmitted from parents to children (Bandura, 1997; Dweck, 2008). The children are then more likely to avoid cognitive engagement with unfamiliar or challenging problem solving situations (Ahmavaara & Houston, 2007). Such constraints on cognitive development are said to explain the link between SES and cognitive and brain development (Hackman et al., 2010; Loughnan et al., 2011).

The point is that the sociocognitive mode of adaptation evolved in humans means that differential development of individuals may reside, not so much in intrinsic (personal) limitations, as in the social structures that control access to cultural tools, although it has ramifications at all, including biological, levels. This has implications for our strategies for the promotion of cognitive development in humans, and explains why simple "compensatory" measures with respect to a few "environmental" variables may not be enough.

REFERENCES

Ahmavaara, A., & Houston, D. M. (2007). The effects of selective schooling and self-concept on adolescents' academic aspiration: an examination of Dweck's self-theory. *British Journal of Educational Psychology, 77*, 613–632.

Albright, T. D., & Stoner, G. R. (2002). Contextual influences on visual processing. *Annual Review of Neuroscience, 25*, 339–379.

Ball, P. (2008). Cellular memory hints at the origins of intelligence. *Nature, 451*, 385.

Bandura, A. (1997). *Self-efficacy: The exercise of control*. New York: Freeman.

Becks, L., Hilker, F. M., Malchow, H., Jürgens, K., & Arndt, H. (2005). Experimental demonstration of chaos in a microbial food web. *Nature, 435*, 1226–1229.

Beggs, J. M. (2008). The criticality hypothesis: how local cortical networks might optimize information processing. *Philosophical Transactions of the Royal Society A, 366*, 329–343.

Bertalanffy, L. von (1968). *General systems theory*. New York: Braziller.

Bertschinger, N., & Natschläger, T. (2004). Real-time computation at the edge of chaos in recurrent neural networks. *Neural Computation, 16*, 1413–1436.

Berzat, A., & Hall, A. (2010). Cellular responses to extracellular guidance cues. *EMBO Journal, 29*, 2734–2745.

Bird, R. J. (2003). *Chaos and life: Complexity and order in evolution and thought*. New York: Columbia University Press.

Bjorklund, D. F., Ellis, B. J., & Rosenberg, J. S. (2007). Evolved probabilistic cognitive mechanisms: an evolutionary approach to gene × environment × development interactions. *Advances in Child Development and Behavior, 35*, 1–39.

Bjorklund, D. F., & Pelligrini, A. (2001). *The origins of human nature: Evolutionary developmental psychology*. New York: American Psychological Association.

Bollati, V., & Baccarelli, A. (2010). Environmental epigenetics. *Heredity, 105*, 105–112.

Braun, S., & Madhani, H. D. (2012). Shaping the landscape: mechanistic consequences of ubiquitin modification of chromatin. *EMBO Reports, 13*, 619–630.

Bromley, S. K., Mempel, T. R., & Luster, A. R. (2008). Orchestrating the orchestrators: chemokines in control of T cell traffic. *Nature Immunology, 9*, 970–980.

Buescher, J. M., Leibermeister, W., Jules, M., Uhr, M., Muntel, J., Botella, E., et al. (2012). Global network reorganization during dynamic adaptations of *Bacillus subtilis* metabolism. *Science, 335*, 1099–1103.

Buss, D. M., & Reeve, H. K. (2003). Evolutionary psychology and developmental dynamics: response to Lickliter and Honeycutt. *Psychological Bulletin, 129*, 848–853.

Chen, L., Zhang, S., & Srinivasan, M.V. (2003). Global perception in small brains: topological pattern recognition in honeybees. *Proceedings of the National Academy of Sciences, 100*, 6884–6889.

Coen, E. (1999). *The art of genes. How organisms make themselves*. Oxford: Oxford University Press.

Cole, M., & Cagigas, X. E. (2010). Cognition. In M. Bornstein (Ed.), *Handbook of developmental science*. London: Psychology Press.

Confer, J. C., Easton, J. A., Fleischman, D. S., Goetz, C. D., Lewis, D. M.G., Perilloux, C., et al. (2010). Evolutionary psychology. Controversies, questions, prospects and limitations. *American Psychologist, 65*, 110–126.

Conway, C. M., & Christiansen, M. H. (2005). Modality-constrained statistical learning of tactile, visual, and auditory sequences. *Journal of Experimental Psychology: Learning, Memory, and Cognition, 31*, 324–390.

Cosmides, L., Barrett, H. C., & Tooby, J. (2010). Adaptive specializations, social exchange, and the evolution of human intelligence. *Proceedings of the National Academy of Sciences, 107*, 9007–9014.

Cosmides, L., & Tooby, J. (1994). Origins of domain-specificity: evolution of functional organization. In L. A. Hirschfeld & S. A. Gelman (Eds.), *Mapping the mind: Domain specificity in cognition and culture*. Cambridge: Cambridge University Press.

Crespi, E. J., & Denver, R. J. (2005). Ancient origins of human developmental plasticity. *American Journal of Human Biology, 17*, 44–54.

Deal, R. B., & Henikoff, S. (2010). Capturing the dynamic epigenome. *Genome Biology, 11*, 218.

Dean, L. G., Kendal, R. L., Schapiro, S. J., Thierry, B., & Laland, K. N. (2012). Identification of the social and cognitive processes underlying human cumulative culture. *Science, 336*, 1114–1118.

del Moral, R., González, M., Navarro, J., & Marijuán, P. C. (2011). From genomics to scientomics: expanding the bioinformation paradigm. *Information, 2*, 651–671.

Dennett, D. (2011). Homonculi rule: reflections on Darwinian populations and natural selection. *Biology and Philosophy, 26*, 475–488.

Dweck, C. (2008). *Mindset: The new psychology of success*. New York: Random House.

Edelman, G. M. (2006). Synthetic neural modeling and brain-based device. *Biological Theory, 1*, 1–12.

Ellis, B. J., & Bjorklund, D. F. (2005). *Origins of the social mind: Evolutionary psychology and child development*. New York: Guildford Press.

Feil, R., & Fraga, M. F. (2012). Epigenetics and the environment: emerging patterns and implications. *Nature Reviews Genetics, 13*, 97–109.

Flinn, M. V., Geary, D. C., & Ward, C. V. (2005). Ecological dominance, social competition, and coalitionary arms races: why humans evolved extraordinary intelligence. *Evolution and Human Behavior, 26*, 10–46.

Fodor, J. (2000). *The mind doesn't work that way: The scope and limits of computational psychology*. Cambridge, MA: MIT Press.

Fogel, A., King, B. J., & Shanker, S. (Eds.), (2008). *Human development in the twenty-first century: Visionary ideas from systems scientists*. Cambridge: Cambridge University Press.

Freeman, W. F. (1995). *Societies of brains*. Hillsdale, NJ: Erlbaum.

Freeman, W. F. (2000). A proposed name for aperiodic brain activity: stochastic chaos. *Neural Networks, 13*, 11–13.

Geary, D. C. (2007). Educating the evolved mind: conceptual foundations for an evolutionary educational psychology. In J. S. Carlson & J. R. Levin (Eds.), *Educating the evolved mind. Psychological perspectives on contemporary educational issues.* (Vol. 2). Greenwich, CT: Information Age.

Gershenson, C., & Lenaerts, T. (2008). Evolution of complexity. *Artificial Life, 14*, 241–243.

Gilbert, S. F. (2001). Ecological developmental biology: developmental biology meets the real world. *Developmental Biology, 233*, 1–12.

Giurfa, M. (2003). Cognitive neuroethology: dissecting non-elemental learning in a honeybee brain. *Current Opinion in Neurobiology, 13*, 726–735.

Goldberger, A. L., Amaral, L. A. N., Hausdorff, J. M., Ivanov, P. Ch., Peng, C-K., & Stanley, H. E., (2002). Fractal dynamics in physiology: Alterations with disease and aging. *Proceedings of the National Academy of Sciences, 99* (Suppl. 1), 2466–2472.

Gottlieb, G. (1991). Experiential canalization of behavioral development: theory. *Developmental Psychology, 27*, 4–13.

Gough, N. R. (2012). A coincidence detector with a memory. *Science Signalling, 5*, ec48.

Greener, M. (2007). Taking on creationism. Which arguments and evidence counter pseudoscience? *EMBO Reports, 12*, 1107–1109.

Greenspan, R. J., & van Swinderen, B. (2004). Cognitive consonance: complex brain functions in the fruit fly and its relatives. *Trends in Neurosciences, 27*, 707–711.

Grubb, M. S., & Thompson, I. D. (2004). The influence of early experience on the development of sensory systems. *Current Opinion in Neurobiology, 14*, 503–512.

Guerin, S., & Kunkle, D. (2004). Emergence of constraint in self-organizing systems. *Nonlinear Dynamics, Psychology, and Life Sciences, 8*, 131–147.

Hackman, D. A., Farah, M. J., & Meaney, M. J. (2010). Socioeconomic status and the brain: mechanistic insights from human and animal research. *Nature Reviews Neuroscience, 11*, 651–659.

Hartenstein, V. (2006). The neuroendocrine system of invertebrates: a developmental and evolutionary perspective. *Journal of Endocrinology, 190*, 555–570.

Helikar, T., Kochi, N., Konvalina, J., & Rogers, J. A. (2011). Boolean modeling of biochemical networks. *The Open Bioinformatics Journal, 5*, 16–25.

Kaneko, K. (2006). *Introduction to complex systems biology.* New York: Springer-Verlag.

Kauffman, S. (1995). *At home in the universe: The search for the laws of self-organization and complexity.* Oxford: Oxford University Press.

Kholodenko, B. N., Hancock, J. F., & Kolch, W. (2010). Signalling ballet in space and time. *Nature Reviews Molecular Cell Biology, 11*, 414–426.

Krain, L. P., & Denver, R. J. (2004). Developmental expression and hormonal regulation of glucocorticoid and thyroid hormone receptors during metamorphosis in Xenopus laevis, *Journal of Endocrinology, 181*, 91–104.

Langton, C. (Ed.), (1989). *Artificial life.* Redwood City, CA: Addison-Wesley.

Le Hir, H., Nott, A., & Moore, M. J. (2003). How introns influence and enhance eukaryotic gene expression. *Trends in Biochemical Sciences, 28*, 215–220.

Lee, D. H., Granja, J. R., Martinez, J. A., Severin, K., & Ghadiri, M. R. (1996). Emergence of symbiosis in peptide self-replication through a hypercyclic network. *Nature, 382*, 525–528.

Lee, T. S., Stepleton, T., Potetz, B., & Samonds, J. (2010). Neural encoding of scene statistics for surface and object inference. In S. Dickinson, A. Leonardis, B. Schiele & M. Tarr (Eds.), *Object categorization: Computer and human vision perspectives.* Cambridge: Cambridge University Press.

Lehrer, M., Srinivasan, M., Zhang, S. W., & Horridge, G. A. (1988). Motion cues provide the bee's visual world with a third dimension. *Nature, 332*, 356–357.

Leising, K. J., & Blaisdell, A. P. (2009). Associative basis of landmark learning and integration in vertebrates. *Comparative Cognition and Behavior Reviews, 4,* 80–102.

Levine, M., & Tjian, R. (2003). Transcription regulation and animal diversity. *Nature, 424,* 147–151.

Lickliter, R., & Honeycutt, H. (2003). Developmental dynamics and contemporary evolutionary psychology: status quo or irreconcilable views? *Psychological Bulletin, 129,* 866–872.

López-Maury, L., Marguerat, S., & Bähler, J. (2008). Tuning gene expression to changing environments. *Nature Reviews Genetics, 9,* 583–594.

Loughnan, S., Kuppens, P., Allick, J., Balazs, K., de Lemus, S., Dummont, K., et al. (2011). Economic inequality is linked to biased self-perception. *Psychological Science, 22,* 1254–1258.

Luscombe, N. M., Babu, M. M., Yu, H., Snyder, M., & Teichmann, S. A. (2004). Genomic analysis of regulatory network dynamics reveals large topological changes. *Nature, 431,* 308–312.

Maguire, E. A., Gadian, D. G., Johnsrude, I. S., Good, C. D., Ashburner, J., Frackowiak, R. S. J., et al. (2000). Navigation-related structural change in the hippocampi of taxi drivers. *Proceedings of the National Academy of Sciences, 97,* 4398–4403.

Manu, S. S., Spirov, A. V., Gursky, V. V., Janssen, H., Kim, A.-R., Radelescu, O., et al. (2009). Canalization of gene expression in the *Drosophila* blastoderm by gap gene cross regulation. *PLoS Biology, 7,* e1000049.

Marijuán, P. C., Navarro, J., & del Moral, R. (2010). On prokaryotic intelligence: strategies for sensing the environment. *Biosystems, 99,* 94–103.

Martin, W., & Russell, M. (2002). On the origins of cells: a hypothesis for the evolutionary transitions from abiotic geochemistry to chemoautotrophic prokaryotes, and from prokaryotes to nucleated cells. *Philosophical Transactions of the Royal Society, Series B, 358,* 59–85.

May, A. (2011). Experience-dependent structural plasticity in the adult human brain. *Trends in Cognitive Sciences, 15,* 475–482.

Mazur, S. (Ed.), (2008). *The Altenberg 16: An exposé of the evolution industry.* New York: Random House.

Meaney, M. J. (2010). Epigenetics and the biological definition of gene × environment interactions. *Child Development, 81,* 41–79.

Miller, G. (2009). On the origin of the nervous system. *Science, 325,* 24–26.

Minugh-Purvis, N., & McNamara, K. J. (2002). *Human evolution through developmental change.* New York: Johns Hopkins University Press.

Neisser, U. (2006). Foreword. In M. J. Spivey (Ed.), *Continuity of mind.* Oxford: Oxford University Press.

O'Regan, J. K., & Noë, A. (2001). A sensorimotor account of vision and visual consciousness. *Behavioral and Brain Sciences, 24,* 939–1011.

Odling-Smee, F. J., Laland, K. N., & Feldman, M. W. (2003). Niche construction: the neglected process in evolution. *Monographs in population biology.* (Vol. 37). Princeton: Princeton University Press.

Pauli, A., Rinn, J. L., & Schier, A. F. (2011). Non-coding RNAs as regulators of embryogenesis. *Nature Reviews Genetics, 12,* 136–149.

Paz-y-Miño, C. G., Bond, A. B., Kamil, A. C., & Balda, R. P. (2004). Pinyon jays use transitive inference to predict social dominance. *Nature, 430,* 778–788.

Peisajovich, S. G., Garbarino, J. F., Wei, P., & Lim, W. A. (2010). Rapid diversification of cell signalling phenotypes by modular domain recombination. *Science, 328,* 368–372.

Piaget, J. (1988). Piaget's theory. In K. Richardson & S. Sheldon (Eds.), *Cognitive development to adolescence.* Hove: Erlbaum.

Pinker, S. (1997). *How the mind works.* London: Penguin.

Pinker, S. (2006). The blank slate. *The General Psychologist, 41,* 1–8.

Pinker, S. (2010). The cognitive niche: coevolution of intelligence, sociality, and language. *Proceedings of the National Academy of Sciences, 107*(Supplement 2), 8993–8999.

Prokopenko, M., Boschetti, F., & Ryan, A. (2009). An information theoretic primer on complexity, self-organization and emergence. *Complexity, 15,* 11–28.

Ramanathan, S., & Broach, J. (2007). Do cells think? *Cellular and Molecular Life Sciences, 64,* 1801–1804.

Richardson, K. (1998). *Models of cognitive development.* Hove: Psychology Press.

Richardson, K. (2002). What IQ tests test. *Theory and Psychology, 12,* 283–314.

Richardson, K. (2010). *The evolution of intelligent systems: how molecules became minds.* Basingstoke: Palgrave Macmillan.

Richerson, P. J., & Boyd, R. (2005). *Not by genes alone: How culture transformed human evolution.* Chicago: University of Chicago Press.

Rogoff, B., Moore, L., Najafi, B., Dexter, A., Correa-Chávez, M., & Solís, J. (2007). Children's development of cultural repertoires through participation in everyday routines and practices. In J. E. Grusec & P. D. Hastings (Eds.), *Handbook of socialization.* NY: Guilford.

Rose, C. R. (2005). Integrating ecology and developmental biology to explain the timing of frog metamorphosis. *Trends in Ecology and Evolution, 20,* 129–135.

Rose, H., & Rose, S. (Eds.), (2001). *Alas poor Darwin.* London: Vintage.

Salthe, S. N. (2008). Natural selection in relation to complexity. *Artificial Life, 14,* 363–374.

Shamma, S. (2001). On the role of space and time in auditory processing. *Trends in Cognitive Sciences, 5,* 340 348.

Shenhav, S., Solomon, A., Lancet, D., & Kafri, R. (2005). Early systems biology and prebiotic networks. In C. Priami (Ed.), *Transactions on computational systems biology.* Berlin: Springer-Verlag.

Siri, B., Quoy, M., Delord, B., Cessac, B., & Berry, H. (2007). Effects of Hebbian learning on the dynamics and structure of random networks with inhibitory and excitatory neurons. *Journal of Physiology (Paris), 101,* 136–148.

Skarda, C., & Freeman, W. J. (1987). How brains make chaos in order to make sense of the world. *Behavioral and Brain Sciences, 10,* 161–195.

Slonim, N., Atwal, G. S., Tkačik, G., Bialek, W. (2005). *Estimating mutual information and multi–information in large networks.* Arxiv preprint: cs/0502017.

Smith, P. K., Jostmann, N. B., & Galinsky, A. D. (2008). Lacking power impairs executive functions. *Psychological Science, 19,* 441–447.

Soujik, V., & Armitage, J. P. (2010). Spatial organization in bacterial chemotaxis. *EMBO Journal, 29,* 2724–2733.

Spencer, J., Thomas, M. S.C., & McClelland, J. L. (Eds.), (2009). *Toward a new grand theory of development: Connectionism and dynamical systems theory re-considered.* Oxford: Oxford University Press.

Sporns, O., & Kötter, R. (2004). Motifs in brain networks. *Public Library of Science: Biology, 2,* 1910–1918.

Stearns, S. C. (1989). The evolutionary significance of phenotypic plasticity. *BioScience, 39,* 436–447.

Tagkopoulos, I., Liu, Y.-C., & Tavazoie, S. (2008). Predictive behavior within microbial genetic networks. *Science, 320,* 1313–1317.

Taylor, R. J., Falconnet, D., Niemisto, A., Ramsey, S. A., Prinz, S., Shmulevich, I., et al. (2009). Dynamic analysis of MAPK signaling using a high-throughput microfluidic single-cell imaging platform. *Proceedings of the National Academy of Sciences, 106,* 3758–3763.

Tomasello, M., & Warnekan, F. (2008). Human behaviour: share and share alike. *Nature, 454,* 1057–1058.

Tooby, J., & Cosmides, L. (1995). The psychological foundations of culture. In J. H. Barkow, L. Cosmides & J. Tooby (Eds.), *The adapted mind: Evolutionary psychology and the generation of culture.* Oxford: Oxford University Press.

Tschacherm, W., & Haken, H. (2007). Intentionality in non-equilibrium systems? the functional aspects of self-organized pattern formation. *New Ideas in Psychology, 25,* 1–15.

Visscher, P. M., Smith, D., Hall, S. J.G., & Williams, J. L. (2001). A viable herd of genetically uniform cattle. *Nature, 409,* 303.

Vygotsky, L. S. (1981). The genesis of higher mental functions. In J. V. Wertsch (Ed.), *The concept of activity in Soviet psychology.* New York: Sharpe.

Waddington, C. H. (1957). *The strategy of the genes.* London: George Allen & Unwin.

Wagner, A., & Wright, J. (2007). Alternative routes and mutational robustness in complex regulatory networks. *BioSystems, 88,* 163–172.

Wang, X. (2004). The unexpected consequences of a noisy environment. *Trends in Neurosciences, 27,* 364–366.

Warren, W. H., & Shaw, R. E. (Eds.), (1985). *Persistence and change: Proceedings of the first international conference on event perception.* Hillsdale NJ: Lawrence Erlbaum.

White, L. A., Coppola, D. M., & Fitzpatrick, D. (2004). The contribution of sensory experience to the maturation of orientation selectivity in ferret visual cortex. *Nature, 411,* 1049–1053.

Wiedemann, C. (2010). Neuronal activation: the symphony of transcription. *Nature Reviews Neuroscience, 11,* 372.

Wilkins, A. S. (2008). Canalisation: a molecular genetic perspective. *BioEssays, 19,* 257–262.

Wilkinson, R., & Pickett, K. (2009). *The spirit level.* London: Penguin.

Wimpenny, J. H., Weir, A. A.S., Clayton, L., Rutz, C., & Kacelnik, A. (2009). Cognitive processes associated with sequential tool use in New Caledonian crows. *PLoS One, 4,* e6471.

Woollett, K., & Maguire, E. A. (2011). Acquiring "the knowledge" of London's layout drives structural brain changes. *Current Biology, 21,* 2109–2114.

Yang, Y. (2009). Growth and patterning in the limb: signaling gradients make the decision. *Science Signalling, 2.* pe3.

CHAPTER SIX

Embodiment and Agency: Toward a Holistic Synthesis for Developmental Science

David C. Witherington*,[1], Shirley Heying[†]

*Department of Psychology, University of New Mexico, Albuquerque, NM, USA
†Department of Anthropology, University of New Mexico, Albuquerque, NM, USA
[1]Correspondence author: E-mail: dcwither@unm.edu

Contents

Abstract

Relational, systems-oriented approaches are strongly positioned to advance theory and research in developmental science and to cement a process orientation to development at all levels of organization—from the biological to the psychological and sociocultural—despite continued prominence in the field of biologically reductionist explanatory accounts. However, the inclusive, explanatorily pluralistic ontological framework involved in adopting a relational perspective on developing systems is not always fully appreciated, explicitly articulated or even followed by devotees of the perspective. In this chapter, we highlight the importance of holistically couching interlevel relations—those that obtain vertically between levels of organization, such as between the biological and psychological levels—in terms of wholes and parts and of recognizing the different modes of causal explanation that obtain depending on whether the relations move from parts-to-whole or whole-to-parts. This, in turn, yields an explanatory pluralism under which all living systems, at any level of organization, exist as both subjects and objects. We ground this discussion by examining the ontological compatibility with a relational developmental systems perspective of two systems-oriented approaches to embodiment: the dynamic systems approach of Thelen and Smith (1994, 2006) and the enactive approach of Varela, Thompson, and Rosch (1991).

1. INTRODUCTION

Steady growth over the years 1991–2010 in the prominence and interdisciplinary influence of neuroscience has thoroughly ensconced biological analysis in mainstream discussions of psychological development and functioning, fueling, in turn, questions of how to conceptualize the role of biological processes in psychological explanation (Johnson, 2010; Marshall, 2009; Miller, 2010). Unfortunately, conceptualization of biological–psychological interplay lags far behind neuroscience's methodological and technological advances and all too commonly trades on the traditional discourse of biological reductionism, wherein lower order, biological levels of analysis are viewed as causally privileged and foundational relative to higher order levels, such as the psychological (Miller, 2010; Spencer, Blumberg, et al., 2009). In the wake of such discourse, psychological processes—thought, memory, emotion, and intentionality—become either isomorphic to or, at best, epiphenomenal byproducts of the brain's neurochemistry. Under these circumstances, brains themselves become agentive overseers of organismic functioning, assuming causal primacy relative to peripheral components of the nervous system by processing sensory information from the periphery and generating motor programs for enacting the musculature. Genes set the whole process in motion by innately endowing organisms with prefigured specifications for gene–environment interaction and its developmental consequences. Despite substantial progress in applying the mechanics of neuroscience to psychology, biological reductionism of this kind still carries theoretical weight in psychological and even developmental science (Marshall, 2009; Overton, 2006; Shimamura, 2010).

Fortunately, for proponents of developmental science, biological reductionism's recent resurgence in the middle of what Miller (2010) has called "the Decades of the Brain" (p. 720) has been paralleled by an equally strong resurgence of systems approaches to development and biological–psychological interplay (Gottlieb, Wahlsten, & Lickliter, 2006; Lerner, 2006). These approaches are emblematic of what Overton (1998, 2006) terms *relational metatheory*, within which classic polarities such as structure–function and process–pattern are recognized as distinct, alternative yet equally legitimate and interdependent perspectives taken on the same whole. Some of the earliest formalizations of a systems orientation to development arose in biological circles through the work of von Bertalanffy (1933) and Weiss (1939) and extended through the work of developmental psychobiologists such as Schneirla (1957, 1960), Kuo (1967) and Lehrman (1970). Central to the classic developmental theorizing of Piaget (1952, 1970, 1971) and Werner (1948, 1957), the systems orientation to development has spawned numerous modern incarnations, from the transactional approach (Sameroff, 1983), the developmental psychobiological systems view (Gottlieb, 1991, 1992; Gottlieb et al., 2006), and the developmental systems approach (Ford & Lerner, 1992; Lerner, 2002, 2006) to holistic interactionism (Magnusson, 1995; Magnusson & Stattin, 2006) and the dynamic systems (DS) approaches, such as Thelen and Smith (1994, 2006) and van Geert (1991, 1994).

All these systems approaches to development unite around antireductionist stances and process orientations. Rather than explaining organismic complexity through recourse to foundational biological elements and their reductive summation, systems approaches appeal to the active, reciprocal, interpenetrating processes of interaction that transpire at all levels of organismic organization (Ford & Lerner, 1992; Gottlieb, 1992). This relational ontology establishes that components of any system, at any level of analysis, are what they are by virtue of their relationships to other components. In other words, components do not house an elemental identity independently of the relations in which they engage, for the relations involve an "interpenetration among parts" (Overton, 2003, p. 359) and are consequently essential to the components' very nature (Kitchener, 1982; Lerner, 1978; von Bertalanffy, 1968).

Yet the full implications of this relational ontology are not always realized across different systems approaches. Just as components of a system are what they are by virtue of their relations to other components, so too the relations among components of any system are what they are by virtue of the organizational structure of the components involved (Bickhard & Campbell, 2000; Witherington, 2011). Components of any given system not only interrelate horizontally in part-to-part interactions—which alter the nature of the parts themselves—to give rise vertically, in *upward* fashion, to wholes (new structures/patterns/levels of organization), but part-to-part interactions also derive vertically *downward* from higher order wholes (O'Connor, 1994; Silberstein, 2006; Sperry, 1986). The *holistic* nature of these interlevel interactions—in which the organization of the system as a whole necessarily conditions the very parts and relationships that comprise it—critically informs our understanding of the interplay between biological and psychological levels of organization. Yet systems approaches to development are not entirely unified with regard to their conceptualization of interlevel interactions, with this lack of unification ultimately revolving around whether wholes are accorded explanatory status in their own right or simply viewed as emergent, irreducible objects of part–part relations (Witherington, 2007, 2011).

As we will discuss in this chapter, such variation in systems approaches has its metatheoretical underpinnings in fundamentally different conceptualizations of holism and the nature of explanation itself. After providing an overview of the metatheoretical issues that play out in both reductionist and systems-oriented approaches, we will argue that fully embracing the holism integral to a relational developmental systems approach (Ford & Lerner, 1992; Overton, 2006, 2010) entails framing all living systems, at any level of organization—whether that system is a cell, an organ, an organism, or an organism–environment unity—as both *subjects* and *objects* (Levins & Lewontin, 1985), as *both* integrated wholes and top-down subjects of their own organization *and* bottom-up emergent objects of the relations that characterize their component parts. We conclude by contrasting two systems-oriented approaches to embodiment—the DS approach of Thelen and Smith (1994, 2006), Smith (2005) and Thelen, Schoner, Scheier, and Smith (2001) and the enactive approach of Varela, Thompson, Rosch (1991), Di Paolo, Rohde, & De Jaegher (2010), Stewart (2010) and Thompson (2007)—to highlight ontological differences among systems thinkers in adherence to relationism's tenet of holism, with critical ramifications for how we view the biological from the vantage point of psychological and developmental science.

2. THE MECHANISTIC HOMOGENEITY OF REDUCTIONISM

Current resurrections of biological reductionist thought hold all the earmarks of a classic mechanistic ontology (Pepper, 1942), emblematic of what Overton (1998, 2006) terms *split metatheory*. In the mechanistic world view, organisms are machine-like collections of discrete, independent parts; those parts establish the elemental bedrock out of which all organismic complexity arises and to which all complexity ultimately reduces. The organism itself, taken as a whole, is nothing more than the summation of its elemental parts and their additive interactions. It is the passive recipient or consequent of efficient causal forces—temporally antecedent propelling or initiating forces—emanating either from within, via, for example, neural or genetic activity, or from without, via environmental stimulation (Lerner, 1978, 2002; Pepper, 1942; Reese & Overton, 1970).

For the biological reductionist, like all mechanistically oriented thinkers, genes and environment—or brains and bodies—constitute independent receptacles of information for the specification of form and serve as antecedent causal forces acting on the organism to implement transfer of said information, much as the kinetic energy of one billiard ball is imparted to another by spatiotemporal contact. Given the specific formative priority that biological reductionism assigns to those forces of "biological" origin, development arises primarily from the organism's "original nature"—that is, the innately given genetic code and the prespecified "information" it contains—and secondarily from the environmental circumstances to which the organism is exposed, which lead either to facilitation and elaboration or to inhibition of the blueprint set forth in the genome (Derksen, 2010; Ingold, 2004; Schneirla, 1960). The biological reductionist tendency to prefigure the developmental emergence of structures and functions in the informational potential of the genome applies equally to the prefiguring of real-time behavior generation in the information processing and pattern generating properties of the brain (Thompson, 2007).

In its mechanistic focus, biological reductionism, like any other kind of reductionism, disembeds the organism from its environment and divides the total organism into a collection of independent parts. Its uniqueness resides in its nativist assignment of causal primacy to parts endogenous to the organism, but reductionism can alternatively assign causal primacy to parts of the environment, as in classic empiricist accounts when forces of environmental origin impress preexistent information upon organisms (Lerner, 1978, 2002; Overton, 2010). Irrespective of which informational source does the heavy lifting or carries the potential, reductionist thinking construes both real- and developmental-time process as the part-to-part transfer or transmission of preexistent information, with some parts (e.g., the genes or the brain) more foundational to the process than others by dint of their "informed" status and causal prowess (Oyama, 1985). The organism qua organism—as an embodied whole—is regarded as simply a byproduct of preexistent information transferred from more elemental causal sources—"the passive object of autonomous internal and external forces" (Levins & Lewontin, 1985, p. 89)—reflecting, in Oyama's (1985) words, a "preoccupation with organisms as material objects whose design and functioning must be imparted to them" (p. 12).

The logical pitfalls of explaining developmental emergence and real-time action generation through the mechanistic appeal to preexistent information, privileged sources, and processes of transmission have been well documented (e.g., Anastasi, 1958; Oyama, 1985; Smith & Thelen, 1993). In the case of biological reductionism, if information from the genome establishes a plan for development, prefiguring the developmental process itself, then what accounts for the development of that information in its own right? Similarly, if information from the brain establishes a plan for organismic action, prefiguring the real-time process of action generation itself, then what accounts for the generation of that information in its own right? Arguments that explain action generation and developmental emergence through design must eventually address the generation and development of the design itself without invoking another level of design lest they succumb to infinite regress (Oyama, 1985; von Bertalanffy, 1933).

The appeal to preexistent information effectively negates the idea that *new* actions, structures and functions truly do emerge, as *irreducible wholes*, in real time and over the course of development. Though the biological reductionist recognizes that developing structures and functions themselves do not literally preexist their epigenetic "realization" in an individual's ontogeny, the genetic information for building those structures does preexist the epigenetic processes that ultimately render that information manifest, meaning that the emergent structures and functions themselves are ultimately reducible to that information and its transmission. Similarly, although the actual actions organism perform do not literally preexist their real-time "realization" in an individual's performance—every action is unique, given the need to flexibly adapt to the intricacies of ever-changing intra- and extraorganismic contextual circumstances—information in the brain for building those acts does preexist the real-time processes that ultimately render that information manifest, meaning that the basic form of the act is reducible to that antecedent information and its transmission (Gottlieb et al., 2006; Oyama, 1985; Thelen & Smith, 2006). Biological reductionism's information transmission process essentially prescribes developmental end-products before they develop and ultimately transfers explanation for origins from ontogenetic to phylogenetic sources (Oyama, 1985; Thelen & Smith, 1994). This argument by design, in turn, severely marginalizes developmental process in the generation of structure and function and actively conflicts with the ontological stance of emergence and self-organization in epigenesis to which developmental science has long been committed: how do novel levels of organization emerge from precursor levels of organization, absent planning or prescription (Gottlieb, 1991; van Geert, 1998)?

3. SYSTEMS APPROACHES AND EXPLANATORY PLURALISM

Systems proponents search for the key to the developmental emergence of structure and function and to the real-time emergence of organismic activity not in preexistent instructions, privileged levels of analysis, and processes of transmission but in the constructive activity of real- and developmental-time processes themselves. Contrary to reductionism, living organisms are complex, multilevel/multicomponent systems of organization—from biological levels, such as genes, cells, tissues and organs, to different levels of psychological structure, such as sensorimotor, cognitive and metacognitive—and themselves are embedded within and stand in dynamic relation to multiple levels of extraorganismic organization—from physical and sociocultural environments to broader historical contextual circumstances (Ford & Lerner, 1992; Gottlieb, 1992; von Bertalanffy, 1968). It is the mutual, dynamic relations, or what Gottlieb (1991) terms "coactions," among various intra- and extraorganismic components—not the components themselves—that ultimately engender the real-time production of behavior and the developmental-time emergence of new structures and functions, with no component or level (e.g., genes, brain, or the biological level of organization) of this overarching organism–environment system assuming formative privilege or foundational status relative to any other component or level (Lerner, 2006; Thelen & Smith, 1994, 2006).

Any specific action an organism performs in context (e.g., a reach for a cup) is not generated, via information or otherwise, by any one of the components (e.g., the brain) that comprise the organism or its environment, but instead arises, unscripted, from a reciprocal confluence of components that comprise the organism–environment system (e.g., brain–body–environment relations), with all of the components coequal partners and none holding privileged causal status in the real-time construction of the action pattern. Such is the hallmark of the embodiment paradigm (Clark, 1997; Thelen et al., 2001; Varela et al., 1991; Wheeler, 2005). Similarly, new patterns of organization in an organism's development (e.g., the emergence of reaching ability) cannot be reduced to instructions or information in any one of the components (e.g., the genes) that comprise the developing system, but instead spontaneously emerge from a reciprocal confluence of components that comprise the organism–environment system, without being instructionally prefigured in or causally driven by any of those coequal components. Such is the hallmark of a holistic, or in Gottlieb's (1970, 1991) terms "probabilistic," epigenesis paradigm (Overton, 2010). As Oyama (1985) writes, "It is in this ontogenetic crucible that form appears and is transformed, not because it is immanent in some interactants and nourished by others, or because some interactants select from a range of forms present in others, but because any form is created by the precise activity of the system" (p. 34).

In place of biological reductionism's unidirectional, bottom-up causal flow from biological to psychological processes, systems approaches substitute bidirectional, reciprocal flow among all components of the organism–environment system, from biological and psychological to sociocultural components (Ford & Lerner, 1992; Gottlieb, 1992). The bidirectionality and mutuality of interactions among a system's components receives its fullest treatment in the developmental systems approaches of Lerner, Gottlieb and colleagues (Ford & Lerner, 1992; Gottlieb, 1992; Gottlieb et al., 2006; Lerner, 2002), which Overton (2006, 2010) has identified as most illustrative of the relational metatheory's inclusive ontological framework. Developmental systems approaches stress that the multiple levels of organization in living systems, from the cellular to the organism–environment levels, involve a *hierarchical*—or as Ford and Lerner (1992) prefer, *heterarchical*—structuring in the sense that these levels are vertically organized with respect to one another, invoking part–whole relationships. Genes comprise parts of the whole that is the cellular level of organization, cells comprise parts of the whole that is the tissue level of organization, organs comprise parts of the whole that is the organ system level of organization (e.g., the cardiovascular system), and organisms comprise parts of the whole that is the organism–environment level of organization. So even though genes, cells, tissues, and organs are all components/parts of the organism as a whole, each of these parts is also a whole, or level of organization, in its own right relative to its lower order parts (e.g., the cell is a whole relative to genes but a part relative to tissues). And, as bidirectionality characterizes all interactions in a system, interlevel relationships among the components of any system play out in both vertical directions—parts-to-whole *and* whole-to-parts interaction—across all levels of organization, from the biological to the psychological and sociocultural.

When framing the interactional process in bottom-up, *parts-to-whole* terms, systems approaches replace reductionism's part-to-part transmission of preexistent information with parts-to-whole emergence of truly novel wholes. Given that parts are "internally" related such that their identity depends on their relation to other parts (Kitchener, 1982), each part-to-part interaction involves simultaneous influence between parts (e.g., part A is affecting, while simultaneously being affected by, part B, and vice versa), effectively eliminating absolutist notions of antecedent and consequent, cause and effect at the part-to-part level (Ford & Lerner, 1992). However, from the bottom–up vantage point, the new whole—to which these part–part relations give rise—can be framed as a consequent relative to the push-from-behind, antecedent forces of part–part relations; in effect, an efficient causal framing of sorts still prevails at the level of parts-to-whole. Since the whole is multiply determined—with each part of the whole a necessary but not a sufficient interactant in the joint production of the whole—efficient cause becomes distributed across all the parts that make up the whole such that each part equally influences the emergence of the whole without determining it (Turvey, Shaw, & Mace, 1978). This distribution of efficient cause essentially amounts to a singular confluence of multiple, bidirectional part-to-part relations among components of the system spontaneously and nonlinearly giving rise to a new level of organization in the system, both at the level of real-time activity and at the level of developmental transformation.

When framing the interactional process in top-down, *whole-to-parts* terms, however, a fundamentally different mode of causal explanation— a "systematic causation" (Silberstein, 2006)—needs to be invoked, one that endows the emergent organization of a system (its structure, form, and pattern) with causal significance. This necessitates moving beyond the antecedent–consequent, force-like process orientation of efficient causality, distributed or otherwise, because conceptualizing the vertically downward influence of higher order wholes on their lower order constituents in efficient causal terms essentially amounts to a revival of vitalistic dualism, through appeal to some vital force or entelechy (Campbell, 1974; Emmeche, Koppe, & Stjernfelt, 2000; Juarrero, 1999; Sperry, 1986). In downward or systematic causation, a system's organization—the unitary structure of the whole—constitutes a subject of explanation in its own right and establishes an *organizational* explanatory grounding for the very part–part relations that give rise to it.

Thus, viewing living organisms as multilevel systems of organization supporting bidirectional causal influence within the same level of organization (e.g., relations between cells within the biological or between representation subsystems within the psychological) and between levels of organization (e.g., relations between the psychological and biological levels of organization or between the tissue and cellular levels of organization within the biological) requires an explanatory pluralism that moves beyond the temporal framing of efficient cause. Such an explanatory pluralism is central to the relational developmental systems perspective that Overton (2006, 2010) articulates and follows from the "levels of organization" frameworks endorsed in the developmental systems approaches of Lerner, Gottlieb, and colleagues (though it often receives little *explicit* focus in these specific systems approaches). To fully articulate the nature of whole-to-parts systematic causality and the integrative focus of explanatory pluralism, we must first elaborate the metatheoretical foundations for the relational developmental systems perspective: the synthesis of organismic and contextualist world views. This metatheoretical elaboration will illustrate why the full implications of relational ontology are not always realized in systems approaches.

3.1. Organismic–Contextualist Synthesis

A relational metatheory for understanding developmental systems relies on what Overton (2010) terms a "principled synthesis" (p. 11) of the contextualist and organismic world views (see also Overton, 1984, 2007; Overton & Ennis, 2006). The organismic world view or organicism (Pepper, 1942) takes as its basic metaphor the organism as active constructor of reality through interaction with the world (Reese & Overton, 1970). By this perspective, any living, organized system constitutes an irreducible, integrated whole, and its development is marked by irreversible, progressive and qualitative changes in the formal properties of that whole (Overton, 1984; Pepper, 1942; Scholnick, 1991). Organicism grounds its explanatory framework in the abstraction of formal properties from real-time system activity in the ever-changing present (Lerner & Kauffman, 1985).

Explanation in organicism involves formulations of what Aristotle called *formal* and *final* causes. Formal and final causes differ from efficient cause. Neither formal nor final cause operates as a temporal antecedent or propelling force; rather, these causes are atemporal, organizational levels of explanation that lend *meaning* to the temporal, cause–effect sequences of efficient cause (Rychlak, 1988). Efficient cause necessarily presupposes a temporal context, but formal and final causes explain without recourse to the flow of time by invoking abstraction itself as a means of explanation (Rychlak, 1988). Formal causes specifically abstract a pattern—form, structure, and organization—from the particular, real-time content of a phenomenon and employ that pattern to explain the phenomenon. Final causes abstract a function, future-end, or purpose from the particular, real-time content of a phenomenon to explain the phenomenon. In both cases, the formal and teleological patterning has explanatory value because "it introduces order and organization into the domain under investigation" (Overton, 1991, p. 220).

Formal and final causes are commonplace in scientific explanation, though frequently not acknowledged as such. We invoke formal causes in psychology, for example, when explaining real-time organismic activity by means of psychological constructs such as emotions, cognitive, and personality structures—constructs that capture organismic functioning as a whole, across specific actions and contexts—or through appeals to a particular stage or developmental level of organization. We invoke final causes when explaining an organism's actions by means of goals or intentions or, more generally, by means of the function served by the action, conscious or otherwise. At the level of developmental time, we invoke final causes when we posit ideal endpoints and directional sequences of organizational change as meaningful contexts in which to embed the understanding of a phenomenon at any given time, such as explaining real-time behavior by means of Werner's (1957) orthogenetic principle by which development proceeds in terms of increasing differentiation and hierarchic integration.

But far too rarely do psychologists acknowledge the holistic, higher order nature of these kinds of explanation. Instead most routinely misconstrue them as efficient causes, reifying them as concrete parts of the system in the process, and thus incorrectly framing formal and final causes as antecedent "forces" initiating the consequent activity from which they are abstracted (Lourenco & Machado, 1996; Rychlak, 1988). Whereas efficient cause serves to contextualize—in bottom-up fashion—our understanding of an organism's specific activity in terms of the physicochemical events and conditions that temporally and regularly precede that activity, formal and final causes are not meant to explain the incessant variability in the real-time content of activity or to identify the antecedent circumstances for such content. Instead, formal and final causes serve to contextualize our understanding of an organism's specific activity in terms of the organization and directional purpose that activity evinces as a whole (Overton, 1991; Tolman, 1991). Their explanation is one of the atemporal embeddedness, not the temporal precedence.

With its focus on explanation by way of abstraction, organicism largely neglects issues of timing, intra- and interindividual variability and the particularities of action-in-context (Pepper, 1942), all of which constitute key concerns for the relational ontology of developmental systems approaches (Lerner, 2006). The contextualist world view effectively counters organicism's neglect by highlighting the particularities of time and context in the ontological framework it offers for understanding action and its development. "The event alive in its present" aptly captures the fundamental metaphor for this framework (Pepper, 1942, p. 232). Rather than appeal to the abstract, generalizable forms of organicism's higher order holistic focus, contextualism's holism grounds itself in the present, real-time activities of organisms in specific settings and contexts (Overton, 1991; Pepper, 1942). As Overton (2007) writes, "the holism of organicism is about parts–whole relations of self-organizing systems, while the holism of contextualism is about parts–whole relations of the adaptive act" (p. 158). Change and novelty mark the core foci of contextualism; Pepper (1942) writes that "nothing is more empirically obvious to a contextualist than the emergence of a new quality in every event" (p. 256). Contextualism starts with the present event of specific action in a specific context and proceeds to other events immediately past and in the immediate future. However, it is to the immediacy of the present that contextualism steadfastly clings.

In the eyes of the contextualist, the abstractions to which organicism appeals lose sight of the process unfolding in the particularities of action adapted to local context (Chandler, 1997). Understanding development requires a grounding of analysis in the variability that action demonstrates during real-time, adaptive encounters with everyday contexts, for development is continuous with such real-time change, moving from particular to particular (Overton, 1991). In contrast to organicism's integrative, organizational approach, contextualism's outlook is dispersive, characterizing the world in terms of "multitudes of facts rather loosely scattered about and not necessarily determining one another to any considerable degree" (Pepper, 1942, pp. 142, 143). Espousing a "horizontal cosmology," the contextualist world view dispenses with hierarchy, valuing instead the flattening of all systems of thought that rely on multiple, vertically structured levels of analysis or organization (Kendler, 1986; Pepper, 1942). Organization is consequently devalued as mere "appearance" and regarded as explanatorily epiphenomenal relative to the "reality" of the real-time, temporal dynamics that give rise to it.

Overall, in the realm of explanatory focus, organicism's neglect of the lower order dynamics of particulars is matched by contextualism's neglect of the higher order organizational properties of the whole. An organismic–contextualist synthesis, couched in terms of a relational developmental systems perspective (Overton, 2006, 2010), preserves the distinct explanatory foci of each world view while obviating their neglectful weaknesses. Within the inclusive ontology of relational metatheory, both organicism and contextualism operate not as absolute modes of truth but as *interdependent*, complementary frames—each legitimate in its own right, neither privileged as modes of explanation—whose individual meanings are necessarily contextualized one within the other and within the inclusive whole of which they are differentiated, relational parts. Both perspectives reflect organizationally distinct but relationally unified, indissociable lines of sight (Overton, 2006, 2010).

With these metatheoretical foundations in place, the issue of lower order parts-to-whole interaction is addressed through contextualism's focus on the "grass-roots" particularities of real-time acts adapting to their contextual settings—a dynamic interplay of organismic system components constrained instrumentally by local context—yielding, in turn, the continuous emergence of new action qualities, or wholes. Similarly, the issue of higher order whole-to-parts systematic causality, or downward causation, is addressed through organicism's formal and final cause outlook. As formal and final cause, downward causation operates through whole-to-part organizational *constraint*, in which "the whole at least partially determines what contributions are made by its parts" (van Gulick, 1993, p. 251). Constraint involves the selective effect that higher order pattern has on its lower order foundations (Campbell, 1974; Moreno, 2008). Downward causation does not operate by generating new physicochemical "forces" or by altering existing forces but through the "selective activation of physical powers" (van Gulick, 1993, p. 252). The higher order form and function of a system topologically constrain the distributed efficient causal interactions among the system's components by selecting among a wide set of interactional possibilities. That is, the nature of local action-in-context interactions cannot be fully understood in the absence of the self-organizing whole, in which they are embedded (El-Hani & Pereira, 2000; Moreno, 2008; Murphy, 2009). Efficient cause becomes meaningful only in the presence of formal and final cause via the organizational complexity that characterizes the system qua system (El-Hani & Pereira, 2000).

Thus, higher order forms, reflecting the abstracted organization of a system that characterizes functioning across a variety of contexts, necessarily frame the lower order local dynamics on which they depend. Form abstracted from the here-and-now is as integral to explanation as real-time contextual factors. Such is the contribution of an organismic focus on formal wholes. The embedded real-time system activity in various real-time, concrete contexts of adaptation provides an equally essential explanatory framework. This taps into the dynamics involved in the real-time emergence and consolidation of action content as well as the developmental emergence of new system forms. Such is the contribution of a contextualist focus on real-time particularities (Witherington, 2007). In the explanatory pluralism of a relational developmental systems perspective, both organicism and contextualism—as different but simultaneous levels of explanation—constitute necessary lines of sight for our understanding of system activity and development, but with neither assuming ontological privilege nor precedence (Overton, 2007). A relational ontology is as much about global-to-local organizational constraint as it is about local-to-global dynamic construction, as much about whole-to-parts relations as about parts-to-whole relations via the interdependence and interpenetration of multiple levels of reality, from the micro to the macro (Emmeche et al., 2000; Finnemann, 2000; Thompson, 2007). All parts-to-whole relations presuppose whole-to-parts relations, just as all whole-to-parts relations presuppose parts-to-whole relations, at all levels of living system organization.

Modern systems approaches in developmental science have primarily targeted the local-to-global side of explanation, especially in light of two decades worth of increased foregrounding of issues pertaining to intra- and interindividual variability, temporality, and context specificity in developmental science (Lerner, 2006). Consequently, system organization is typically characterized as something to be explained—as an *object* of inquiry. But just as critical is the characterization of system organization as a *subject* of explanation in its own right. Through downward causation, systems qua *systems* coordinate and control the properties of their lower order components and the interactions of those components with one another, preserving their very organization despite flux, renewal and turnover in the lower order components that dynamically engender them (Thompson, 2007). Systems are simultaneously objects and subjects. What does it mean to view any living system as a subject of its own self-organization? In the realm of psychological explanation characterizing humans as active, meaning-making agents is commonplace enough, though such characterizations are often inappropriately framed in efficient—rather than formal and final—causal terms, leading to the structural reification and antecedent–consequent misinterpretation of systematic causality. But how does the view of living systems as subjects play out broadly at the level of biological organization, from the cell to the organ system to the organism as biologically integrated totality? To answer this question, we now turn to the biological theorizing of Hans Jonas and Francisco Varela.

3.2. Biological Systems as Subjects

Jonas' (1966) philosophical writings on biology and teleology point to the centrality of purposive existence at all levels of living organization. Rather than draw clear and unassailable boundaries between the psychological properties of human existence (e.g., thinking, feeling, intentionality) and the biological properties of life (e.g., construed in physical, objective terms), Jonas asserts that the fundamental qualities of human experience "have their rudimentary traces in even the most primitive forms of life" (p. xxiii). Basic selfhood and continuity of identity across time—which in turn yield an organized, meaningful perspective from which any living system engages its surround and makes sense of that surround via rudimentary processes of assimilation—comprise life in all of its organized guises, meaning that the "organic, even in its lowest forms prefigures mind, and…mind, even on its highest reaches, remains part of the organic" (p. 1). All systems of life, whatever the level of complexity, actively organize their relation to their surrounding as subjects of their own activity, as much as they are organized by their relation to their surround.

Jonas (1966) points out that this fundamental purposiveness of life, in which system organization maintains its identity in the face of ceaseless exchange of matter and energy with its surround, sets living systems apart from nonliving systems. Organization in nonliving systems is merely a consequence of the material constituents, or parts, that give rise to it. A change in any of its parts will necessarily result in a change in the nonliving system's overall form; as Jonas writes, "In the realm of the lifeless, form is no more than a changing composite state, an accident, of enduring matter" (p. 80). In any living system, however, the reverse holds true—the living system, as a whole, actively unites an entire network of material constituents in virtue of itself, for the sake of itself and in order to continually sustain itself. In Jonas' words, "Here wholeness is self-integrating in active performance, and form for once is the cause rather than the result of the material collections in which it successively subsists" (p. 79). The unity of the living system is a *self-unifying* unity rather than simply a *unified* unity, and as a self-unifying system, teleology is intrinsic to the living system (Jonas, 1966).

Thus, all biological systems as wholes have a purposive existence through maintenance of their identity, an active maintenance that both establishes their independence, as organized wholes, from the material flux on which they are built and grounds them incessantly in ceaseless flow of material and energy exchange with their surround (Jonas, 1966). For Jonas, metabolism—a process core to all organic existence and the defining quality of life—provides a poignant example of the living system's purposive existence and consequent subject orientation. In order to maintain its very existence, a living system must incessantly exchange matter with its surroundings via metabolism. Through metabolic functioning, the living system builds up originally and replaces continually the very material components, or parts, that constitute it as a whole. Yet the constant renewal of the living system's internal material components does not result in constant change of its overall form. Rather, it is precisely through the incessant renewal of its internal matter that a living system maintains sameness at the level of its organization, illustrating that form transcends matter. Thus, a living system "is never the same materially and yet persists as its same self, *by* not remaining the same matter" (Jonas, 1966, p. 76).

A living system's active maintenance of self—a formal whole—in the middle of continuous metabolic constituent renewal allows it to dynamically establish an identity that is distinct from its surround (Jonas, 1966; Thompson, 2007, p. 152). If the system did not maintain its own organization—its identity—it would cease to exist; thus, in the living system's drive to continue its own existence, it maintains a self-centered identity in which its very being is its own doing and that necessarily places it in contraposition to the external world (Barbaras, 2010; Jonas, 1966; Thompson, 2007). Jonas explains:

> *Profound singleness and heterogeneousness within a universe of homogeneously interrelated existence mark the selfhood of organism. An identity which from moment to moment reasserts itself, achieves itself, and defies the equalizing forces of physical sameness all around, it truly pitted against the rest of things (p. 83).*

The living system's active maintenance of its own identity in relation to the external world establishes an internal–external boundary that distinguishes the system from its material surround, ultimately yielding the relational pair "organism–environment" (Jonas, 1966).

Imminent in the relation between any living system qua system and its surround is the system's perpetual need to reach beyond itself into its surround to obtain foreign matter essential for its own internal renewal. This outward reach for external matter to satisfy its own internal needs results in a peculiar relatedness of dependence and possibility with the environment that requires the system's continual openness to its surround (Jonas, 1966). For living systems such as plants, contact with the external source of supply necessary for self-maintenance is continuous as their roots sustain an invariant contact with the system's food supply; Jonas writes that "through their continuous contact with the source of supply, the organism–environment relation functions automatically and no further apparatus for adaptation to short-term changes is necessary" (p. 103). Consequently, plants are fully and permanently integrated with their metabolic surround, generating a direct and stable openness to the environmental constituents necessary for their metabolic needs. Animals, in contrast, are not perpetually contiguous with such metabolic aliment and must typically reach well beyond themselves to access the foreign matter they need to sustain themselves. In place of osmotic interchange, animals "depend on the unassured presence of highly specific and nonpermanent organic bodies" to meet their metabolic needs (Jonas, 1966, p. 103). The active movement required of animals to obtain matter from a noncontiguous environment of metabolic resources distinguishes animality from other living forms in its mediated quality. Thus, as living systems move from plant to animal life, there is an increasing disclosure of the world and an increasing individuation of self (Jonas, 1966).

At whatever level of organization and complexity, the living system, in purposively striving to continue its being by seeking matter outside of itself for continuous inward self-renewal, is both cause and effect of itself, revealing its subjectivity. Jonas (1966) emphasizes that the living system's openness to its surround "is grounded in the fundamental transcendence of organic form relative to its matter, for it is this which constitutionally refers it beyond its given material composition to foreign matter as needed and potentially its own" (p. 84). The living system's transcendence over its own material constituents illustrates that as a living form, the system is not a mere consequence of the renewal of its matter, or parts, but is the cause (Barbaras, 2010). Thus, as a self-organizing subject that has as its purpose the very desire to continue its being, any living system must be understood as intrinsically teleological, necessitating the examination of all living systems in formal and final—not just efficient—causal terms.

Extending Jonas' phenomenological framing of living form's transcendence over its matter through its relation to its surround is the notion of living systems as autopoietic, developed in the writings of Varela (1979) and Weber and Varela (2002). In Varela's framework, living systems, as autopoietic, construct themselves by generating the very boundary conditions necessary for creation and maintenance of their *self*-organization as living subjects; autopoiesis thus presupposes a phenomenological framing of living matter. Weber and Varela (2002) elaborate:

autopoiesis proposes an understanding of the radical transition to the existence of an individual, a relation of an organism with it-self, and the origin of 'concern' based on its ongoing self-produced identity...Thus, autopoiesis is a singularity among self-organizing concepts in that it is on the one hand close to strictly empirical grounds, yet provides the decisive entry point into the origin of individuality and identity, connecting it, through multiple mediation with human lived body and experience, into the phenomenological realm (p. 116).

An autopoietic system produces the very components that give rise to it, establishing its own self-maintaining processes (Thompson, 2007; Weber & Varela, 2002). This is not to suggest that autopoietic systems construct themselves independently from their surround. The autopoietic system exists as thermodynamically open, engaging in incessant exchange of energy and matter with its surround, yet establishes organizational closure in that it metabolically maintains itself as an individual unity, as an invariant organization of patterning, in the face of continuous turnover and renewal of its material and energetic constituents (Moreno & Umerez, 2000; Varela, 1979; Weber & Varela, 2002). Autopoietic systems actively regulate the very external boundary conditions that produce them—by regulating the flow of energy and matter—thereby incorporating external boundary conditions into their own dynamics (Thompson, 2007; Juarrero, 2009).

For Varela, the living cell exemplifies autopoiesis, functioning as a self-producing, metabolic system wherein "every molecular reaction in the system is generated by the very same system that those molecular reactions produce" (Thompson, 2007, p. 92; Varela, 1979). The cell dynamically and continuously constructs its own membrane—the semipermeable boundary that establishes self-other distinction, the divide between itself and its surround—through metabolic processes that exist and operate only by means of the membrane they construct (Thompson, 2007). Cells maintain themselves through incessant generation and dissolution of their material constituents, both those that comprise the interior of the cell as well as those that make up its membrane. The products of cellular process are necessary conditions for the enactment of the process itself (Juarrero-Roque, 1985).

Autopoiesis, like Jonas' construal of metabolism, establishes agency—and consequently downward causation—as a central property of living systems and requires the framing of living phenomena in final causal, teleological terms, wherein "purposiveness" or "that for the sake of which" provides a key component of scientific explanation. Autopoietic systems construct themselves by generating the very constraints that establish far-from-equilibrium conditions, demonstrating an agency through self-determination. Such systems cannot be adequately explained through sole reliance on the temporal dynamics of efficient cause. As Kant (1790/2007) classically established, something that organizes itself is a "natural purpose," meaning that it is "both cause and effect of itself" (p. 199). The temporal, antecedent–consequent explanation from parts-to-wholes under which efficient causality operates dissolves from the vantage point of the system as a whole, necessitating instead a teleological explanation in terms of end or purpose because "every part is thought as *owing* its presence to the *agency* of all the remaining parts, and also as existing *for the sake of the others* and of the whole" (Kant, 1790/2007, pp. 201, 202). As Juarrero-Roque (1985) and Weber and Varela (2002) have persuasively argued, Kant's insistence on a teleological account of the organism as self-organizing and self-producing finds ontological validation in the empirically grounded theory of autopoiesis. The language of purposiveness and teleology is critical to the nature of whole-to-part relations (Thompson, 2007).

Self-organization in autopoietic systems requires a self—the system's organizational integrity—within which the local dynamics of the system are contextualized to explain system patterning. Juarrero (2009) writes that "by reversing the exergonic direction of classically thermodynamic processes and bringing a measure of control inside the systems, the appearance of endergonic processes creates an integrity and self-direction that were previously absent" (p. 92). Agency thus characterizes the autopoietic system qua *system*, controlling its own organizational integrity in the face of material and energetic turnover. As downward causation, agency involves formal and final—not efficient—causes. Agency does not consist of a central executive housed within the system dictating what the organism does in homunculus fashion. Neither it is imposed on the organism from outside. Rather, agency captures system functioning as a whole—it captures the relational dynamics of the system as a unifying unity—and is causally foundational to the study of self-organization in life, not an epiphenomenal byproduct of it. As a foundation to all life, such agency should not be confused with the deliberative, reflective agency associated with higher order, emergent forms of psychological functioning, such as consciousness (Gentile, 2008; Martin, Sugarman, & Hickinbottom, 2009). Agency exists at different levels of organization, and the basic agency of self-production, as a formative characteristic of living systems, organizationally and developmentally precedes deliberative, reflective forms of agency. But agency, in a rudimentary form, pervades all life.

4. FALLING SHORT OF A RELATIONAL PERSPECTIVE: ONTOLOGICAL DIVERGENCE INSTANTIATED THROUGH THE CONCEPT OF EMBODIMENT

Within a relational developmental systems perspective, all living phenomena are explicable only when viewed in *both* subject and object terms. As subjects, living systems construct themselves by generating the boundary conditions necessary for the creation and maintenance of their self-organization, demonstrating, in the process, an agency through self-determination; consequently, as Thompson (2007) writes, life is "a self-affirming process that brings forth or enacts its own identity and makes sense of the world from the perspective of that identity" (p. 153). Agentive self-maintenance and sense-making—as final and formal causes—apply whenever relationships in any living system are viewed from the global-to-local vantage point, whether that system is a cell, a tissue, an organ, or an entire organism. This "interiority of selfhood and sense-making" (Thompson, 2007, p. 225) becomes framed in terms of specifically *psychological* levels of organization— thought, emotion, intention, consciousness—when the entire organism, as an integrated whole, becomes the focus of inquiry. Prereflective psychological levels of organization in agency and sense-making involve the organism's immediate, symbiotic immersion in the world, where knowing inheres in the act of doing as an embodied and embedded activity absent any self-referential awareness of self as subject standing in relation to the world as object. In contrast, reflective psychological levels involve a clear subject–object structure characterized by a detached "Cartesian" stance in which organisms as subjects take the world, their actions in it, and their felt experience as objects of contemplation.

Thus, from the most complex to the most rudimentary levels of system organization, living systems entail both global-to-local organizational constraint and local-to-global dynamic construction, requiring, in turn, the explanatory pluralism of efficient, material, formal, and final causes via the complementary merging of organismic and contextualist world views. However, not all of the developmentally oriented systems approaches that are routinely included under the ontological umbrella of relational metatheory entertain the explanatory pluralism of an organismic–contextualist merger (Overton, 2007, 2010; Witherington, 2007). In fact, one of the most prominent systems approaches in the years 1991–2010—the DS approach of Thelen, Smith and their colleagues[1] (Spencer, Perone, & Buss, 2011; Spencer, Perone, & Johnson, 2009; Thelen et al., 2001; Thelen & Smith, 1994, 2006)—explicitly rejects the explanatory significance of organization, viewing instead all system patterning as an end product of more fundamental system process dynamics (Witherington, 2007, 2011; Witherington & Margett, 2011). Through this decidedly antistructuralist stance, the DS approach of Thelen and Smith eschews the organismic world view—and its focus on explanation through abstraction via formal and final cause—in favor of promoting an exclusive contextualism, which, as Overton (2007) has elaborated, essentially operates as a mechanism–contextualism blend that promotes a *split* rather than relational metatheoretical approach to the study of development.

The seeds of split ontology rest in contextualism's dispersive character. Pepper (1942) writes that contextualism is "constantly on the verge of falling back upon underlying mechanistic structures, or of resolving into the overarching implicit integrations of organicism" (p. 235). When contextualism merges with mechanism, the contextualist focus on local-to-global dynamic construction becomes ontologically foundational, with the local, temporally bound dynamics of particular actions in real-time context— captured via distributed efficient causality—constituting the only basis for explanation. The organization that emerges from the ceaseless flux of these real-time dynamics exists as little more than a momentary, epiphenomenal byproduct of bottom-up causal forces. Insofar as holism is concerned, although higher order organization in the system is irreducible to lower order organization, all organization is, in the final analysis, causally reducible to the local processes that give rise to it. The "appearance" of organization ultimately reduces to the "reality" of ceaseless flux, with pattern (the whole) serving as consequent to the antecedent of process (part–part relations).

[1] Two ontologically distinct camps of dynamic systems approaches to development exist. The version of dynamic systems that Thelen, Smith and colleagues espouse—referred to as the contextualist DSP (Witherington, 2007) or the Bloomington approach (van Geert & Steenbeek, 2005)—endorses a split metatheoretical framework and is the focus of this chapter. However, an alternative approach to dynamic systems—revealed in the work of Paul van Geert and Marc D. Lewis, among others, and referred to as the organismic–contextualist DSP (Witherington, 2007) or the Groningen approach (van Geert & Steenbeek, 2005)—endorses a fully relational metatheoretical framework and stands as a model instantiation of the relational developmental systems perspective. We focus selectively on Thelen and Smith's DS approach because its conceptualization of embodiment has achieved much more prominence in embodied cognition circles and has been much more widely promulgated as characteristic of the dynamic systems approach to embodiment (e.g., Shapiro, 2011).

The manner in which Thelen and Smith's DS approach operates within this split ontology and engenders new forms of reductionism is vividly illustrated by examining a concept key to the relational developmental systems perspective: embodiment (Overton, 1994, 2006, 2008). For Overton (2006, 2008), embodiment involves *both* the idea that our physical bodies condition and are constitutive of the way we psychologically engage the world *and* the idea that we, as embodied wholes, are agentive subjects of our own activity. In fact, Overton (2006) explicitly highlights the necessary framing of the body in psychological subject, not just physical object terms: "as a relational concept, embodiment includes not merely the physical structures of the body but the body as a form of lived experience, actively engaged with the world of sociocultural and physical objects" (p. 48). In other words, from one vantage point within a relational ontology, embodiment means viewing organismic functioning as the joint product, or object, of continuously coupled relations among coequal components of brain, body, and environment. But from an equally important vantage point, embodiment means viewing organismic activity as the meaningful engagement of a bodily subject—the organism as an integrated whole, an "embodied center of agency and action" (Overton, 2006, p. 48)—with its world. Thelen and Smith's DS approach, as we now demonstrate, treats embodiment exclusively from the former vantage point while denying the explanatory value of the latter, splitting Overton's relational framework of embodiment in the process.

4.1. Thelen and Smith's DS Approach to Embodiment

Thelen and Smith's DS approach to embodiment resolves to the question of how psychological processes, such as cognition, are coupled to the real-time dynamics of physical activity in a physical world (Smith, 2005; Spencer, Perone, et al., 2009; Thelen et al., 2001). Cognition, in this light, concerns the real-time adaptive acts of organisms in relation to specific task contexts; cognition as an abstract, context-general characterization of a living system's organization is wholly absent from consideration. From the standpoint of Thelen and Smith's DS approach, the only explanatory value of cognition (or of psychological constructs in general) lies in its being "bound to the real-time bodily processes through which we act in a physical world" (Smith, 2005, p. 288), for such constructs are "only useful if linkable in context to...real-time sensorimotor activity" (Spencer, Perone, et al., 2009, p. 88). Rendering psychological constructs meaningful, in other words, involves casting these processes in the same continuous, spatiotemporal terms as physical body movements and neurological functioning (Thelen et al., 2001).

Work on the A–not–B error from Thelen, Smith and colleagues (e.g., Clearfield, Dineva, Smith, Diedrich, & Thelen, 2009; Schutte, Spencer, & Schoner, 2003; Smith, Thelen, Titzer, & McLin, 1999) illustrates both this casting process and its ontological implications. When framed in traditional, Piagetian terms, the A–not–B error serves as a criterion for a certain level of organization in the development of infants' objectification of self and world— i.e., the object concept—via progressive differentiation of their activity from the objects they act upon. The A–not–B error is typically made by infants in Piaget's sensorimotor stage 4 in the development of the object concept. The error is made after infants successfully retrieve an object completely hidden under an occluder at location A and continue to search at location A after the object is subsequently hidden under an occluder in a new location, B. Piaget characterized the formal properties of this activity—its formal cause— in terms of partial but not complete phenomenological objectification of self and world: objects, though now clearly differentiated for the infant from her/ his activity, nonetheless remain for the infant wed to a particular spatial location and are therefore not fully objectified (Piaget, 1954).

Thelen et al. (2001) summarily reject Piaget's formal explanation for the A–not–B error and substitute what they consider an "embodied" conceptualization of infants' performance in terms of "the coupled dynamics of looking, planning, reaching, and remembering within the particular context of the task" (p. 5). Absent from their account is any semblance of formal explanation via an organizational construct, such as object concept (Smith, 2005); instead, Thelen et al. ground explanation of the A–not–B error solely in terms of the concrete, real-time dynamics of visually guided reaching, decomposed into the practical processes of looking, sensorimotor remembering, motor planning, and motor execution. Thelen et al. further argue that shifting relationships among these processes—unfolding in real-time within particular task circumstances—account both for errors and successes in reaching for occluded objects and that the same dynamics apply irrespective of level of developmental organization, meaning that the same dynamics explain both the perseverative reaching errors of young infants and the absent-minded errors of adults (e.g., automatically reaching for your keys where you normally keep them even when you see they are not in that location). By reducing more abstracted constructs, such as object concept— that capture the formal, organizational aspects of real-time activity—to more concrete, temporally unfolding, continuous activities, replete with graded activation and decay, Thelen and Smith's DS approach couches its conceptualization of cognition firmly in the realm of *content*, not form, with their focus of explanation targeting the particulars of infant in-the-moment behaviors embedded within a real-time, task-specific context.

In itself, the content focus of Thelen, Smith, and colleagues' DS approach yields a "dynamic field model" that successfully captures and predicts the extensive, intra- and interindividual variability of responding across context variations of the A-not-B task. Dynamic field models have also been extended to other domains of functioning, such as habituation, working memory, and visuospatial cognition (Spencer et al., 2011; Spencer, Perone, et al., 2009). These models exemplify embodiment at the level of local-to-global focus, wherein relations among brain, body, and environment jointly produce patterning in organismic functioning. However, within the ontological framework of Thelen and Smith's DS approach to embodiment, *only* a local-to-global focus constitutes explanation. The global-to-local vantage point of embodiment, focusing on the organism as an agentive, integrated whole in meaningful engagement as a bodily subject with its world, holds no explanatory value in its own right; only the real-time, task-specific dynamics of activity in context count as cause. In other words, from Thelen and Smith's DS approach, explanation arrives not by abstracting higher order organizational properties or by embedding organismic action in a developmental sequence, but by embedding action in a task-specific context. Bottom-up, distributed efficient cause— *not formal and final causes*—operates as the only viable means of explanation in this framework.

That Thelen and Smith's DS approach to embodiment operates through the exclusive lens of local-to-global, distributed efficient cause is revealed in Thelen et al.'s (2001) (see also Thelen & Smith, 1994) conceptualization of Piagetian formal explanation. Thelen et al. make the all-too-common mistake of construing Piaget's object concept account in functional, antecedent–consequent terms rather than in organizational, constraint terms (Muller & Carpendale, 2001; Staddon, Machado, & Lourenco, 2001; Witherington, 2007). They view infants' stage 4 object concept as a concrete "entity" or process that temporally precedes, exists independent of, and serves as the efficient causal force for the real-time performance of an organism in context. Given that such an abstracted entity cannot readily account for the enormous context-dependent variability in infant responding to the task and that no reasonable account exists for how this abstracted entity could actually motivate real, physical activity in the real, physical world, Thelen et al. (2001) argue that the object concept is explanatorily vacuous; psychological concepts such as object concept "do not offer us any help in understanding the mechanisms and processes involved in succeeding or failing at the A-not-B....there is a gap between invoking such constructs and specifying how they actually operate to motivate real life behavior" (pp. 71, 72).

In response to commentaries by Muller and Carpendale (2001) and Staddon et al. (2001) that underscore Piagetian explanation as formal, not functional in nature, Thelen et al. (2001) expose the explanatory monism of their DS approach by asking "what is the real use of a mental structure with only formal properties?" (p. 71) and "if Piaget meant for mental structures to be only formal descriptions, our foundational issue remains: where are the mechanisms that produce behavior?" (p. 72). These questions reflect Thelen et al.'s assumption that structural accounts offer explanation only insofar as they involve temporally based, efficient causal mechanisms, concretely grounded in real-time movement dynamics. Since structures, as abstractions, cannot cause in that fashion, structural explanation from Thelen et al.'s perspective cannot serve as true explanation. Thelen et al. fail to recognize that structure and organization can explain in their own right and on their own terms, as frameworks, à la formal and final cause, within which the particular, real-time dynamics of interest to Thelen et al. are meaningfully contextualized. Their analysis at the level of content serves not as a complement to, but as a substitute for, the organizational explanation offered in Piaget's account. By reducing the organism as a whole to its activity in context, Thelen et al. lose sight of the organizational whole that characterizes organismic functioning in its totality. Their account splits Overton's (2006, 2008) relational framework of embodiment by privileging the physical object and repudiating the phenomenological subject.

5. ENACTION AND EMBODIMENT: A HOLISTIC FOCUS

Thelen and Smith's DS approach to embodiment reveals the insidious ease with which reductionist framing and foundationalism can work their way into systems thinking, undermining the relational ontology of organismic–contextualist synthesis as a consequence. Given the particular focus of many modern incarnations of developmentally oriented systems perspectives on the local-to-global side of process orientation and on issues of variability and context specificity, lapses into split metatheoretical thinking become more and more likely in developmental systems thinking unless balance is struck through equal focus on the global-to-local, "structured whole" side of the relational matrix. Truly relational systems thinking broadens the notion of a "process" orientation to conceptually envelop organization, establishing an orientation of *organized process* such that organization informs our understanding of process as much as process informs our understanding of organization (Bickhard & Campbell, 2000; Witherington, 2011).

Only an approach to embodiment that values both the physical object and psychological subject poles as equally necessary and legitimate lines of sight fulfills the stipulations of a relational developmental systems perspective. Varela et al.'s (1991) enactive approach, which has enjoyed increasing prominence of late in the realm of cognitive science (Di Paolo et al., 2010; Stewart, 2010; Thompson, 2007), fits these stipulations. Our prior discussions of Jonas' philosophical writings on biology and teleology and of Varela's writings on autopoiesis have already established the core themes of self-organization, agency, and sense-making in the enactive approach; as Di Paolo et al. (2010) elaborate:

> For the enactivist the body is the ultimate source of significance; embodiment means that mind is inherent in the precarious, active, normative, and worldful process of animation, that the body is not a puppet controlled by the brain but a whole animate system with many autonomous layers of self-constitution, self-coordination, and self-organization (p. 42).

The autonomy of the living system qua system exists precisely because of its being inextricably bound to its surround. All living systems—from cells to organisms—are both independent and interdependent, subject and object, constructing their own autonomous organization through active, dynamic exchange with their surround and sustaining such organization in the face of flux in their material constituents. Through their dynamically maintained organization, living systems—as agentive centers of activity—regulate their exchange with the environment and "cast a web of significance on their world" (Di Paolo et al., 2010, p. 39), constructing a unique perspective that organizationally frames such exchanges and in turn imbues them with meaning. Living systems as wholes—as subjects of their own activity—constrain the very dynamics of interaction that give rise to them.

Without an acknowledgment of the global-to-local processes of constraint that define a system's organizational causality, the embodied meaning of a living system's activity remains obscure. In Thelen and Smith's DS approach to embodiment, the task-specific context preexists an organism's encounter with it and plays a privileged role in organizing the organism's activity (Witherington, 2007, 2011). Such an approach regards as superfluous the characterization of organisms and their development in organized whole, context-general terms. But the task-specific context alone is ill-suited to fully convey the individualized meaning of a context for an organism, as the organism itself, taken as a whole, establishes an equally important context of meaning for its own activity. Any system qua system constitutes a critical mode of explanation in its own right, as its own context of whole-to-parts.

The organism, as an active agent—in Weber and Varela's (2002) words, "an autonomous being who does not suffer passive world encounter but fashions a world of meaning from within" (p. 115)—imparts meaning to its action through its organization. The distributed efficient causal relations among the dynamic processes of visually guided reaching in a task-specific context to which Thelen et al. (2001) appeal themselves gain meaning against a backdrop of the organism as a whole and are in the service (for the sake of) of the organism as a whole. The very nature of the task itself and the functional significance of an organism's action in the task-context derive meaning from that organism's particular level of developmental organization. As Levins and Lewontin (1985) assert, "organisms determine what is relevant" (p. 99). Thus, every action of the organism speaks both to task-specific adaptation and to organism-specific organization (Piaget, 1952), and both vantage points inform our understanding of embodiment.

We must stress that, within the inclusive ontological stance of the enactive approach, the local-to-global dynamic construction of system organization of Thelen and Smith's DS approach stands as an important explanatory perspective complementing the global-to-local perspective of living systems enacting their own worlds. Relational ontology fragments when *either* the subject *or* the object pole of inquiry assumes privileged and exclusive status; as distinct yet indissociable lines of sight, both vantage points are critical to explanation in developmental science. As biological analysis pervades more and more of psychological and developmental science, seeing both the biological and the psychological as integrated levels of organization, involving distinct but complementary local-to-global and global-to-local processes all the way up and all the way down, is critical to encouraging a relational developmental systems perspective. Avoiding the snares of reductionism—biological or otherwise—demands an explanatory pluralism wherein all living systems exist as both subjects and objects.

REFERENCES

Anastasi, A. (1958). Heredity, environment, and the question 'how?'. *Psychological Review, 65,* 197–208.

Barbaras, R. (2010). Life and exteriority: the problem of metabolism. In J. Stewart, O. Gapenne & E. A. Di Paolo (Eds.), *Enaction: Toward a new paradigm for cognitive science* (pp. 89–122). Cambridge, MA: MIT Press.

Bickhard, M. H., & Campbell, D. T. (2000). Emergence. In P. B. Anderson, C. Emmeche, N. O. Finnemann & P. V. Christiansen (Eds.), *Downward causation: Minds, bodies and matter* (pp. 322–348). Aarhus: Aarhus University Press.

Campbell, D. T. (1974). 'Downward causation' in hierarchically organized biological systems. In F. J. Ayala & T. Dobzhansky (Eds.), *Studies in philosophy of biology: Reduction and related problems* (pp. 179–186). London: Macmillan.

Chandler, M. (1997). Stumping for progress in a post-modern world. In E. Amsel & K. A. Renninger (Eds.), *Change and development: Issues of theory, method and application* (pp. 1–26). Mahwah, NJ: Erlbaum.

Clark, A. (1997). *Being there: Putting brain, body and world together again.* Cambridge, MA: MIT Press.

Clearfield, M. W., Dineva, E., Smith, L. B., Diedrich, F. L., & Thelen, E. (2009). Cue salience and infant perseverative reaching: tests of the dynamic field theory. *Developmental Science, 12,* 26–40.

Derkson, M. (2010). Realism, relativism, and evolutionary psychology. *Theory & Psychology, 20,* 467–487.

Di Paolo, E. A., Rohde, M., & De Jaegher, H. (2010). Horizons for the enactive mind: values, social interaction, and play. In J. Stewart, O. Gapenne & E. A. Di Paolo (Eds.), *Enaction: Toward a new paradigm for cognitive science* (pp. 33–87). Cambridge, MA: MIT Press.

El-Hani, C. N., & Pereira, A. M. (2000). Higher-level descriptions: why should we preserve them? In P. B. Andersen, C. Emmeche, N. O. Finnemann & P. V. Christiansen (Eds.), *Downward causation: Minds, bodies and matter* (pp. 118–142). Aarhus: Aarhus University Press.

Emmeche, C., Koppe, S., & Stjernfelt, F. (2000). Levels, emergence, and three versions of downward causation. In P. B. Anderson, C. Emmeche, N. O. Finnemann & P. V. Christiansen (Eds.), *Downward causation: Minds, bodies and matter* (pp. 13–34). Aarhus: Aarhus University Press.

Finnemann, N. O. (2000). Rule-based and rule-generating systems. In P. B. Anderson, C. Emmeche, N. O. Finnemann & P. V. Christiansen (Eds.), *Downward causation: Minds, bodies and matter* (pp. 278–302). Aarhus: Aarhus University Press.

Ford, D. H., & Lerner, R. M. (1992). *Developmental systems theory: An integrative approach.* Newbury Park, CA: Sage.

Gentile, J. (2008). Agency and its clinical phenomenology. In R. Frie (Ed.), *Psychological agency: Theory, practice, and culture* (pp. 117–135). Cambridge: MIT Press.

Gottlieb, G. (1970). Conceptions of prenatal behavior. In L. R. Aronson, E. Tobach, D. S. Lehrman & J. S. Rosenblatt (Eds.), *Development and evolution of behavior: Essays in memory of T. C. Schneirla* (pp. 111–137). San Francisco, CA: W. H. Freeman.

Gottlieb, G. (1991). Experiential canalization of behavioral development: theory. *Developmental Psychology, 27,* 4–13.

Gottlieb, G. (1992). *Individual development and evolution: The genesis of novel behavior.* New York: Oxford University Press.

Gottlieb, G., Wahlsten, D., & Lickliter, R. (2006). The significance of biology for human development: a developmental psychobiological systems view. In R. M. Lerner (Ed.), & W. Damon & R. M. Lerner (Editors-in-chief). *Handbook of child psychology. Theoretical models of human development* (Vol. 1, 6th ed., pp. 210–257). Hoboken, NJ: Wiley.

Ingold, T. (2004). Beyond biology and culture: the meaning of evolution in a relational world. *Social Anthropology, 12,* 209–221.

Johnson, M. H. (2010). Functional brain development during infancy. In J. G. Bremner & T. D. Wachs (Eds.), *The Wiley-Blackwell handbook of infant development* (Vol. 1, 2nd Ed., pp. 294–319). West Sussex, UK: Wiley-Blackwell.

Jonas, H. (1966). *The phenomenon of life: Toward a philosophical biology.* Evanston, IL: Northwestern University Press.

Juarrero, A. (1999). *Dynamics in action: Intentional behavior as a complex system.* Cambridge, MA: MIT Press.

Juarrero, A. (2009). Top-down causation and autonomy in complex systems. In N. Murphy, G. F. R. Ellis & T. O'Connor (Eds.), *Downward causation and the neurobiology of free will* (pp. 83–102). Berlin: Springer.

Juarrero-Roque, A. (1985). Self-organization: Kant's concept of teleology and modern chemistry. *Review of Metaphysics, 39,* 107–135.

Kant, I. (1790/2007). *Critique of judgement.* Oxford: Oxford University Press.

Kendler, T. S. (1986). World views and the concept of development: a reply to Lerner and Kauffman. *Developmental Review, 6,* 80–95.

Kitchener, R. F. (1982). Holism and the organismic model in developmental psychology. *Human Development, 25,* 233–249.

Kuo, Z. Y. (1967). *The dynamics of behavior development.* New York: Random House.

Lehrman, D. S. (1970). Semantic and conceptual issues in the nature-nurture problem. In L. R. Aronson, E. Tobach, D. S. Lehrman & J. S. Rosenblatt (Eds.), *Development and evolution of behavior: Essays in memory of T. C. Schneirla* (pp. 17–52). San Francisco, CA: W. H. Freeman.

Lerner, R. M. (1978). Nature, nurture, and dynamic interactionism. *Human Development, 21,* 1–20.

Lerner, R. M. (2002). *Concepts and theories of human development* (3rd ed.). Mahwah, NJ: Erlbaum.

Lerner, R. M. (2006). Developmental science, developmental systems, and contemporary theories of human development. In R. M. Lerner (Ed.), & W. Damon & R. M. Lerner (Editors-in-chief). *Handbook of child psychology. Theoretical models of human development* (Vol. 1, 6th ed., pp. 1–17). Hoboken, NJ: Wiley.

Lerner, R. M., & Kauffman, M. B. (1985). The concept of development in contextualism. *Developmental Review, 5,* 309–333.

Levins, R., & Lewontin, R. (1985). *The dialectical biologist.* Cambridge, MA: Harvard University Press.

Lourenco, O., & Machado, A. (1996). In defense of Piaget's theory: a reply to 10 common criticisms. *Psychological Review, 103,* 143–164.

Magnusson, D. (1995). Individual development: a holistic integrated model. In P. Moen, G. H. Elder & K. Lusher (Eds.), *Linking lives and contexts: Perspectives on the ecology of human development* (pp. 19–60). Washington, DC: APA Books.

Magnusson, D., & Stattin, H. (2006). The person in context: a holistic-interactionistic approach. In R. M. Lerner (Ed.), & W. Damon & R. M. Lerner (Editors-in-chief). *Handbook of child psychology. Theoretical models of human development* (Vol. 1, 6th ed., pp. 400–464). Hoboken, NJ: Wiley.

Marshall, P. J. (2009). Relating psychology and neuroscience: taking up the challenges. *Perspectives on Psychological Science, 4,* 113–125.

Martin, J., Sugarman, J. H., & Hickinbottom, S. (2009). *Persons: Understanding psychological selfhood and agency.* New York: Springer.

Miller, G. A. (2010). Mistreating psychology in the decades of the brain. *Perspectives on Psychological Science, 5,* 716–743.

Moreno, A. (2008). Downward causation requires naturalized constraints: a comment on Vieira and El-Hani. *Cybernetics and Human Knowing, 15,* 135–144.

Moreno, A., & Umerez, J. (2000). Downward causation at the core of living organization. In P. B. Andersen, C. Emmeche, N. O. Finnemann & P. V. Christiansen (Eds.), *Downward causation: Minds, bodies and matter* (pp. 99–117). Aarhus: Aarhus University Press.

Muller, U., & Carpendale, J. I.M. (2001). Objectivity, intentionality, and levels of explanation. *Behavioral and Brain Sciences, 24*, 55–56.

Murphy, N. (2009). Introduction and overview. In N. Murphy, G. F. R. Ellis & T. O'Connor (Eds.), *Downward causation and the neurobiology of free will* (pp. 1–28). Berlin: Springer.

Overton, W. F. (1984). World views and their influence on psychological theory and research: Kuhn-Lakatos-Laudan. In H. W. Reese (Ed.), *Advances in child development and behavior* (Vol. 18, pp. 191–226). Orlando: Academic Press.

Overton, W. F. (1991). The structure of developmental theory. In P. van Geert & L. P. Mos (Eds.), *Annals of theoretical psychology* (Vol. 7, pp. 191–235). New York: Plenum Press.

Overton, W. F. (1994). The arrow of time and the cycle of time: concepts of change, cognition and embodiment. *Psychological Inquiry, 5*, 215–237.

Overton, W. F. (1998). Developmental psychology: philosophy, concepts, and methodology. In W. Damon (Series Ed.), & R. M. Lerner (Vol. Ed.), *Handbook of child psychology. Theoretical models of human development* (Vol. 1, 5th ed., pp. 107–188). New York: Wiley & Sons.

Overton, W. F. (2003). Metatheoretical features of behavior genetics and development. *Human Development, 46*, 356–361.

Overton, W. F. (2006). Developmental psychology: philosophy, concepts, methodology. In W. Damon & R. M. Lerner (Editors-in-chief) & R. M. Lerner (Vol. Ed.), *Handbook of child psychology. Theoretical models of human development* (Vol. 1, 6th ed., pp. 18–88). Hoboken, NJ: Wiley.

Overton, W. F. (2007). A coherent metatheory for dynamic systems: relational organicism–contextualism. *Human Development, 50*, 154–159.

Overton, W. F. (2008). Embodiment from a relational perspective. In W. F. Overton, U. Muller & J. L. Newman (Eds.), *Developmental perspectives on embodiment and consciousness* (pp. 1–18). New York: Psychology Press.

Overton, W. F. (2010). Life-span development: concepts and issues. In R. M. Lerner (Series Ed.), & W. F. Overton (Vol. Ed.), *Cognition, biology, and methods The handbook of life-span development* (Vol. 1, pp. 1–29). Hoboken, NJ: John Wiley & Sons.

Overton, W. F., & Ennis, M. D. (2006). Cognitive-developmental and behavior-analytic theories: evolving into complementarity. *Human Development, 49*, 143–172.

Oyama, S. (1985). *The ontogeny of information: Developmental systems and evolution*. Cambridge: Cambridge University Press.

O'Connor, T. (1994). Emergent properties. *American Philosophical Quarterly, 31*, 91–104.

Pepper, S. C. (1942). *World hypotheses: A study in evidence*. Berkeley, CA: University of California Press.

Piaget, J. (1952). *The origins of intelligence in children*. New York: International Universities Press.

Piaget, J. (1954). *The construction of reality in the child*. New York: Basic Books.

Piaget, J. (1970). Piaget's theory. In P. H. Mussen (Ed.), *Carmichael's manual of child psychology* (3rd ed., pp. 703–732). New York: John Wiley & Sons.

Piaget, J. (1971). *Biology and knowledge*. Chicago, IL: University of Chicago Press.

Reese, H. W., & Overton, W. F. (1970). Models of development and theories of development. In L. R. Goulet & P. B. Baltes (Eds.), *Life-span developmental psychology: Research and theory* (pp. 115–145). New York: Academic Press.

Rychlak, J. F. (1988). *The psychology of rigorous humanism* (2nd ed.). New York: New York University Press.

Sameroff, A. J. (1983). Developmental systems: contexts and evolution. In W. Kessen (Ed.), *Handbook of child psychology History, theory, and methods* (Vol. 1, pp. 237–294). New York: Wiley & Sons.

Schneirla, T. C. (1957). The concept of development in comparative psychology. In D. B. Harris (Ed.), *The concept of development: An issue in the study of human behavior* (pp. 78–108). Minneapolis, MN: University of Minnesota Press.

Schneirla, T. C. (1960). Instinctive behavior, maturation-experience and development. In B. Kaplan & S. Wapner (Eds.), *Perspectives in psychological theory: Essays in honor of Heinz Werner* (pp. 303–334). New York: International Universities Press.

Scholnick, E. K. (1991). The development of world views: towards future synthesis? In P. van Geert & L. P. Mos (Eds.), *Annals of theoretical psychology* (Vol. 7, pp. 249–259). New York: Plenum Press.

Schutte, A. R., Spencer, J. P., & Schoner, G. (2003). Testing the dynamic field theory: working memory for locations becomes more spatially precise over development. *Child Development, 74,* 1393–1417.

Shapiro, L. (2011). *Embodied cognition.* London: Routledge.

Shimamura, A. P. (2010). Bridging psychological and biological science: the good, bad, and ugly. *Perspectives on Psychological Science, 5,* 772–775.

Silberstein, M. (2006). In defence of ontological emergence and mental causation. In P. Clayton & P. Davies (Eds.), *The re-emergence of emergence: The emergentist hypothesis from science to religion* (pp. 203–226). Oxford: Oxford University Press.

Smith, L. B. (2005). Cognition as a dynamic system: principles from embodiment. *Developmental Review, 25,* 278–298.

Smith, L. B., & Thelen, E. (Eds.), (1993). *A dynamic systems approach to development: Applications.* Cambridge, MA: MIT Press.

Smith, L. B., Thelen, E., Titzer, R., & McLin, D. (1999). Knowing in the context of acting: the task dynamics of the A-not-B error. *Psychological Review, 106,* 235–260.

Spencer, J. P., Blumberg, M. S., McMurray, B., Robinson, S. R., Samuelson, L. K., & Tomblin, J. B. (2009). Short arms and talking eggs: why we should no longer abide the Nativist-Empiricist debate. *Child Development Perspectives, 3,* 79–87.

Spencer, J. P., Perone, S., & Buss, A. T. (2011). Twenty years and going strong: a dynamic systems revolution in motor and cognitive development. *Child Development Perspectives, 5,* 260–266.

Spencer, J. P., Perone, S., & Johnson, J. S. (2009). The dynamic field theory and embodied cognitive dynamics. In J. P. Spencer, M. S. C. Thomas & J. L. McClelland (Eds.), *Toward a unified theory of development: Connectionism and dynamic systems theory re-considered* (pp. 86–118). Oxford: Oxford University Press.

Sperry, R. W. (1986). Macro-versus micro-determinism. *Philosophy of Science, 53,* 265–270.

Staddon, J. E. R., Machado, A., & Lourenco, O. (2001). Plus ca change, Jost, Piaget, and the dynamics of embodiment. *Behavioral and Brain Sciences, 24,* 63–65.

Stewart, J. (2010). Foundational issues in enaction as a paradigm for cognitive science: from the origin of life to consciousness and writing. In J. Stewart, O. Gapenne & E. A. Di Paolo (Eds.), *Enaction: Toward a new paradigm for cognitive science* (pp. 1–32). Cambridge, MA: MIT Press.

Thelen, E., Schoner, G., Scheier, C., & Smith, L. B. (2001). The dynamics of embodiment: a field theory of infant perseverative reaching. *Behavioral and Brain Sciences, 24,* 1–86.

Thelen, E., & Smith, L. B. (1994). *A dynamic systems approach to the development of cognition and action.* Cambridge, MA: MIT Press.

Thelen, E., & Smith, L. B. (2006). Dynamic systems theories. In W. Damon & R. M. Lerner (Editors-in-chief) & R. M. Lerner (Vol. Ed.), *Handbook of child psychology. Theoretical models of human development* (Vol. 1, 6th ed., pp. 258–312). Hoboken, NJ: Wiley.

Thompson, E. (2007). *Mind in life: Biology, phenomenology, and the sciences of mind.* Cambridge, MA: Belknap Press of Harvard University Press.

Tolman, C. W. (1991). For a more adequate concept of development with help from Aristotle and Marx. In P. van Geert & L. P. Mos (Eds.), *Annals of theoretical psychology* (Vol. 7, pp. 349–362). New York: Plenum Press.

Turvey, M.T., Shaw, R. E., & Mace, W. (1978). Issues in the theory of action: degrees of freedom, coordinative structures, and coalitions. In J. Requin (Ed.), *Attention and performance* (Vol. 7, pp. 557–595). Hillsdale, NJ: Erlbaum.

van Geert, P. (1991). A dynamic systems model of cognitive and language growth. *Psychological Review, 98*, 3–53.

van Geert, P. (1994). *Dynamic systems of development: Change between order and chaos.* New York: Harvester Wheatsheaf.

van Geert, P. (1998). We almost had a great future behind us: the contribution of non-linear dynamics to developmental-science-in-the-making. *Developmental Science, 1*, 143–159.

van Geert, P., & Steenbeek, H. (2005). Explaining after by before: basic aspects of a dynamic systems approach to the study of development. *Developmental Review, 25*, 408–442.

van Gulick, R. (1993). Who's in charge here? and who's doing all the work? In J. Heil & A. Mele (Eds.), *Mental causation* (pp. 233–256). Oxford: Oxford University Press.

Varela, F. J. (1979). *Principles of biological autonomy.* New York: Elsevier/North Holland.

Varela, F. J., Thompson, E., & Rosch, E. (1991). *The embodied mind: Cognitive science and human experience.* Cambridge, MA: MIT Press.

von Bertalanffy, L. (1933). *Modern theories of development: An introduction to theoretical biology.* London: Oxford University Press.

von Bertalanffy, L. (1968). *General systems theory: Foundations, development, applications.* New York: George Braziller.

Weber, A., & Varela, F. J. (2002). Life after Kant: natural purposes and the autopoietic foundations of biological individuality. *Phenomenology and the Cognitive Sciences, 1*, 97–125.

Weiss, P. (1939). *Principles of development: A text in experimental embryology.* New York: Henry Holt & Company.

Werner, H. (1948). *Comparative psychology of mental development.* New York: International Universities Press.

Werner, H. (1957). The concept of development from a comparative and organismic point of view. In D. B. Harris (Ed.), *The concept of development: An issue in the study of human behavior* (pp. 125–148). Minneapolis, MN: University of Minnesota Press.

Wheeler, M. (2005). *Reconstructing the cognitive world: The next step.* Cambridge, MA: MIT Press.

Witherington, D. C. (2007). The dynamic systems approach as metatheory for developmental psychology. *Human Development, 50*, 127–153.

Witherington, D. C. (2011). Taking emergence seriously: the centrality of circular causality for dynamic systems approaches to development. *Human Development, 54*, 66–92.

Witherington, D. C., & Margett, T. E. (2011). How conceptually unified is the dynamic systems approach to the study of psychological development? *Child Development Perspectives, 5*, 286–290.

CHAPTER SEVEN

The Origins of Variation: Evolutionary Insights from Developmental Science

Robert Lickliter
Department of Psychology, Florida International University, Miami, FL, USA
E-mail: lickliter@gmail.com

Contents

Abstract

Evidence from contemporary epigenetic research indicates that it is not biologically meaningful to discuss genes without reference to the molecular, cellular, organismal, and environmental context within which they are activated and expressed. Genetic and nongenetic factors, including those beyond the organism, constitute a dynamic relational developmental system. This insight highlights the importance of bringing together genetics, development, and ecology into one explanatory framework for a more complete understanding of the emergence and maintenance of phenotypic stability *and* variability. In this Chapter, I review some examples of this integrative effort and explore its implications for developmental and evolutionary science, with a particular emphasis on the origins of phenotypic novelty. I argue that developmental science is critical to this integrative effort, in that evolutionary explanation cannot be complete without developmental explanation. This is the case because the process of development generates the phenotypic variation on which natural selection can act.

This work was supported in part by the NSF grant BCS 1057898.

Advances in Child Development and Behavior, Volume 44
ISSN 0065-2407, http://dx.doi.org/10.1016/B978-0-12-397947-6.00007-6

"At the present time there is hardly any question in biology of more importance than this of the nature and causes of variability ...."
(Darwin, 1882, p. vi)

1. INTRODUCTION

It has been over 150 years since Charles Darwin and Alfred Russell Wallace's papers outlining their remarkably similar theories of evolution were read to the Linnean Society (1858). Their theories of evolution rested on two basic ideas: (1) the common descent of all living organisms and (2) natural selection as the major agent of evolutionary change. Darwin's more detailed theory of evolution laid out the following year in *The Origin of Species* (1859) was founded on the notion of descent with modification (changes occur over generations producing new species over time). While Darwin's theory of evolution certainly set biology on a new course, offering an explanation for the remarkable biological diversity observed across plants and animals, it did not explain the origins of phenotypic modifications or novelties, a fact pointed out by St. George Mivart in his book, *On the Genesis of Species* (1871).

Mivart was one of Darwin's most prominent critics (documented by the fact that Darwin devoted almost an entire Chapter in the final sixth revision of the *Origin* to address Mivart's criticisms of his theory). Mivart was among the first to point out that natural selection can account for the preservation and increase in frequency of phenotypic traits and characters within a population, as Darwin proposed, but could not account for their origin (see also Cope, 1886). Harris (1904, p. 401) captured Mivart's insight succinctly when he wrote that "natural selection may explain the survival of the fittest, but it cannot explain the arrival of the fittest." For Mivart, natural selection was not a creative force in evolution, but rather was the eliminator of the unfit. Mivart (1871, p. 240) noted that "natural selection favors and develops useful variations, though it is impotent to originate them." His insight that natural selection can only change the frequency or range of phenotypic variations already present in a population led Mivart to relegate natural selection to a more minor role in evolution than Darwin had proposed and to argue that other factors that were capable of generating the variations upon which selection can then act must be at play. Mivart thought that these factors must somehow be based on the united action of internal and external forces that modified individual development, but he was necessarily vague as to how this might work, referring to this process as "obscure and mysterious". Of course, Mivart and his contemporaries of nineteenth century biology knew relatively little about the details of development or the internal and external forces at play in evolutionary change. Nevertheless, Mivart's insight that the origins of new forms or traits must come about through changes in the process of development was a shift in thinking about the mechanism of evolution, one that has received serious attention within the biological sciences only relatively recently (e.g., Arthur, 1997; Carroll, 2005; Gerhart & Kirschner, 1997; Hall, 1999; Müller & Wagner, 1991; Raff, 1996; West-Eberhard, 1989, 2003).

The lack of concern with the role of development in evolution over much of the past century was not due to Darwin, who came to increasingly appreciate the importance of development to heredity over the course of his career (see Winther, 2000). Rather, the dismissal of development can be traced to the pervasive influence of the "Modern" evolutionary synthesis of Darwinian and Mendelian principles forged by Theodosius Dobzhansky, Ronald Fisher, Julian Huxley, Ernst Mayr, Sewell Wright, and others in the 1930s and 1940s (Amundson, 2005; Gottlieb, 1992; Lickliter & Honeycutt, 2009; Robert, 2004). This view of evolution held that variation in populations was the result of random genetic mutation and recombination, and as a result, proposed that populations evolve by changes in gene frequencies due to genetic drift, gene flow, and natural selection. The "Modern Synthesis" of the twentieth century thus promoted population genetics as key to understanding evolutionary change and dismissed the importance or relevance of development in understanding evolutionary issues.

Due in large part to the prominence of population genetics in evolutionary theory over much of the twentieth century, it is only within the past several decades that developmental and evolutionary scientists have focused their theoretical and empirical efforts on exploring the nature of the links between development and evolution (see Gottlieb, 1992; Hall, 1999; Sanson & Brandon, 2007; West-Eberhard, 2003 for examples). A concern with how development is involved in evolutionary change is now evident among biologists and psychologists working in formally diverse areas of research, including genomics, cellular and molecular biology, developmental biology, evolutionary theory, ecology, and comparative and developmental psychology, as well as philosophers of biology (e.g., Arthur, 2004; Bateson & Gluckman, 2011; Davidson, 2001; Gilbert & Epel, 2009; Hall, Pearson, & Müller, 2004; Lickliter & Honeycutt, 2003; Moore, 2008; Nijhout, 2003; Richardson, 1998; Robert, 2004).

As we gain a deeper appreciation of the importance of the process of development to the production of phenotypic variation, new questions are being raised about how and to what extent developmental change contributes to evolutionary change (e.g., Bjorklund, 2006; Blumberg, 2008; Johnston & Gottlieb, 1990; Lickliter & Schneider, 2006; Moore, 2002; Oyama, 1985; Oyama, Griffiths, & Gray, 2001; West-Eberhard, 2003), topics I review in this Chapter. My overarching goal is to explore how an integration of genetic, epigenetic, and environmental levels of analysis can provide a conceptual framework for understanding how developmental systems can generate novel phenotypes. More generally, I argue that evolutionary explanation cannot be complete without developmental explanation. This is because the process of development generates the phenotypic variation on which natural selection can act.

2. SOURCES OF PHENOTYPIC VARIATION

A long-standing problem of both developmental and evolutionary theory has been how to account for the sources of phenotypic variability (as well as phenotypic stability) observed within and across generations. In other words, how is it that some anatomical, physiological, and behavioral traits remain the same across time and some are modified or change across time? As all students of introductory biology know, variation in phenotypic traits and characters evident across individuals of a species was a crucial component of Darwin's theory of evolution by natural selection. He realized that for natural selection to act, individuals must vary in terms of their anatomy, physiology, or behavior. For Darwin (as well as for Alfred Russell Wallace), such variations provided the opportunity for natural selection to filter out those anatomical structures, physiological capabilities, and behavioral forms that are less successful and promote those that offer some reproductive advantage, thereby providing the engine for evolutionary change. Despite the importance of phenotypic variation to Darwin's theory, the sources of phenotypic variation were not well understood in nineteenth or even much of twentieth century biology. Darwin himself admitted "our ignorance of the laws of variation is profound" (1859, p. 167). As it turned out, it was not until the twenty-first century that most biologists finally began to consider phenotypic variation as an important area of study rather than as noise or nuisance to their experimental designs (e.g., Bateson & Gluckman, 2011; DeWitt & Scheiner, 2004; Hallgrímsson & Hall, 2005; Piersma & van Gils, 2010; Pigliucci, 2001;

West-Eberhard, 2003; but see Baldwin, 1902; Bateson, 1894; Brooks, 1883 for notable exceptions).

As a result, a new perspective on the sources of phenotypic variation has taken shape over the past several decades. This perspective is based on a relatively simple but fundamental insight: given that all phenotypes arise during ontogeny as products of development, it follows that the primary basis for phenotypic variation within and across generations must be the patterns and processes of development. *Development is thus critical to evolution because it is the process that provides the phenotypic variation on which natural selection can act.* The thread of this insight can be traced back to several embryologists and developmental biologists working in the first half of the twentieth century, including Walter Garstang (1922), Edward Russell (1930), Gavin de Beer (1930), Richard Goldschmidt (1940), Conrad Waddington (1942), and Ivan Schmalhausen (1949). Although each of these biologists had a distinctive perspective on how to characterize the links between development and evolution, they all promoted the notion that changes in individual development were a potentially important basis for evolutionary change. For example, Waddington (1942) was highly critical of evolutionary models in which genes were portrayed as directly causing development or were directly acted upon by natural selection. Schmalhausen (1949) was likewise critical of genocentric models of development and evolution and, similar to Waddington, emphasized the importance of the environment in inducing changes in development. As I review later, these views were well outside mainstream twentieth century thinking about evolution, but are now being reconsidered across the biological and psychological sciences. This shift has involved moving beyond the notion of genes as the fundamental cause of phenotypic traits, thereby allowing for a reconsideration of a variety of extragenetic factors at play in the emergence, maintenance, and modification of phenotypes within and across generations.

As a result of this conceptual shift, there is growing recognition of the necessity of considering and defining the complex transactions among genetics, development, and ecology in order to understand the range of morphological structures, shifts in behavioral repertoires, and other instances of phenotypic variation observed across plant and animal species (e.g., Bateson & Gluckman, 2011; Gilbert, 2005; Gilbert & Epel, 2009; Nijhout, 2003; Piersma & van Gils, 2010; Schlichting & Pigliucci, 1998; West-Eberhard, 2003). This *relational* approach views the novelty-generating aspects of evolution as being the result of the developmental dynamics of living organisms, situated and competing in specific ecological contexts, and not simply the result of random genetic mutations, genetic drift, or genetic recombination. This is a paradigmatic shift in emphasis, as genetic factors were argued to be the only evolutionary relevant source of phenotypic variation by the neo-Darwinian or "Modern Synthesis" school of evolutionary biology that dominated twentieth century life sciences (e.g., Dobzhanksy, 1937; Huxley, 1942; Mayr, 1963; Williams, 1966).

3. TWENTIETH CENTURY PERSPECTIVES ON DEVELOPMENT AND EVOLUTION

Following the rediscovery of Mendel's work in 1900 and the growing influence of Mendelian genetics during the next several decades, evolutionary biology came to distance itself from its earlier concerns with embryology and developmental biology and embrace the new science of population genetics (see Amundson, 2005; Gilbert, 1994; Gottlieb, 1992 for useful overviews). Unlike experimental embryology, population genetics was not focused on the development of the individual organism, rather it focused on organisms as members of a breeding population and on how best to calculate the probabilities of changing gene frequencies in this population of breeding organisms under this or that set of circumstances over generations. In particular, population geneticists were interested in understanding the role of genetic mutation, genetic recombination, and selection in the changes in gene frequencies found within a population and how these genetic changes resulted in evolutionary change (Provine, 1971). This approach concentrated on the traits of adults in populations and virtually ignored questions about how these traits were actually realized during the course of individual development. The noted evolutionary biologist Maynard-Smith (1985) argued that attempting explanations of evolution in terms of individual development was an "error of misplaced reductionism".

The Modern Synthesis (see Mayr & Provine, 1980) was able to effectively sidestep concerns with the role of development in evolution by proposing that there were two relatively independent classes of causal factors responsible for an individual's phenotypic traits: (1) *ultimate causes*, those that derive from internal or intrinsic factors (e.g., genes), molded over evolutionary time by natural selection and (2) *proximate causes*, everything else that interacts with these internal factors during development to provide the materials or experiences necessary to trigger the expression of form and function thought to be encoded in the genes (e.g., the environment).

This causal dichotomy for explaining phenotypic outcomes was grounded on the assumption that development is primarily internally determined, set on course at conception by genetic programs (*ultimate causes*) that had been designed and selected over evolutionary time. In contrast, *proximate causes* were defined as those factors involved in "decoding the genetic program" (Mayr, 1974). Developmental factors were thus seen as proximate causes, making development essentially irrelevant to the understanding of evolution (Lickliter & Berry, 1990). Widespread acceptance of this proximate–ultimate distinction effectively kept development and evolution as separate scientific concerns during most of the second half of the past century. Since genotypes were thought to be the direct causes of phenotypes, evolutionary biology had no causal or explanatory need for the process of development.

Watson and Crick's discovery of the structure and function of DNA in 1953 served to reaffirm the genocentric position of the Modern Synthesis: if genes are DNA and copying errors from DNA to RNA to protein is the source of genetic variation, then evolution must indeed be "changes in gene frequencies in populations" (Dobzhansky, 1937). Development was thus increasingly viewed as merely the reading out of genetic programs that were assumed to be the products of natural selection. This gene-centered perspective had at its core an underlying false premise that went unquestioned by many psychologists and biologists over the course of the twentieth century: the bodily forms, physiological processes, and behavioral patterns of organisms could be specified *in advance* of the organism's development.

However, this assumption of prespecification is a profoundly *nondevelopmental* point of view. Adult traits are seen to be the result of genetic instructions or programs, with little concern for the intervening resources, relations, and causes that transforms the adult from the zygote. This view has several serious shortcomings, not the least being that it assumes *as a given* the developmental outcomes that actually require a causal developmental analysis (Gottlieb, 1997; Kuo, 1967; Robert, 2004). Although nativists have continued to apply the notion of prespecification to both developmental and evolutionary issues (e.g., Buss, 2005; Carruthers, Laurence, & Stich, 2007; Pinker, 2002; Spelke & Newport, 1998), it has become increasingly clear that perspectives that favor the notion of prespecified phenotypic traits are not up to the task of making sense of the dynamics of the developmental process and its varied outcomes (Lerner, 1991, 2006; Lewkowicz, 2011; Overton, 2006; Oyama, 1985; Spencer et al., 2009).

4. TAKING A DEVELOPMENTAL POINT OF VIEW

Any successful theory integrating development and evolution must ultimately account for (1) the emergence of complexity of organization by differentiation, (2) the stability of form and function across generations, and (3) the origin and range of variability across individuals of a species (Laubichler & Maienschein, 2007; Lickliter, 2013). Attempts at this intellectual synthesis have engaged (and frustrated) scientists for centuries. Indeed, much of the content of the eighteenth and nineteenth centuries theorizing about development and evolution focused on explaining the possible mechanisms for these three phenomena (see Depew & Weber, 1995; Gould, 1977; Mayr, 1982 for useful overviews). As briefly reviewed above, in the twentieth century, biologists eventually converged on a bottom-up approach to account for the similarities and differences observed across individuals, holding that genes were the key to understanding the fundamental characteristics of development. Genes came to be seen as the cause for an organism's growth and development, as well as the cause for the intergenerational stability and variability of traits and qualities observed within species (see Keller, 2000; Sapp, 2003). Widespread application of this gene-centered framework resulted in significant advances in molecular and cellular biology and fostered the growth and popularity of fields such as sociobiology, behavioral genetics, and evolutionary psychology in the twentieth century. However, recent advances in several fields of biology, including what Conrad Waddington (1957) termed *epigenetics* and what Brian Hall (1992) and others have termed *evolutionary developmental biology*, have made it clear that gene-centered approaches to developmental and evolutionary issues minimize or simply overlook the wide range of factors, transactions, and contingencies at play in both development and evolution.

A key insight contributing to these advances in the biological sciences is that changes in development (brought on by changes in the context of development) are necessary in order to explain the types of variation that can be filtered by natural selection. Recent research with Darwin's finches (genus *Geospiza*), famous for their role in Darwin's formulation of the principle of natural selection, provides a useful example of how the complex interplay of molecular, cellular, and ecological factors contribute to relatively rapid and dramatic phenotypic change (in this case, the variety of beak shapes observed across these 13 species of finches distributed across the Galapagos Islands). Such developmental plasticity provides a potent pathway for organisms to rapidly change structure and function in response to environmental resources and changes (see West-Eberhard, 2003). In the case of Darwin's finches, in the time frame of just 1–2 million years, a founding group of finches from South America generated more than a dozen different finch species in the remote Galapagos Islands, including some with large pliers-like beaks capable of cracking nuts and seeds and some with forceps-like beaks able to extract insects from fruit. Darwin had noted these birds' remarkable differences in beak size and shape on his visit to the Galapagos Islands during his *Beagle* voyage in 1835, but due to the degree of variation across species, he did not realize at that time that they were all finches. Further reflection on this variation after his return to England contributed to Darwin's formulation of the critical role of natural selection in the direction of evolutionary change.

The standard genocentric explanation of the striking variation in beak size and shape seen across these closely related finch species suggests that genetic mutation, recombination, and reassortment of genes in an island's founder population would occasionally result in variant birds that had somewhat smaller and more forceps-like beaks or somewhat larger and more pliers-like beaks than those of the founder population. These individuals would be more likely to explore and exploit different food niches (insects vs. seeds), potentially leading to increasing geographic and behavioral isolation from one another. Morphological change would be gradual in this scenario, but over many generations, differential reproduction (based in part on relative feeding success) would eventually result in the selection of several variations of the original founders' beak type.

In contrast, recent synthesis of molecular, cellular, and ecological research indicates that the pathway to the remarkable variations observed in beak size and shape is more contingent on the context of development *and* more rapid than traditional views of evolutionary change would suggest (e.g., Abzhanov et al., 2006; Abzhanov, Protas, Grant, Grant, & Tabin, 2004; Grant, Grant, & Abzhanov, 2006). Current evidence indicates that the size and shape of the finch beak are determined during development by the growth and differentiation of neural crest cells that settle around the mouth of the developing bird embryo. These neural crest cells produce a growth factor protein called bone morphogenetic protein 4 (*Bmp4*), which stimulates the deposition of bone and beak materials during embryogenesis. This protein is produced earlier in embryonic development and at higher levels in the finch species with larger and wider beaks than in the closely related finch species with longer and narrower beaks (Abzhanov et al., 2004). Further, when *Bmp4* is experimentally introduced into the beak neural crest cells of chicken embryos, they develop broader and larger beaks than control chicks.

The introduction of other growth factors did not have this effect. Related work has found that a protein that mediates calcium signaling and plays a role in cell and tissue differentiation (calmodulin or CaM) is expressed at higher levels in finch species with longer and narrower beaks than in those with the longer wider beaks (Abzhanov et al., 2006). It appears that a variety of inter-related factors, including the number of neural crest cells, the level of signaling that stimulates or inhibits the production of growth factor protein and calmodulin, and the types of signals that induce cell death of the neural crest cells, are all at play in generating the beak shape variation seen across these finch species. How these various factors and their relations are regulated by the birds' experience and ecology (particularly the type of food sources available) is not fully understood, but given the wide adaptability of neural crest cells, it seems that relatively large modifications in beak size and shape have been accomplished with relatively few changes in the developmental process. This potential for rapid phenotypic adjustment to the contextual features of development has important implications for evolutionary change, in that it would increase the likelihood that members of the population could quickly take advantage of new or changing resources and habitats (Gottlieb, 2002).

The finch beak example illustrates how a focus on the complexities of the dynamics of development and evolution is bringing together genetics; molecular, cellular, and developmental biology; neuroscience; and evolutionary biology to construct a more comprehensive explanation of the ways and means of the stability and variability of phenotypic development (e.g., Lickliter & Harshaw, 2010; Müller & Newman, 2003; Neumann-Held & Rehmann-Sutter, 2006; Overton, 2006). This integrative approach is providing important opportunities for developmental science to contribute to evolutionary theory, a dramatic shift from the state of affairs just several decades ago. As Robert (2008) has pointed out, taking a developmental point of view requires understanding that there is more to development than differential gene expression, that development is not a genes-*plus*-environment phenomenon, and that the causal analysis of development is required to understand evolution.

5. DEVELOPMENTAL SOURCES OF PHENOTYPIC STABILITY AND VARIABILITY

The morphologist Pere Alberch (1982) pointed out over 30 years ago that development contributes to the evolutionary process in at least two key ways, one *regulatory* and the other *generative*. First, the process of development constrains phenotypic diversity by limiting the "range of the possible" in terms of both form and function. This robustness of development, despite genetic or environmental perturbations, is the *regulatory* function of development (Siegal & Bergman, 2002; Wimsatt, 1986). Some years ago, Maynard Smith and colleagues (1985, p. 266) defined developmental constraint as "a bias on the production of variant phenotypes or a limitation on phenotypic variability caused by the structure, character, composition, or dynamics of the developmental system". In the most general sense, developmental constraints result from the physical properties of biological materials and the temporal and spatial limitations on the relations among internal and external factors at play in developmental processes. These constraints effectively bias the course of evolution, limiting the type and range of variation available to natural selection.

The process of development also introduces phenotypic variation and novelties of potential evolutionary significance. This is the *generative* function of development. For example, many phenotypes show graded responses to factors or events that occur along natural environmental gradients (e.g., temperature, pH levels) and dichotomous responses (polyphenisms) to factors or events that occur in a dichotomous manner (e.g., the presence or absence of predators or particular food items, see Nijhout, 2003). This flexibility of phenotypic outcomes in response to variations in or modifications of genetic and environmental factors is referred to as *phenotypic plasticity*. West-Eberhard (2003, p. 33) defined phenotypic plasticity as "the ability of an organism to react to an internal or external environmental input with a change in form, state, movement, or rate of activity". She argues that such plasticity in response to changed environmental conditions is an important basis of evolutionary novelties (see also Frankino & Raff, 2004; Johnston & Gottlieb, 1990; van der Weele, 1999).

For example, new or novel behaviors brought on by alterations in normal prenatal and/or postnatal rearing environments can lead to new organism–environment relationships, including changes in diet, habitat use, and/or social and reproductive behavior. These behavioral shifts can be maintained across generations if such changes or alterations in the developmental rearing environment persist over time, promoting a cascade of possible changes in morphology and physiology over time (Gottlieb, 2002; Johnston & Gottlieb, 1990; Kuo, 1967).

The phenomenon of domestication, the process by which organisms change in terms of morphology, physiology, or behavior as a result of the human control of their breeding, feeding, and care (Hale, 1969), provides an informative and often overlooked illustration of the complex dynamics involved in this transgenerational shift in phenotypes (Belyaev, 1979; Lickliter & Ness, 1990; Price, 1999). The variance of phenotypes among wild and domestic strains of a single species has long been appreciated. Darwin (1859, 1868), for example, documented the wide array of alterations in size, shape, coloration, productivity, and behavior evident in domesticated animals and speculated on their possible sources. Following the neo-Darwinian synthesis of the first half of the twentieth century and its emphasis on population genetics, most students of domestication assumed that the morphological, physiological, and behavioral differences observed between wild and domestic strains of animals could be explained by random and nonrandom genetic mechanisms associated with captive rearing. These genetic mechanisms include natural and artificial selection, inbreeding, genetic drift, and genetic mutation (Price & King, 1968).

Although the importance of genes as sources of phenotypic variation in both wild and domestic animals is indisputable, domestication is certainly not simply a matter of changing gene frequencies. The transition from free living to captivity is accompanied by many and varied changes in an animal's physical, biological, and social environments and we know that these changes can bring about significant modifications in phenotypic development. For example, Clark and Galef (1981) have shown that specific differences in the morphology, physiology, and behavior of wild and domestic strains of gerbils (*Meriones unguiculatus*) can be traced to relatively minor changes in the developmental resources available in their early rearing experiences. Gerbils reared in standard laboratory cages without access to shelter show accelerated eye opening following birth, earlier sexual maturity, increased docility, and reduced reactivity to humans when compared to gerbils reared in laboratory conditions that allow free access to shelter, as would normally occur in the wild. Of course, the change from free living to captivity for most species is typically accompanied by changes in the availability of not only shelter but also space, food and water, predation, and possibilities for social interaction (Price, 1999). The influence of such changes on the nature and range of phenotypic change under domestication remains relatively unexplored (Lickliter & Ness, 1990).

One research program that has attempted to address these changes is that of Dmitri Belyaev (1979) (see also Trut, Plyusnina, & Oskina, 2004) on the domestication of silver foxes (*Vulpes vulpes*). Selection for tame behavior in silver foxes began in the 1950s and continues to the present. Selection was based solely on behavioral criteria. It is important to note that such selective breeding (common in cases of domestication) is selecting for "developmental outcomes" (in this case, tameness), not genes. In addition to becoming more doglike in their behavior over the course of more than 40 generations, the silver foxes quickly showed a number of other phenotypic modifications, including changes in the skeleton (shortened legs, tail, and snout and a widened skull), hormonal changes, altered tail and ear posture, and decreased sexual dimorphism. Belyaev (1979) proposed that the experiential conditions of domestication led to neural and hormonal changes that in turn activated dormant genes, thereby revealing hidden genetic potentials previously undetected in wild silver foxes. This idea remains speculative, but Belyaev's interpretation that certain genes were able to switch from dormant to active states in response to changes in environmental conditions is certainly plausible in the light of recent advances in epigenetics (Hallgrímsson & Hall, 2011) and would help explain the rapid rate of phenotypic changes observed across only a few generations. The more we learn about the mechanisms by which the environment (both internal and external, see Stotz, 2006) can influence the activation and expression of genes (see Bateson & Gluckman, 2011; Gilbert, 2010; Hallgrímsson & Hall, 2011), the more it becomes clear that gene/environment coaction has to be a cornerstone of explanations of phenotypic variation.

6. THE ECOLOGY OF DEVELOPMENT AND EVOLUTION

The growing acknowledgment of the dynamics of development by biologists and psychologists over the past several decades has fueled a renewed interest in how the developmental process contributes to evolutionary change (e.g., Arthur, 2004; Bjorklund, 2006; Gilbert, Opitz, & Raff, 1996; Gottlieb, 1992, 2002; Lickliter & Honeycutt, 2009; Pigliucci, 2007; Robert, 2004; West-Eberhard, 2003). Of particular importance in this concern is the recognition that variations in morphology, physiology, and/or behavior arising from modifications to the developmental process can place organisms in different ecological or functional relationships with their environments. If these phenotypic variations provide even slight advantages in survival and reproduction, then competitors without the novel phenotype will eventually decrease in frequency in the population, thereby contributing to evolutionary change.

For example, a European passerine bird, the blackcap (*Sylvia atricapilla*) has shown changes in its migratory behavior over the past several decades that have resulted in changes in wing shape, beak size, mating behavior, size of egg clutches, and success at fledging young (Bearhop, Fiedler, Furness, Newton, Votier, & Waldron et al., 2005). Many passerine birds are seasonal migrants and the timing of spring migration constrains when breeding starts each year. Until recently, all European blackcaps migrated back and forth together, spending summers in northern Europe and the British Isles and winters in Portugal, Spain, and North Africa before gathering in mating grounds in southern Germany and Austria to breed. Blackcaps were typically seen in the British Isles only during the summer months, but the number of them wintering in Britain and Ireland has increased dramatically over the past 40 years. This change is thought to be due to the increased availability of winter provisioning provided by bird feeders, landscapers, and other related human activities, as well as an increase in winter temperatures. The resulting shift in migratory patterns has allowed northern-wintering blackcaps to be exposed some 10 days earlier than their southern-wintering counterparts to the critical photoperiods that contribute to the initiation of migration and the onset of gonadal development. Even though all blackcaps continue to gather each year at the same mating sites in Germany and Austria, isotopic data indicate that northern blackcaps that winter in the United Kingdom arrive earlier at the breeding grounds and establish territories and mate with other earlier-arriving birds; southern-wintering blackcaps arrive at the same mating sites some 2 weeks later and are more likely to mate with each other, serving to reproductively isolate northern-wintering birds from the later-arriving southern-wintering population. This shift in migratory patterns appears to confer an advantage to the northern blackcaps, which lay more eggs per season than do their later-arriving cohorts from the south (Bearhop et al., 2005).

The blackcap provides a compelling example of how a change in behavior (in this case, a change in migratory patterns brought on by changes in food availability) can lead to changes in the timing of breeding, which in turn can lead to the effective reproductive isolation of populations and ultimately, divergence and even sympatric speciation. Contrary to the assumptive base of the neo-Darwinian synthesis of the past century, the introduction of phenotypic variation on which natural selection acts is not simply the result of random genetic mutations. Rather, variations in phenotypes and the resulting opportunities for evolutionary change are the result of a wide range of epigenetic processes occurring at different timescales and involving factors internal *and* external to the developing individual.

The blackcap example thus suggests that that understanding the limits and the possibilities of developmental systems is crucial for furthering our understanding of evolution (Lickliter & Harshaw, 2010; Lickliter & Honeycutt, 2009; Oyama, 1985; Oyama, Griffiths, & Gray, 2001). This task will require both description and experimentation, with the goal of explaining how each generation sets up the necessary developmental conditions and resources for the next and how specific changes in developmental conditions lead to specific changes in behavior, anatomy, physiology, and gene expression. For example, differences in physical (body weight, endocrine responses) and behavioral (fearfulness) measures have long been observed between groups of rats whose mothers (Dennenberg & Whimbey, 1963; Whimbey & Dennenberg, 1967) or grandmothers (Dennenberg & Rosenburg, 1967) were handled or not handled as infants. Despite its obvious importance to both developmental and evolutionary concerns, these types of transgenerational effects on both physiological responsiveness (for example, the development of the hypothalamic adrenocortical system) and behavioral responsiveness (including curiosity, novelty seeking, and emotional regulation) remain poorly understood. We do know, however, that aspects of maternal behavior such as licking and grooming influence gene expression in rat pups, as measured by increases in mRNA coding for proteins involved in behaviors known to be affected by differences in maternal care, such as hypothalamic pituitary adrenal axis (HPA) stress response and spatial learning (reviewed by Meaney, 2001, 2010). For example, rat pups that receive relatively high levels of maternal licking and grooming following birth have more hippocampal glucocorticoid receptors. These receptors serve as a brake on the HPA stress response and as a result these pups show less physiological and behavioral response to stress throughout the life span than do those which received lower levels of maternal grooming (Champagne, Francis, Mar, & Meaney, 2003).

Drawing on decades of work by developmental psychobiologists (see Michel & Moore, 1995), we know that the conditions that best favor the expression of modified or novel phenotypes are species–atypical alterations in environmental conditions and contingencies that occur early in ontogeny (e.g., Blumberg, 2008; Denenberg, 1969; Gottlieb, 1971; Kuo, 1967; Levine, 1956; Renner & Rosenzweig, 1987). Of course, shifts in behavior brought about by alterations to the developmental system can arise at any stage of the life cycle, but are generally more likely to occur earlier in individual development. This point was highlighted by several evolutionary theorists over the past century (e.g., de Beer, 1930; Garstang, 1922; Goldschmidt, 1940; Waddington, 1975), who despite their different backgrounds and perspectives realized the significance of embryonic and neonatal periods of development for the generation of phenotypic novelties. These early periods of development are a time of rapid morphological, physiological, and behavioral change, and modifications to an individual's developmental system during this time can initiate a host of physical and behavioral changes, and in some cases (given the availability of appropriate developmental conditions) persist across subsequent generations. Developmental science has much to contribute in this area, particularly to explore how previous developmental outcomes and current experiences in specific contexts combine to influence these processes.

For example, during the later stages of prenatal development the precocial avian embryo is oriented in the egg such that its left eye is occluded by the body and yolk sac, whereas the right eye is exposed to diffuse light passing through the egg shell when the brooding hen is intermittently off the nest during the incubation period. This differential prenatal visual stimulation resulting from the embryo's invariant postural orientation in the egg has been shown to facilitate the development of the left hemisphere of the brain in advance of the right hemisphere. Further, this light-induced developmental advantage for the left hemisphere has been shown to influence the direction of hemispheric specialization for a variety of postnatal behaviors, including visual discrimination, spatial orientation, feeding behavior, and various visual and motor asymmetries (reviewed in Rogers, 1995). Altering the normal pattern of light stimulation available during prenatal development can modify this typical pattern of brain and behavioral development (Deng & Rogers, 2002). For example, a left spatial turning bias is seen in the large majority of quail chicks following hatching (>85%, Casey & Lickliter, 1998). Research has shown that this species-typical turning bias can be reversed by occluding the right eye and stimulating the left eye with light prior to hatching. Further, the induction of such lateralization can be prevented by incubating eggs in darkness or by providing the same level of light stimulation to both eyes in the period prior to hatching (Casey & Lickliter, 1998). These findings suggest an equipotentiality for hemispheric specialization and indicate that late prenatal experience can have a powerful influence on the stability and the variability of functional lateralization.

Similar findings from birds and mammals have demonstrated that the features of available prenatal and early postnatal sensory stimulation (such as amount, intensity, or the timing of presentation and the sources of stimulation) coact with specific organismic characteristics (such as the stage of organization of the sensory systems, previous history with the given properties of stimulation, and the current state of arousal of the young organism) to contribute to the developmental course of species-typical perceptual biases and preferences, learning, and memory (Harshaw & Lickliter, 2011; Lickliter, 2005; Spear, 1984; Spear & McKenzie, 1994). Changes in these basic processes can set up a trajectory of experiential events that can result in modifications to typical patterns of species identification, habitat selection, diet preference, and other key aspects of the organism–environment system. Such phenotypic variations provide the opportunity for natural selection to filter out those novelties that are less successful and promote those that provide some reproductive advantage.

Environmental conditions, including social factors, can influence development by a rich interplay of both external and internal signals. Work with desert locusts, a well-known agricultural pest, illustrates the intricate links between internal and external factors contributing to the effects of experience on phenotypic plasticity. The desert locust (*Schistocerca gregaria*) is usually cryptic in color (green), has short wings and a large abdomen, and is solitary. It typically actively avoids other locusts and flies alone at nighttime. However, under certain climatic conditions and the resulting increase in desert vegetation, their numbers can explode, triggering a rapid increase in population density that results in a dramatic transformation of their color (now black and bright yellow) and social behavior (now gregarious). Normally, solitary locusts now molt into adults with longer wings and more slender abdomens and form bands and eventually swarms consisting of billions of locusts, causing catastrophic damage to agricultural crops.

This dramatic transformation includes many morphological, physiological, and behavioral changes involving numerous chemical messengers and changes in the expression of more than 500 genes (Kang, Chen, Zhou, Liu, Zheng, & Li et al., 2004). Anstey, Rogers, Ott, Burrows, and Simpson (2009) have shown that the key agent in this dramatic phenotypic plasticity is the neurotransmitter serotonin, which is synthesized in the locust's thoracic nervous system in response to multiple sensory cues (touch, smell or sight) provided by social contact with other locusts when the population density increases. Within as little as 2 h of proximity to other locusts, elevated serotonin levels (up to three times the levels seen in solitary locusts) can switch behavior from mutual aversion to mutual attraction, recruiting additional chemical messengers and allowing the formation of the enormous locust swarms that wreak havoc on human populations.

Further, the gregarious migratory phenotype can be retained for several generations after the crowding experience that initiated the original transformation of the solitary phenotype. This transgenerational effect is mediated, at least in part, by the neurotransmitter L-dopa, which is introduced by gregarious females into the foam surrounding their eggs during egg laying. If this foam is experimentally transferred from eggs laid by gregarious female locusts onto eggs laid by solitary females, the solitary eggs hatch into gregarious locusts (McCaffery & Simpson, 1998).

Of particular importance to our concern with the links between development and evolution is the fact that such phenotypic variability across individuals (the grist for the mill of natural selection) can be generated by genetic *and* nongenetic means. These means are varied, including random mutation, drift and other well-recognized genetic processes; they also include less well-studied extragenetic components, such as maternal cytoplasmic constituents, mRNA, and chromatin-modifying enzymes (chromatin is the protein–DNA complex involved in fitting the genome into the cell nucleus), all known to influence changes by which the fertilized egg cell transforms into a complex organism during embryonic development and in later life allows cells to respond to hormones, growth factors, and other regulatory molecules (Crews, 2011; Jablonka & Lamb, 1998).

7. EPIGENETICS: BRIDGING THE GAP BETWEEN GENOTYPE AND PHENOTYPE

Recent advances in genetics and molecular and developmental biology have converged to demonstrate that the expression of genes is affected or modified not only by other genes but also by the local cellular as well as the extracellular environment of the developing organism, including cell cytoplasmic factors, hormones, and sensory, motor, and social stimulation provided by the external environment (reviewed in Davidson, 2001; Jablonka & Raz, 2009; Johnston & Edwards, 2002; Gottlieb, 1998). This advance in our understanding of genomics and its links to ecology has fueled the growth of *epigenetics*, the study of emergent properties in the origin of the phenotype in development and in the modification of phenotypes in evolution (Hallgrímsson & Hall, 2011).

Conrad Waddington first described epigenetics early in the 1940s as the branch of biology that studies the causal interactions of genes with their environment that bring the phenotype into being (Waddington, 1942). Of course, the genetic, molecular and cellular details of phenotypic development were poorly understood at that point in time. Indeed, in the first half of the twentieth century, the gene was largely a theoretical concept without a physical identity (Crews & McLachlan, 2006). Nevertheless, based in part on his experimental work with fruit flies, Waddington came to question the canonical view that there was a simple correspondence between genes and phenotypic traits and proposed that only an understanding of the interaction of genes with each other and with the internal and external environment of the organism could successfully account for phenotypic development. Waddington was advocating a new conceptual framework for the study of development and evolution, one that emphasized changes in what he termed "developmental systems". From this view, the contribution of the genome always depends on the influence of the features of its surrounding contexts, beginning with the cytoplasmic environment provided by the mother's egg at conception. Waddington's efforts to integrate genetics, development, and evolution was well ahead of the prevailing consensus of his time and was motivated by what he viewed as the inability of population genetics to provide a workable model of the operation of genes in development and evolution (Hall, 2001).

The epigenetic framework of development and evolution outlined by Waddington, as well as a handful of other biologists and psychologists working in mid-twentieth century (e.g., Gottlieb, 1971; 1987; Ho & Saunders, 1979; Kuo, 1967; Løvtrup, 1974; Matsuda, 1987), went relatively ignored for the next several decades. This was due in large part to the widespread acceptance of the gene-centered framework of the Modern Synthesis across the life sciences during this time. However, a growing body of evidence drawn from genetics, cellular and developmental biology; neuroscience; and developmental psychology has converged in recent years to provide substantial support for the validity of the epigenetic framework (e.g., Bjorklund, 2006; Hallgrímsson & Hall, 2011; Gottlieb, Wahlsten, & Lickliter, 2006; Ho, 1998; Jablonka & Lamb, 1995, 2005; Meaney, 2010; Michel & Moore, 1995; Szyf, Weaver, & Meaney, 2007). Epigenetics-oriented approaches to development and evolution have the phenotype rather than the genotype as their primary focus of interest and in the general sense, are concerned with aspects of the process of development that lead to flexibility or plasticity when the environment or genome changes (West-Eberhard, 2003). Epigenetic research is now documenting the intricate relational regulatory networks involved in the developmental process, as well as pointing to the need to revise several enduring ideas regarding development and heredity over the past century. These include the notions that (1) genes contain specific programs or instructions for building organisms, (2) genes are the exclusive means by which these instructions are reliably transmitted from one generation to the next, and (3) there can be no meaningful feedback from the environment to the genes (see Lickliter & Honeycutt, 2009 for discussion).

A wide range of recent findings in epigenetics (e.g., Crews, 2011; Meaney, 2010; Zhang & Ho, 2011) have made clear that gene activity or expression is determined by the developmental system as a whole, with positive and negative feedback loops between genes, cells, organs, body, and environment. For example, in rats, higher levels of nurturing maternal behavior leads to postnatal remodeling of the hippocampal glucocorticoid receptor gene (GR, known to be involved in glucocorticoid feedback sensitivity and response to stress). These changes in GR emerge over the first week of life, are reversed with cross-fostering to less-nurturing females, and persist into adulthood (Weaver et al., 2004). The epigenetic framework emerging in contemporary biology highlights the developmental contingency of gene expression and thus focuses research attention on dynamic developmental processes at many levels of organization, and in its best forms, without an implicit bias as to what factors or levels are driving the process. Further, in keeping with a relational developmental systems perspective, epigenetics emphasizes that all developmental processes take place in context. Larsen and Attalah (2011) provide a poignant example of this idea, pointing out that a fertilized egg, once cracked open and stirred, can no longer produce an embryo, even though its entire genome remains intact.

As the philosopher Richard Burian (2005, p. 177) has described it, "the context-dependence of the effects of nucleotide sequences entails that what a sequence-defined gene does cannot be understood except by placing it in the context of the higher-order organizations of the particular organisms in which it is located and in the particular environments in which those organisms live." A growing number of developmental biologists and psychologists are thus expanding the focus of their research attention to not only the internal features of the developing organism (genes, cells, and hormones) but also the contributions of the physical, biological, and social resources available to the individual in its developmental context (e.g., diet, temperature, social interaction, see Gilbert, 2005). These internal and external resources generate a range of phenotypic outcomes of interest to psychologists, including behavioral profiles or "personalities" (see Groothuis & Trillmich, 2011) observed within and across species.

8. THE IMPORTANCE OF BEHAVIOR TO EVOLUTIONARY CHANGE

In neo-Darwinian views of evolution, changes in phenotype are seen as the outcome of genetic change in a population (presumably produced by natural selection). However, as discussed earlier, phenotypic change must come before natural selection is possible. In other words, variation must exist in a population before selection among the variants can occur. In this light, a number of biologists and psychobiologists have proposed that behavior can potentially play a leading role in evolutionary change (e.g., Bateson, 1988; Ho, 1998; Johnston & Gottlieb, 1990; Oyama, Griffths, & Gray, 2001; Plotkin, 1988; Stamps, 2003; Wcislo, 1989). While the importance of behavior as an agent of evolutionary change is not a new idea (Baldwin, 1896; Hardy, 1965; Lloyd Morgan, 1896; Wyles, Kunkel, & Wilson, 1983), it has yet to be fully incorporated into evolutionary theory. The work of Gilbert Gottlieb (1987, 1991, 1992) provided an innovative framework for making sense of how changes in behavior could contribute to the evolutionary process. In Gottlieb's view of evolution, enduring transgenerational phenotypic change can occur at the behavioral, anatomical, and physiological levels before it occurs at the genetic level. His theory proposes that changes in development that result in a novel behavioral shift that recurs across generations can facilitate new organism–environment relationships. These new relationships, which can include "invasion" of novel environments, can bring out latent possibilities for morphological or physiological change. Eventually, a change in gene frequencies may also occur as a result of geographically or behaviorally isolated breeding populations. Thus, changes in behavior can be the first step in creating new phenotypic variants on which natural selection can act (Gottlieb, 1992, 2002).

In this view of evolutionary change, genetic change is often a secondary or tertiary consequence of enduring transgenerational behavioral changes brought about by alterations of normal or species-typical development. These developmental modifications often put individuals in new relations with their local environments, subjecting them to new selection pressures and increasing the likelihood of eventual change in the genetic composition of the population. This perspective introduces a plurality of possible pathways to evolutionary change, complementing genetic factors such as mutation, recombination, and drift.

In a seminal paper exploring the nature of the links between development and evolution, Johnston and Gottlieb (1990) provided an example of how new phenotypes may arise due to an enduring change in behavior before changes in gene frequencies. They describe a scenario in which a population of rodents whose normal diet consists of soft vegetation encounters a new food source of relatively hard but highly nutritious seeds. As the animals learn to sample and eventually increase the representation of seeds in their diet, a number of developmental effects of their new diet become evident, including possible changes in body size and composition, fecundity, age of sexual maturation, and indirect changes in morphology. For example, as the diet changes from soft vegetation to harder seed items, the mechanical stresses exerted on growing jaw tissues during development will change. Given that patterns of bone growth are known to be determined, in part, by forces exerted on the growing bone (Frost, 1973), the skeletal anatomy of the jaw will likely be different in the animals that experience hard vs. soft diets early in life. Such changes in diet have, in fact, been shown to affect the jaw and skull of rats (Bouvier & Hylander, 1984). In this example, behavioral change in members of a population (a preference for a new diet of hard seeds) leads to specific anatomical changes (modification of the jaw and teeth). Such changes can endure across generations, and as long as the new diet remains available, may eventually lead to changes in gene frequency as a result of long-term behavioral or geographic isolation among variants within the population. Following this line of thinking, West-Eberhard (2005, p. 6547) has argued that "genes are probably more often followers that leaders in evolutionary change".

Developmental scientists are well suited to provide systematic investigation of the developmental and ecological dynamics contributing to such behavioral plasticity or malleability. Starting with the pioneering work of Hymovitch (1952), Levine (1956), and Cooper and Zubek (1958), a large body of research has explored the influence of early experiential alterations on later exploratory behavior and problem-solving abilities (see Renner & Rosenzweig, 1987 for a review). Expanding on this research tradition is critical to a more complete understanding of the mechanisms responsible for shifts in behavior and psychological functioning due to changes in species-typical environments, resulting in changes in the activity of the organism and leading to potential variations within and across generations in anatomy, morphology, or physiology (see Gariepy, Rodriguiz, & Jones, 2002). The example of Darwin's finches discussed earlier highlights how, in a relatively short time, birds arriving from the mainland were able to occupy many different habitats. Finches' observed behavioral flexibility and adaptability likely played an important role in their successful colonization of new environmental settings (Tebbich, Sterelny, & Teschke, 2010).

The observed behavioral changes reported in enriched rearing and early handling experiments (i.e., enhanced exploratory behavior, increased problem-solving abilities, and resistance to stress) are the types of behaviors that could promote the seeking out and utilization of new habitats, leading to a host of other potential phenotypic novelties and setting the stage for possible evolutionary change. In this light, Gottlieb (1997) pointed out that animals that have had considerable variation in social and physical experiences early in life are more likely to seek out variation later in life, showing greater levels of exploratory behavior and novelty seeking than animals having more limited early experience. This sort of behavioral plasticity, the willingness to approach and explore novel objects, places, or situations (termed "neophilia" by Thorpe, 1956), can increase the likelihood of particular individuals utilizing or migrating to new habitats, where they could encounter different types of functional demands. Although many changes in functional demands would be transient, others, including diet, day length, and climate, could be long lasting and persist across generations, revealing latent morphological or physiological variability not expressed in the original environment. How these processes work and the underlying biology involved remains poorly understood (but see Duckworth, 2009) and psychology can provide a developmental and ecological perspective to the ways and means of transgenerational processes and their effects.

A related task for developmental science is to provide more detailed empirical evidence for the role of behavior as a leading edge in the evolutionary process. Changes in behavior brought about by changes in prenatal and postnatal rearing environments have been well documented in comparative psychology (e.g., Kuo, 1967; Lickliter, 2005; Michel & Moore, 1995; Renner & Rosenzweig, 1987) but how such changes are significant to evolutionary issues has received little empirical or conceptual attention. Experimental demonstrations of how novel behavioral phenotypes lead to genetic, morphological, or physiological alterations can help further explicate the specific biological and psychological mechanisms involved in the behavioral initiation of evolutionary change. This approach can investigate behavior as the *product* of development and also as a component of the *process* by which development takes place (Stotz, 2008). This type of approach will require a multidimensional process-oriented methodology that includes a variety of levels of analysis beyond the behavioral level, including the environmental regulation of gene expression and cellular function, the effects of sensory stimulation on neural and hormonal responsiveness, and the direct and indirect effects of a developing organism's biological, ecological, and social organization (see Meaney, 2010; West-Eberhard, 2005).

9. PATHS INTO THE FUTURE

We have not yet arrived at a comprehensive synthesis of development and evolution, but it seems to be underway. Whether the relational developmental approach to evolution outlined in this Chapter is up to this task remains to be seen, but it does provide a perspective that has moved many biologists and psychologists beyond outdated notions of genetically *or* environmentally determined phenotypic development. This shift, long overdue, is raising new questions for developmental and evolutionary scientists about the importance of activity and experience, the nature and extent of heredity, and the sources of phenotypic stability and variability (see Charney, 2012; Harper, 2005; Dworkin, Foster, Ledon-Rettig, Moczek, Nijhout, & Sultan et al., 2011; Overton & Müller, 2012; Reid, 2007; West-Eberhard, 2003). In addition, conceptual reformulations and empirical evidence emerging from new disciplines and topics of research (genomics, epigenetics, evolutionary developmental biology, ecological developmental biology, and systems biology) are providing developmental science increasing opportunities to contribute to a fuller understanding of the evolutionary process. This scholarship should ultimately provide a real synthesis of development, evolution, and heredity. My focus in this Chapter has been to explore how a developmental point of view is helping forge a more complete explanation for the origin of novel phenotypic traits, one of the most enduring challenges of evolutionary biology. Much work remains to be done on this project, but we can now say with confidence that understanding evolutionary change will require an understanding of development.

REFERENCES

Abzhanov, A., Kuo, W., Hartmann, C., Grant, R. B., Grant, P., & Tabin, C. J. (2006). The calmodulin pathway and evolution of elongated beak morphology in Darwin's finches. *Nature, 442*, 563–564.

Abzhanov, A., Protas, M., Grant, B. R., Grant, P., & Tabin, C. J. (2004). Bmp4 and morphological variation of beaks in Darwin's finches. *Science, 305*, 1462–1465.

Alberch, P. (1982). The generative and regulatory roles of development in evolution. In D. Mosakowski & G. Roth (Eds.), *Environmental adaptation and evolution* (pp. 19–36). Stuttgart, Germany: Fischer-Verlag.

Amundson, R. (2005). *The changing role of the embryo in evolutionary thought: Roots of evo-devo.* Cambridge, UK: Cambridge University Press.

Anstey, M., Rogers, S. M., Ott, S. R., Burrows, M., & Simpson, S. J. (2009). Serotonin mediates behavioral gregarization underlying swarm formation in desert locusts. *Science, 323,* 627–630.

Arthur, W. (1997). *The origin of animal body plans: A study in evolutionary developmental biology.* Cambridge, UK: Cambridge University Press.

Arthur, W. (2004). *Biased embryos and evolution.* New York: Cambridge University Press.

Baldwin, J. M. (1896). A new factor in evolution. *American Naturalist, 30,* 441–451.

Baldwin, J. M. (1902). *Development and evolution.* New York: Macmillan.

Bateson, W. (1894). *Materials for the study of variation treated with especial regard to discontinuity in the origin of species.* London: Macmillan.

Bateson, P. P. G. (1988). The active role of behavior in evolution. In M. W. Ho & S. W. Fox (Eds.), *Evolutionary processes and metaphors* (pp. 191–207). New York: Wiley.

Bateson, P. P. G., & Gluckman, P. (2011). *Plasticity, robustness, development and evolution.* Cambridge, UK: Cambridge University Press.

Bearhop, S., Fiedler, W., Furness, R., Votier, S., Waldron, S., Newton, J., et al. (2005). Assortative mating as a mechanism for rapid evolution of a migratory divide. *Science, 310,* 502–504.

de Beer, G. (1930). *Embryology and evolution.* Oxford, UK: Clarendon Press.

Belyaev, D. (1979). Destabilizing selection as a factor in domestication. *Journal of Heredity, 70,* 301–308.

Bjorklund, D. F. (2006). Mother knows best: epigenetic inheritance, maternal effects, and the evolution of human intelligence. *Developmental Review, 26,* 213–242.

Blumberg, M. S. (2008). *Freaks of nature.* New York: Oxford University Press.

Bouvier, M., & Hylander, W. L. (1984). The effect of dietary consistency on gross and histologic morphology in the craniofacial region of young rats. *American Journal of Anatomy, 170,* 117–126.

Brooks, W. K. (1883). *The law of heredity: A study of the cause of variation and the origin of living organisms.* Baltimore, MD: John Murphy.

Burian, R. M. (2005). *The epistemology of development, evolution, and genetics.* New York: Cambridge University Press.

Buss, D. M. (2005). *The handbook of evolutionary psychology.* New York: Wiley.

Carroll, S. (2005). *Endless forms most beautiful.* New York: W. W. Norton.

Carruthers, P., Laurence, S., & Stich, S. (2007). *The innate mind: Foundations and the future.* New York: Oxford University Press.

Casey, M., & Lickliter, R. (1998). Prenatal visual experience facilitates the development of spatial turning bias in bobwhite quail chicks. *Developmental Psychobiology, 32,* 327–338.

Champagne, F., Francis, D., Mar, & Meaney, M. (2003). Variations in maternal care in the rat as a mediating influence for the effects of environment on development. *Physiology and Behavior, 79,* 359–371.

Charney, E. (2012). Behavior genetics and post genomics. *Behavioral and Brain Sciences, 35,* 331–358.

Clark, M., & Galef, B. (1981). Environmental influences on development, behavior, and endocrine morphology of gerbils. *Physiology and Behavior, 27,* 761–765.

Cooper, R. M., & Zubek, J. P. (1958). Effects of enriched and restricted early environments on the learning ability of bright and dull rats. *Canadian Journal of Psychology, 12,* 159–164.

Cope, E. D. (1886). *The origin of the fittest: Essays on evolution.* London: D. Appleton.E.

Crews, D. (2011). Epigenetic modifications of brain and behavior: theory and practice. *Hormones and Behavior, 59,* 393–398.

Crews, D., & McLachlan, J. A. (2006). Epigenetics, evolution, endocrine disruption, health, and disease. *Endocrinology, 147,* S4–S10.

Darwin, C. (1859). *On the origin of species by means of natural selection.* London: John Murray.

Darwin, C. (1868). *The variation of animals and plants under domestication.* New York: D. Appleton.

Darwin, C. (1882). Prefatory notice: studies in the theory of descent. In A. Weismann (Ed.), *Studies in the theory of descent: with notes and additions by the author* (pp. v–vi). London: Sampson Low, Marston, Searle, and Rivington.

Davidson, E. H. (2001). *Genomic regulatory systems: Evolution and development*. San Diego: Academic Press.

Deng, C., & Rogers, L. (2002). Prenatal visual experience and lateralization in the visual Wultz of the chick. *Behavioural Brain Research, 134,* 375–385.

Dennenberg, V. (1969). The effects of early experience. In E. S. E. Hafez (Ed.), *The behavior of domestic animals*. Baltimore, MD: Williams & Wilkins.

Dennenberg, V., & Rosenburg, K. M. (1967). Nongenetic transmission of information. *Nature, 216,* 549–550.

Dennenberg, V., & Whimbey, A. E. (1963). Behavior of adult rats is modified by the experiences their mothers had as infants. *Science, 142,* 1192–1193.

Depew, D. J., & Weber, B. H. (1995). *Darwinism evolving: System dynamics and the genealogy of natural selection*. Cambridge, MA: MIT Press.

DeWitt, T. J., & Scheiner, S. M. (2004). *Phenotypic plasticity: Functional and conceptual approaches.* New York: Oxford University Press.

Dobzhansky, T. (1937). *Genetics and the origin of species*. New York: Columbia University Press.

Duckworth, R. A. (2009). The role of behavior in evolution: a search for mechanism. *Evolutionary Ecology, 23,* 513–531.

Frankino, W. A., & Raff, R. A. (2004). Evolutionary importance and pattern of phenotypic plasticity. In T. J. DeWitt & S. M. Scheiner (Eds.), *Phenotypic plasticity: Functional and conceptual approaches* (pp. 64–81). New York: Oxford University Press.

Frost, H. M. (1973). *Bone modeling and skeletal modeling errors*. Baltimore, MD: Thomas.

Gariepy, J. L., Rodriguiz, R. M., & Jones, B. C. (2002). Handling, genetic and housing effects on the mouse stress system, dopamine function, and behavior. *Pharmacology, Biochemistry, and Behavior, 70,* 7–17.

Garstang, W. (1922). The theory of recapitulation: a critical re-statement of the biogenetic law. *Journal of the Linnean Society of London, Zoology, 35,* 81–101.

Gerhart, J., & Kirschner, M. (1997). *Cells, embryos, and evolution: Toward a cellular and developmental understanding of phenotypic variation and evolutionary adaptability*. New York: Wiley.

Gilbert, S. F. (1994). Dobzhansky, Waddington, and Schmalhaussen: embryology and the modern synthesis. In M. B. Adams (Ed.), *The evolution of Theodosius Dobzhansky* (pp. 143–154). Princeton, NJ: Princeton University Press.

Gilbert, S. F. (2005). Mechanisms for the environmental regulation of gene expression: ecological aspects of animal development. *Journal of Biosciences, 30,* 65–74.

Gilbert, S. F. (2010). *Developmental biology* (9th ed.). Sunderland, MA: Sinauer.

Gilbert, S. F., & Epel, D. (2009). *Ecological developmental biology*. Sunderland, MA: Sinauer.

Gilbert, S. F., Opitz, J. M., & Raff, R. (1996). Resynthesizing evolutionary and developmental biology. *Developmental Biology, 173,* 357–372.

Goldschmidt, R. (1940). *The material basis of evolution*. New Haven, CT: Yale University Press.

Gottlieb, G. (1971). *Development of species identification in birds: An enquiry into the prenatal determinants of perception*. Chicago: University of Chicago Press.

Gottlieb, G. (1987). The developmental basis of evolutionary change. *Journal of Comparative Psychology, 101,* 262–271.

Gottlieb, G. (1991). Experiential canalization of behavioral development: Theory. *Developmental Psychology, 27,* 4–13.

Gottlieb, G. (1992). *Individual development and evolution: The genesis of novel behavior*. New York: Oxford University Press.

Gottlieb, G. (1997). *Synthesizing nature-nurture: Prenatal roots of instinctive behavior*. Mahwah, NJ: Erlbaum.

Gottlieb, G. (1998). Normally occurring environmental and behavioral influences on gene activity: from central dogma to probabilistic epigenesis. *Psychological Review, 105,* 792–802.

Gottlieb, G. (2002). Developmental-behavioral initiation of evolutionary change. *Psychological Review, 109,* 211–218.

Gottlieb, G., Wahlsten, D., & Lickliter, R. (2006). The significance of biology for human development. In R. Lerner (Ed.), *Handbook of child psychology Theoretical models of human development* (Vol. 1, pp. 210–257). New York: John Wiley.

Gould, S. J. (1977). *Ontogeny and phylogeny.* Cambridge, MA: Harvard University Press.

Grant, P. R., Grant, B. R., & Abzhanov, A. (2006). A developing paradigm for the development of bird beaks. *Biological Journal of the Linnean Society, 88,* 17–22.

Groothuis, T. G., & Trillmich, F. (2011). Unfolding personalities: the importance of studying ontogeny. *Developmental Psychobiology, 53,* 641–655.

Hale, E. B. (1969). Domestication and the evolution of behavior. In E. S. E. Hafez (Ed.), *The behaviour of domestic animals* (pp. 22–42). London: Bailliere, Tindall, and Cassell.

Hall, B. K. (1992). *Evolutionary developmental biology* (1st ed.). London: Chapman & Hall.

Hall, B. K. (1999). *Evolutionary developmental biology* (2nd ed.). Dordrecht: Kluwer.

Hall, B. K. (2001). Organic selection: proximate environmental effects on the evolution of morphology and behaviour. *Biology and Philosophy, 16,* 215–237.

Hallgrímsson, B., & Hall, B. K. (2005). *Variation: A central concept in biology.* New York: Elsevier.

Hallgrímsson, B., & Hall, B. K. (2011). *Epigenetics: Linking genotype and phenotype in development and evolution.* Berkeley, CA: University of California Press.

Hall, B. K., Pearson, R. D., & Müller, G. (2004). *Environment, development, and evolution: Towards a synthesis.* Cambridge, MA: MIT Press.

Hardy, A. (1965). *The living stream.* London: Collins.

Harper, L. V. (2005). Epigenetic inheritance and the intergenerational transfer of experience. *Psychological Bulletin, 131,* 340–360.

Harris, J. A. (1904). *A new theory of the origin of species.* London: The Open Court Publishing.

Harshaw, C., & Lickliter, R. (2011). Biased embryos: prenatal experience and the malleability of species-typical auditory preferences. *Developmental Psychobiology, 53,* 291–302.

Ho, M. W. (1998). Evolution. In G. Grreenberg & M. H. Haraway (Eds.), *Comparative psychology: A handbook* (pp. 107–119). New York: Garland Publishing.

Ho, M. W., & Saunders, P. (1979). Beyond neo-Darwinism: an epigenetic approach to evolution. *Journal of Theoretical Biology, 78,* 573–591.

Huxley, J. S. (1942). *Evolution, the modern synthesis.* London: Allen & Unwin.

Hymovitch, B. (1952). The effects of experimental variations on problem solving in the rat. *Journal of Comparative and Physiological Psychology, 45,* 313–321.

Jablonka, E., & Lamb, M. J. (1995). *Epigenetic inheritance and evolution.* New York: Oxford University Press.

Jablonka, E., & Lamb, M. (1998). Epigenetic inheritance in evolution. *Journal of Evolutionary Biology, 11,* 159–183.

Jablonka, E., & Lamb, M. J. (2005). *Evolution in four dimensions: Genetic, epigenetic, behavioral, and symbolic variation in the history of life.* Cambridge, MA: MIT Press.

Jablonka, E., & Raz, G. (2009). Transgenerational epigenetic inheritance: prevalence, mechanisms, and implications for the study of heredity and evolution. *Quarterly Review of Biology, 84,* 131–176.

Johnston, T. D., & Edwards, L. (2002). Genes, interactions, and development. *Psychological Review, 109,* 26–34.

Johnston, T. D., & Gottlieb, G. (1990). Neophenogenesis: a developmental theory of phenotypic evolution. *Journal of Theoretical Biology, 147,* 471–495.

Kang, L., Chen, X. Y., Zhou, Y., Liu, B. W., Zheng, W., Li, R. Q., et al. (2004). The analysis of large scale gene expression correlated to the phase changes of the migratory locust. *Proceedings of the National Academy of Sciences of the United States of America, 101,* 17611–17615.

Keller, E. F. (2000). *The century of the gene*. Cambridge, MA: Harvard University Press.

Kuo, Z.-Y. (1967). *The dynamics of behavior development: An epigenetic view*. New York: Random House.

Larsen, E. W., & Attalah, J. (2011). Epigenesis, preformation, and the Humpty Dumpty problem. In B. Hallgrímsson & B. K. Hall (Eds.), *Epigenetics: Linking genotype and phenotype in development and evolution* (pp. 103–115). Berkeley, CA: University of California Press.

Laubichler, M. D., & Maienschein, J. (2007). *From embryology to evo-devo: A history of developmental evolution*. Cambridge, MA: MIT Press.

Lerner, R. M. (1991). Changing organism-context relations as the basic process of development: a developmental contextual perspective. *Developmental Psychology, 27*, 27–32.

Lerner, R. M. (2006). Developmental science, developmental systems, and contemporary theories of human development. In R. M. Lerner (Ed.), *Handbook of child psychology Theoretical models of human development* (Vol. 1, pp. 1–17). New York: Wiley.

Levine, S. (1956). A further test of infantile handling and adult avoidance learning. *Journal of Personality, 25*, 70–80.

Lewkowicz, D. J. (2011). The biological implausibility of the nature-nurture dichotomy and what it means for the study of infancy. *Infancy, 16*, 331–367.

Lickliter, R. (2005). Prenatal sensory ecology and experience: Implications for perceptual and behavioral development in precocial birds. *Advances in the Study of Behavior, 35*, 235–274.

Lickliter, R. (2013). Biological development: theoretical approaches, techniques, and key Findings. In P. D. Zelazo (Ed.), *Oxford handbook of developmental psychology* (pp. 65–90). New York: Oxford University Press.

Lickliter, R., & Berry, T. D. (1990). The phylogeny fallacy: developmental psychology's misapplication of evolutionary theory. *Developmental Review, 10*, 348–364.

Lickliter, R., & Harshaw, C. (2010). Canalization and malleability reconsidered: the developmental basis of phenotypic stability and variability. In K. Hood, C. Halpern, G. Greenberg & R. Lerner (Eds.), *The handbook of developmental science, behavior, and genetics* (pp. 491–525). New York: Wiley Blackwell.

Lickliter, R. L., & Honeycutt, H. G. (2003). Developmental dynamics: toward a biologically plausible evolutionary psychology. *Psychological Bulletin, 129*, 819–835.

Lickliter, R., & Honeycutt, H. (2009). Rethinking epigenesis and evolution in light of developmental science. In M. Blumberg, J. Freeman & S. Robinson (Eds.), *Oxford handbook of developmental behavioral neuroscience* (pp. 30–50). New York: Oxford University Press.

Lickliter, R., & Ness, J. (1990). Domestication and comparative psychology: status and strategy. *Journal of Comparative Psychology, 104*, 211–218.

Lickliter, R., & Schneider, S. (2006). The role of development in evolutionary change: a view from comparative psychology. *International Journal of Comparative Psychology, 19*, 150–167.

Lloyd Morgan, C. (1896). *Habit and instinct*. London: E. Arnold.

Løvtrup, S. (1974). *Epigenetics: A treatise on theoretical biology*. New York: John Wiley.

Matsuda, R. (1987). *Animal evolution in changing environments with special reference to abnormal metamorphosis*. New York: Wiley.

Maynard-Smith, J. (1985). Evolution and development. In B. Goodwin, N. Holder & C. C. Wylie (Eds.), *Development and evolution* (pp. 33–45). Cambridge, UK: Cambridge University Press.

Maynard-Smith, J., Burian, R., Kauffman, S., Alberch, P., Cambell, J., Goodwin, B., et al. (1985). Developmental constraints and evolution. *Quarterly Review of Biology, 60*, 265–287.

Mayr, E. (1963). *Animal species and evolution*. Cambridge, MA: Harvard University Press.

Mayr, E. (1974). Behavior programs and evolutionary strategies. *American Scientist, 62*, 650–659.

Mayr, E. (1982). *The growth of biological thought*. Cambridge, MA: Harvard University Press.

Mayr, E., & Provine, W. (1980). *The evolutionary synthesis: Perspectives on the unification of biology.* Cambridge, MA: Harvard University Press.

McCaffery, A., & Simpson, S. (1998). A gregarizing factor in the egg pod foam of the desert locust. *Journal of Experimental Biology, 201,* 347–363.

Meaney, M. J. (2001). Maternal care, gene expression, and the transmission of individual differences in stress reactivity across generations. *Annual Review of Neuroscience, 24,* 1161–1192.

Meaney, M. J. (2010). Epigenetics and the biological definition of gene × environment interactions. *Child Development, 81,* 41–79.

Michel, G., & Moore, C. (1995). *Developmental psychobiology: An integrative science.* Cambridge, MA: MIT Press.

Mivart, S. G. (1871). *On the genesis of species.* London: McMillan.

Moczek, A. P., Sultan, S., Foster, S., Ledon-Rettig, C., Dworkin, I., Nijhout, H. F., et al. (2011). The role of developmental plasticity in evolutionary innovation. *Proceedings of the Royal Society, B, 278,* 2705–2713.

Moore, D. S. (2002). *The dependent gene: The fallacy of nature vs. nurture.* New York: Freeman.

Moore, D. S. (2008). Individuals and populations: how biology's theory and data have interfered with the integration of development and evolution. *New Ideas in Psychology, 26,* 370–386.

Müller, G. B., & Newman, S. A. (2003). *Origination of organismal form: Beyond the gene in developmental and evolutionary biology.* Cambridge, MA: MIT Press.

Müller, G., & Wagner, G. P. (1991). Novelty in evolution: restructuring the concept. *Annual Review of Ecology and Systematics, 22,* 229–256.

Neumann-Held, E. M., & Rehmann-Sutter, C. (2006). *Genes in development and evolution: Re-reading the molecular paradigm.* Durham, NC: Duke University Press.

Nijhout, H. F. (2003). Development and evolution of adaptive polyphenisms. *Evolution and Development, 5,* 9–18.

Overton, W. F. (2006). Developmental psychology: philosophy, concepts, methodology. In R. Lerner (Ed.), *Handbook of child psychology Theoretical models of human development* (Vol. 1, pp. 18–88). New York: John Wiley.

Overton, W. F., & Müller, U. (2012). Metatheories, theories, and concepts in the study of development. In R. Lerner, M. A. Easterbrooks & J. Mistry (Eds.), *Comprehensive handbook of psychology. Developmental psychology.* (Vol. 6). New York: Wiley.

Oyama, S. (1985). *The ontogeny of information: Developmental systems and evolution.* New York: Cambridge University Press.

Oyama, S., Griffith, P. E., & Gray, R. D. (2001). *Cycles of contingency: Developmental systems and evolution.* Cambridge, MA: MIT Press.

Piersma, T., & van Gils, J. A. (2010). *The flexible phenotype: A body-centered integration of ecology, physiology, and behavior.* New York: Oxford University Press.

Pigliucci, M. (2001). *Phenotypic plasticity: Beyond nature and nurture.* Baltimore, MD: Johns Hopkins University Press.

Pigliucci, M. (2007). Do we need an extended evolutionary synthesis? *Evolution, 61,* 2743–2749.

Pinker, S. (2002). *The blank slate: The modern denial of human nature.* New York: Viking.

Plotkin, H. C. (1988). *The role of behavior in evolution.* Cambridge, MA: MIT Press.

Price, E. O. (1999). Behavioral development in animals undergoing domestication. *Applied Animal Behaviour Science, 65,* 245–271.

Price, E. O., & King, J. A. (1968). Domestication and adaptation. In E. S. E. Hafez (Ed.), *Adaptation of domestic animals* (pp. 34–45). Philadelphia, PA: Lea and Febiger.

Provine, W. (1971). *A history of population genetics.* Chicago: University of Chicago Press.

Raff, R. A. (1996). *The shape of life.* Chicago: University of Chicago Press.

Reid, R. G. B. (2007). *Biological emergences: Evolution by natural experiment.* Cambridge, MA: MIT Press.

Renner, M. J., & Rosenzweig, M. R. (1987). *Enriched and impoverished environments*. New York: Springer.

Richardson, K. (1998). *The origins of human potential: Evolution, development, and psychology*. London: Routledge.

Robert, J. S. (2004). *Embryology, epigenesis, and evolution: Taking development seriously*. New York: Cambridge University Press.

Robert, J. S. (2008). Taking old ideas seriously: evolution, development, and human behavior. *New Ideas in Psychology, 26*, 387–404.

Rogers, L. J. (1995). *The development of brain and behaviour in the chicken*. CAB.

Russell, E. S. (1930). *The interpretation of development and heredity*. Oxford, UK: Clarendon Press.

Saap, J. (2003). *Genesis: The evolution of biology*. Oxford, UK: Oxford University Press.

Sanson, R., & Brandon, R. N. (2007). *Integrating evolution and development: From theory to practice*. Cambridge, MA: MIT Press.

Schlichting, C., & Pigliucci, M. (1998). *Phenotypic evolution: A reaction norm perspective*. Sunderland, MA: Sinauer.

Schmalhausen, I. (1949). *Factors of evolution: The theory of stabilizing selection*. Philadelphia, PA: Blakiston.

Siegal, M. L., & Bergman, A. (2002). Waddington's canalization revisited: developmental stability and evolution. *Proceedings of the National Academy of Sciences of the United States of America, 99*, 10528–10532.

Spear, N. (1984). Ecologically determined dispositions control the ontogeny of learning and memory. In R. Kail & N. E. Spear (Eds.), *Comparative perspectives on the development of memory* (pp. 325–358). Hillsdale, NJ: Erlbaum.

Spear, N. E., & McKinzie, D. L. (1994). Intersensory integration in the infant rat. In D. J. Lewkowicz & R. Lickliter (Eds.), *The development of intersensory perception: Comparative perspectives* (pp. 133–161). Hillsdale, N. J: Erlbaum.

Spelke, E. S., & Newport, E. I. (1998). Nativism, empiricism, and the development of knowledge. In R. Lerner (Ed.), *Handbook of child psychology Theoretical models of human development* (Vol. 1, pp. 275–340). New York: John Wiley.

Spencer, J. P., Blumberg, M. S., McMurray, B., Robinson, S. R., Samuelson, L. K., & Tomblin, J. B. (2009). Short arms and talking eggs: why we should no longer abide the nativist-empiricist debate. *Child Development Perspectives, 3*, 79–87.

Stamps, J. (2003). Behavioural processes affecting development: Tinbergen's fourth question comes of age. *Animal Behaviour, 66*, 1–13.

Stotz, K. (2006). With 'genes' like that, who needs an environment? Postgenomic's argument for the 'ontogeny of information'. *Philosophy of Science, 73*, 905–917.

Stotz, K. (2008). The ingredients for a postgenomic synthesis of nature and nurture. *Philosophical Psychology, 21*, 359–381.

Szyf, M., Weaver, I., & Meaney, M. (2007). Maternal care, the epigenome and phenotypic differences in behavior. *Reproductive Toxicology, 24*, 9–19.

Tebbich, S., Sterelny, K., & Teschke, I. (2010). The tale of the finch: adaptive radiation and behavioural flexibility. *Philosophical Transactions of the Royal Society, B, 365*, 1099–1109.

Thorpe, W. (1956). *Learning and instinct in animals*. London: Methuen and Company.

Trut, L. N., Plyusnina, I. Z., & Oskina, I. N. (2004). An experiment on fox domestication and debatable issues of the evolution of the dog. *Russian Journal of Genetics, 40*, 644–655.

van der Weele, C. (1999). *Images of development: Environmental causes of ontogeny*. Albany: State University of New York Press.

Waddington, C. H. (1942). The epigenotype. *Endeavour, 1*, 18–20.

Waddington, C. H. (1957). *The strategy of the genes*. London: Allen & Unwin.

Waddington, C. H. (1975). *The evolution of an evolutionist*. Ithaca, NY: Cornell University Press.

Wcislo, W. (1989). Behavioral environments and evolutionary change. *Annual Review of Ecology and Systematics, 20*, 137–169.

Weaver, I. G., Cervoni, N., Champagne, F. A., D'Alessio, A. C., Sharma, S., Seckl, K. R., et al. (2004). Epigenetic programming by maternal behavior. *Nature Neuroscience, 7*, 847–854.

West-Eberhard, M. J. (1989). Phenotypic plasticity and the origins of diversity. *Annual Review of Ecology and Systematics, 20*, 249–278.

West-Eberhard, M. J. (2003). *Developmental plasticity and evolution.* New York: Oxford University Press.

West-Eberhard, M. J. (2005). Developmental plasticity and the origins of species differences. *Proceedings of the National Academy of Sciences of the United States of America, 102*, 6543–6549.

Whimbey, A. E., & Dennenberg, V. (1967). Experimental programming of life histories. *Behaviour, 29*, 296–314.

Williams, G. C. (1966). *Adaptation and natural selection.* Princeton, NJ: Princeton University Press.

Wimsatt, W. C. (1986). Developmental constraints, generative entrenchment, and the innate-acquired distinction. In W. Bechtel (Ed.), *Integrating scientific disciplines* (pp. 185–208). The Hague: Nijhoff.

Winther, R. (2000). Darwin on variation and heredity. *Journal of the History of Biology, 33*, 425–455.

Wyles, J. S., Kunkel, J. G., & Wilson, A. C. (1983). Birds, behavior, and anatomical evolution. *Proceedings of the National Academy of Sciences of the United States of America, 80*, 4394–4397.

Zhang, X., & Ho, S. M. (2011). Epigenetics meets endocrinology. *Journal of Molecular Endocrinology, 46*, R11–R32.

CHAPTER EIGHT

Cytoplasmic Inheritance Redux

Evan Charney
Sanford School of Public Policy, Duke University, Durham, NC, USA
E-mail: echar@duke.edu

Contents

Abstract

Since the early twentieth century, inheritance was seen as the inheritance of genes. Concurrent with the acceptance of the genetic theory of inheritance was the rejection of the idea that the cytoplasm of the oocyte could also play a role in inheritance and a corresponding devaluation of embryology as a discipline critical for understanding human development. Development, and variation in development, came to be viewed solely as matters of genetic inheritance and genetic variation. We now know that inheritance is a matter of both genetic and cytoplasmic inheritance. A growing awareness of the centrality of the cytoplasm in explaining both human development and phenotypic variation has been promoted by two contemporaneous developments: the continuing elaboration of the molecular mechanisms of epigenetics and the global rise of artificial reproductive technologies. I review recent developments in the ongoing elaboration of the role of the cytoplasm in human inheritance and development.

Advances in Child Development and Behavior, Volume 44
ISSN 0065-2407, http://dx.doi.org/10.1016/B978-0-12-397947-6.00008-8

1. CYTOPLASMIC INHERITANCE

"Omne vivum ex ovo: Every living thing comes from an egg." As implied in William Harvey's famous statement, the developmental fate of an embryo begins in the oocyte. The initial phase of embryonic development takes place during a period of genetic transcriptional silence until the activation of the embryonic genome. Prior to embryonic genome activation (EGA), the embryo depends entirely upon maternal RNAs, maternal DNA (in mitochondria), maternal organelles, proteins, substrates, and nutrients that have been deposited in the cytoplasm of the ovum during oogenesis. These maternal products control almost every aspect of early embryonic development. Collectively, they constitute an extraordinary maternal "cytoplasmic" inheritance. Variations in this cytoplasmic inheritance—in the "quality" of the oocyte—can have profound developmental consequences for offspring, both short and long terms. But with the exception of mitochondrial DNA (mtDNA), the effects of cytoplasmic inheritance are not due to offspring having inherited maternal (or paternal) genes.

I intentionally use the expression "cytoplasmic inheritance" in place of the more common "maternal inheritance," to place the content of this chapter in historical context: in the early twentieth century, the rejection of the existence of cytoplasmic inheritance had important consequences. It represented the triumph of Mendelian inheritance and supported the development of the Modern Synthesis (Amundson, 2005). The view that inheritance is a matter of genes, not cytoplasm, became something of a dogma in genetics. With the denial of cytoplasmic inheritance came a corresponding diminishment of embryology as a scientific discipline (Gilbert, 1998). This dogma persists to this day, most notably in the approach that characterizes contemporary behavioral genetics: inheritance is a matter of the inheritance of alleles, and variation in alleles—generally, single nucleotide polymorphisms—can provide us with clues to understanding variation in complex phenotypes. Hence, embryology can still be ignored because when it comes to inheritance, what matters are genes and genes alone.

It is beyond any doubt today that cytoplasmic inheritance is a key component of human (biological) inheritance. With the growing awareness of the importance of cytoplasmic inheritance has come the resurrection of embryology. At least two contemporaneous forces have led to a growing awareness of the importance of cytoplasmic inheritance/embryogenesis. First is the continued elaboration of the molecular mechanics of epigenetics and the discovery that many of the earliest embryonic processes, those regulated by the maternal cytoplasm, are also epigenetic processes. Second is the global rise of artificial reproductive technologies (ARTs) as a means of conception. Current birth rates of ART-conceived children in a number of developed countries now range from 1% to 3% of all births. Concern about reports of increased risk of negative developmental outcomes among ART-conceived children has focused attention on the early stages of embryonic development prior to implantation. Investigations motivated by this concern have also pointed to the importance of epigenetic mechanisms in preimplantation development (PID). Likewise, research directed at the elaboration of epigenetic mechanisms and possible sources of negative ART outcomes have both pointed to the critical importance of the periconceptual environment(s) in developmental outcomes.

My intent in this chapter is to survey the current state of knowledge of cytoplasmic inheritance. Because what is known about this phenomenon (which is still very little) is enormously complex, I will need both to be selective and to engage in a certain degree of simplification, but hopefully not in a manner that distorts. Part of the elaboration of cytoplasmic inheritance involves a review of some known and conjectured developmental problems associated with ART. My purpose in discussing ART is in no way to pass judgment on the safety of ART procedures or to offer any recommendations regarding their improvement (which I am in no position to do). Rather, I discuss ART because of the insight it provides into PID, i.e., the period prior to the implantation of the blastocyst in the uterus. PID encompasses the period of cytoplasmic regulation of development and the complete activation (by cytoplasmic elements) of the embryonic genome.

2. OOCYTE TO IMPLANTATION (A FEW BASICS)

At the 12–14th week of development, selected immature female germ cells called oogonia begin the first of the two meiotic divisions that will ultimately result in reducing the number of oocyte chromosomes by half. Meiosis I is arrested in the diplotene stage of prophase I, the period when the chromosomes condense along the center of the cell's nucleus in preparation for division. Shortly before birth, all the fetal oocytes in the female ovary have attained this stage and are referred to as *primary oocytes*. A primary oocyte, together with a surrounding layer of flat epithelial cells, is known as a *primordial follicle*. Oocyte maturation inhibitor, secreted by the follicular cells, will keep the primary oocytes in a dormant state for ~12 years or ~1 year prior to the onset of ovulation (Tsafriri & Pomerantz, 1986). Prior to the onset of ovulation, the oocyte is reactivated and experiences a dramatic increase in size concomitant with renewed genomic transcription. Beginning at puberty, periodic hormone secretions induce a few primary oocytes to complete meiosis I, resulting in a small, nonfunctional polar body and a much larger secondary oocyte where most of the cytoplasm from the primary oocyte is concentrated. The secondary oocyte completes growth in the ovary and begins meiosis II, which is again arrested. At ovulation, the mature follicle containing the secondary oocyte ruptures releasing the oocyte. Fertilization triggers the completion of meiosis II with another asymmetric division of the oocyte resulting in the formation of a second polar body.

Fertilization occurs when the head of a single spermatozoon penetrates the zona pellucida, the hard outer membrane of the ovum, and enters the cytoplasm of the oocyte. Shortly after fertilization, the maternal and paternal chromosomes are enclosed in separate nucleic membranes forming a pair of pronuclei. Within 12–18 h postfertilization, the DNA in both the maternal and paternal pronuclei replicate as the two pronuclei approach. Upon contact, the nucleic membranes of both pronuclei dissolve leading to karyogamy, the fusion of the oocyte and sperm haploid nuclei and the formation of a diploid nucleus (i.e., a nucleus containing two sets of chromosomes, one from each parent). Approximately 30 h postfertilization, the first mitotic cell division, or cleavage, occurs, resulting in two cells or blastomeres. After the two-cell stage, the zygote undergoes a series of mitotic divisions without any increase in overall size (i.e., the number of cells increases but the cells become smaller with each cleavage division). Prior to the eight-cell stage, these cells form a loosely arranged clump. At the eight-cell stage, the embryo undergoes a process of compaction in which the individual blastomeres maximize contact, forming a compact ball of cells. Approximately 4 days after fertilization, the compacted embryo forms a 16-cell morula, with the inner cells constituting the inner, and the surrounding cells, the outer cell mass. The inner cell mass (ICM) gives rise to the tissues of the embryo proper, while the outer cell mass forms the trophoblast, that later contributes to the placenta. About the time the morula enters the uterine cavity, fluid begins to penetrate through the zona pellucida into the intercellular spaces of the ICM. With the proliferation of fluid in the morula, the intercellular spaces become confluent and the blastocyst cavity is formed. At days 5–6, the cells of the ICM, now called the embryoblast, move to one pole, and those of the outer cell mass, or trophoblast, flatten and form the epithelial wall. At this stage, the embryo is referred to as a *blastocyst.* Approximately 1 week after fertilization, the zona pellucida disappears allowing implantation of the blastocyst in the uterine wall to begin.

3. MATERNAL EFFECT GENES

Human protein-coding genes typically contain several DNA sequences that code for amino acids, the building blocks of proteins, known as *exons*, interspersed with several introns, noncoding regions. In gene transcription, the introns and exons are first copied to create pre-messenger RNA (pre-mRNA). Pre-mRNA is then assembled to create messenger RNA (mRNA), but the path from pre-mRNA to mRNA to protein is not direct. Alternative splicing is a process in which certain exons are removed or included to produce different mRNA transcripts from the same genomic locus (Chen & Manley, 2009). More than 98% of multiexonic pre-mRNAs are alternatively spliced (Wang, Sandberg, et al., 2008). The distinct proteins produced from alternatively spliced identical pre-mRNAs, called *isoforms*, can have different, even antagonistic activities. Thus, alternative splicing plays a major role in the activity of a wide range of critical cellular processes, and during mammalian embryogenesis, it is a key to generating a viable organism from a single cell (Revil, Gaffney, Dias, Majewski, & Jerome-Majewska, 2010). The process whereby the resulting mRNA is used to form a specific protein is called *translation.*

After the second meiotic arrest, oocyte gene transcription essentially stops and after fertilization, both the maternal and paternal pronuclei are largely transcriptionally silent, as is the zygotic genome after karyogamy (Wong et al., 2010). Transcription does not commence in the embryo until the activation of the embryonic genome, i.e., EGA. While EGA is not an event that occurs at one specific point in zygotic development, but rather in a series of phases or waves, analysis of patterns of de novo embryonic genome transcription indicates a major wave of genome activation at embryonic day 3 (E3), or at approximately the eight-cell stage. However, while genomic transcription is stopped pre-EGA, the translation of oocyte mRNA to synthesize proteins is not.

The oocyte contains a vast array of mRNAs, corresponding to 20–45% of the entire mouse genome (Evsikov et al., 2006; Wang et al., 2004) and their translation, along with the activity of oocyte organelles such as mitochondria, coordinated both temporally and spatially, enables early embryonic development prior to the complete activation of the embryonic genome (Bell, Calder, & Watson, 2008). This coordination includes the timely translation as well as degradation and clearance of oocyte mRNAs. The window of activity of oocyte mRNAs must be carefully regulated inasmuch as their perdurance beyond their appropriate period of activity could interfere with EGA with potentially fatal consequences (Tadros & Lipshitz, 2009).

This reliance on translational control as opposed to transcription is a universal property of oocytes that may allow for the oocyte genome to remain in a more plastic and undifferentiated state in the context of an otherwise highly differentiated cell (Seydoux & Braun, 2006). In order to accomplish this unique regulatory feat, oocytes need to store and then utilize an abundance of factors (Tadros & Lipshitz, 2009). When the oocyte emerges from the dormancy of the first meiotic arrest, it undergoes a period of intense metabolic activity resulting in the synthesis of large amounts of mRNAs, proteins, and macromolecular structures. As opposed to somatic cells, which divide after doubling their volume, the oocyte grows ~200- to 300-fold without dividing (Piko & Clegg, 1982). Those mRNAs destined to play a role in embryogenesis—"maternal mRNAs"—are stored in a form that blocks them from being translated until needed.

The genes that are transcribed to produce maternal mRNAs that play a role in embryogenesis are often referred to as *maternal effect genes* (MEGs) (Mager, Schultz, Brunk, & Bartolomei, 2006). This locution is potentially misleading to the extent that the ultimate effect of any given mRNA should not simply be equated with, or ascribed to, the gene from which it was transcribed. The centrality of the distinction in embryonic development (and in all physiological processes) between transcription and translation and, as we shall see, posttranslational regulatory processes, should make this clear. That said, MEGs impact embryonic development not through genetic inheritance, but through cytoplasmic inheritance of maternal effect transcripts (mRNAs) (Bell et al., 2008; Minami, Suzuki, & Tsukamoto, 2007).

MEGs were first described in *Drosophila* and included genes, the transcripts of which play a critical role in the determination of body axes (Frohnhofer & Nusslein-Volhard, 1986; Nusslein-Volhard, Lohs-Schardin, Sander, & Cremer, 1980; Schupbach & Wieschaus, 1986), but were not described in mammals until 2000 (Christians, Davis, Thomas, & Benjamin, 2000). By definition, a "canonical" or "bona fide" MEG exhibits the following properties (Yurttas, Morency, & Coonrod, 2010). First, the effect of an MEG results not from zygotic inheritance of the gene from the mother or father, but rather from inheritance of the mRNA associated with that gene via the oocyte; second, MEG-knockout (MEG null) embryos gestated by normal MEG (MEG positive) mothers develop normally with the following critical exception. When MEG-null female offspring of MEG-positive mothers reach reproductive maturity, they are able to ovulate normally and their eggs can be fertilized (at least according to the most stringent definition of a MEG), but the resulting embryo arrests early in development. This is because the embryo lacks the necessary cytoplasmic mRNA/protein associated with its mother's MEG. Hence, the phenotype of a MEG-null daughter of a MEG-positive mother will be sterility, while male offspring are able to reproduce normally (and can be, in effect, carriers of the MEG-null mutation). By contrast, neither male nor female zygotes of MEG-negative mothers will survive regardless of their own genotype (Marlow, 2010).

Several different mammalian MEGs have been identified that exhibit these characteristics. For example, Maternal Antigen That Embryos Require (MATER) (given this name because of its initial use as an oocyte antigen in a mouse model of autoimmune premature ovarian failure) is synthesized from the MATER or Nlrp5 gene (Tong et al., 2000). MATER-null daughters gestated by MATER-positive mothers exhibit normal oogenesis, ovulation, and fertilization. When MATER-null daughters conceive, however, the resulting embryos generally do not progress beyond the two-cell stage. Hence, the protein(s) encoded by MATER are necessary for embryogenesis prior to the activation of the embryonic genome. MATER is one of the group of maternally encoded proteins including LOPED, FILIA, and Tle6 that interact with each other to form a protein complex called the *subcortical maternal complex* (Li, Baibakov, & Dean, 2008). Female mice lacking this protein complex can ovulate and their eggs can be fertilized but their embryos do not progress beyond cleavage stage development.

Besides maternally inherited mRNAs that appear to be required only during early embryogenesis, normal embryonic development requires that some mRNAs be both inherited maternally and transcribed from the embryonic genome. The cytoplasmic mRNA provides for a protein necessary prior to EGA, while the embryonic genome provides transcripts for the same protein—or isoforms of that protein—that are necessary post-EGA. An example of this is the cell adhesion protein E-cadherin, coded by the CDH-1 gene (De Vries et al., 2004). Knockout CDH-1 embryos develop to the blastocyst stage but fail to implant in the uterus. Depletion of cytoplasmic E-cadherin prior to fertilization, however, results in the arrest of normal CDH-1 embryos at the two-cell stage. In other words, cytoplasmic E-cadherin is necessary for embryonic development pre-EGA, while embryonic E-cadherin is necessary for development post-EGA.

The number of mammalian genes that have been classified as maternal effect thus far is significantly less than the number of proteins one would expect to be necessary for pre-EGA embryonic development (Yurttas et al., 2010; Zhang et al., 2009). There are potentially many reasons for this, but I will consider five. First, given how much we do not understand about embryonic development, most MEGs have been identified serendipitously, usually when studying their possible involvement in disease phenotypes with no particular relationship to embryogenesis (Tong et al., 2000). Second, the standard research technique for identifying MEGs is a gene knockout experiment or creation of a null mutant strain in mice. The anticipated phenotype for the embryo of a MEG-null mother is either early embryonic arrest or gross deformity. However, the effects may be much more subtle or only manifested later in life, and potential phenotypes of this sort are rarely investigated in the search for MEG genes (Wilkinson, Davies, & Isles, 2007). Third, that a gene knockout gives rise to no observable phenotype can belie its importance in development due to biological robustness. Functional redundancy is one of the mechanisms responsible for biological robustness and involves either functionally redundant duplicate genes (paralogs) (Gu, Steinmetz, Gu, ScharfeDavis, & Li, 2003) or "degeneracy". Degeneracy refers to a situation in which structurally distinct proteins transcribed from distinct genes bear out similar or partially overlapping functions (Edelman & Gally, 2001). Fourth, as a matter of definition, it is often stipulated that in the "strictest sense," mutations in maternal-effect genes do not affect oocyte development, maturation, ovulation, and fertilization, but solely embryonic development (Ma, Zeng, Schultz, & Tseng, 2006). This distinction is problematic inasmuch as the mRNAs involved in embryonic development are synthesized in the oocyte. Fifth, MEGS are usually limited to genes associated with mRNAs that are transmitted in the ooplasm. But as we shall see, there are any number of MEGs associated with the maternal environment beyond the oocyte.

4. EPIGENETICS IN THE PREIMPLANTATION EMBRYO

Epigenetics is the study of heritable changes in gene transcribability and phenotype that occur without changes in DNA sequence (Bollati & Baccarelli, 2010).

A number of MEGs/mRNAs that have been identified thus far play critical roles in epigenetic processes. This is not at all surprising. During mammalian embryogenesis, the maternally and paternally derived genomes must undergo extensive epigenetic remodeling and alteration of their gene transcription patterns to enable embryogenesis. At fertilization, the genomes of male and female gametes possess their own distinctive epigenetic markings. The ovum is unique among cell types in that it has the ability to transition from a differentiated cell fate to one of totipotency, i.e., the ability of a cell to become any differentiated cell type in the human body. To achieve totipotency, the epigenetic states of the germ cells must be reset. To achieve cellular differentiation and development, the embryonic genome must be made accessible to transcription factors in a highly regulated spatiotemporal manner.

4.1. Histone Modification

Within the chromosome, DNA combines with structural proteins called histones to form chromatin, a highly coiled and compact structure. Within chromatin, a repeating structure, the nucleosome, is composed of 146 base pairs of DNA wrapped around a core of eight histone proteins, that maintain the chromatin's shape and structure (Peterson & Laniel, 2004). Five major families of histone proteins have been identified, and these are divided into two groups based upon their location in nucleosomes: histones H2A, H2B, H3 and H4 are known as core histones; histones H1 and H5 are known as linker histones. Core histones form the center of nucleosomes while linker histones bind the nucleosomes to DNA.

Histones are subject to a wide variety of posttranslational biochemical modifications including, but not limited to, lysine acetylation, lysine and arginine methylation, serine and threonine phosphorylation, lysine acetylation, lysine and arginine methylation, serine and threonine phosphorylation, and lysine ubiquitination and sumoylation (Kouzarides, 2007). These modifications occur primarily, but not exclusively, within the histone amino-terminal "tails," the ends of the amino acid chains that make up the histone protein and protrude from the surface of the nucleosome. For a gene to be transcribed, it must be physically accessible to the transcriptional machinery. Modifications to histones can change the structure of the chromatin causing it to wind more or less tightly, making the DNA more or less accessible to transcription factors.

The most widely studied form of histone modification is histone acetylation that loosens histone binding on the DNA structure allowing for a more open chromatin structure and accessibility to transcription factors and RNA. Conversely, histone deacetylation is associated with a closed chromatin structure and inaccessibility to transcription factors. Histone acetylation and deacetylation are regulated by a balance in the activity of two enzymes, histone acetyltransferase (HAT) and histone deacetylase (HDAC). Histone acetylation facilitated by HAT is associated with an open chromatin structure, while histone deacetylation facilitated by HDAC results in a condensed or closed chromatin structure (Wang, Zang, et al., 2008).

The synthesis of histone proteins occurs throughout PID: H3 and H4 are synthesized during the one-cell stage from maternal mRNAs while the synthesis of H2A, H2B, and H1 commences during the late one-cell to two-cell stage. H3.3 is a variant of H3 that correlates with an active transcriptional state, and is observed at the two-cell, four-cell, and blastocyst stages, coinciding with major waves of gene activation during PID (Velker, Michelle, Denomme, & Mann, 2012). The modification of histone proteins on the chromatin of early embryos is very dynamic (Palini, De Stefani, Scala, Dusi, & Bulletti, 2011). Global patterns of histone modifications and chromatin architecture change during the early stages of development (Meshorer et al., 2006). Shortly after fertilization, sperm chromatin undergoes extensive posttranslational modification. Sperm nucleosomes contain protamines in place of histones; postfertilization, the protamines are actively removed and replaced by cytoplasmic histones (Jammes, Junien, & Chavatte-Palmer, 2011).

Genome-wide chromatin analyses suggest that specific combinations of histone marks at DNA promoters and enhancers correlate with the developmental potential and fate of cells (Rada-Iglesias et al., 2011). For example, embryonic stem (ES) cells have a different histone modification landscape than cells with more restricted fates (Hong et al., 2011). In the blastocyst, the very first differentiated cells (those with a more restricted fate) are those that comprise the trophectoderm, which gives rise to the placenta and extraembryonic structures, while cells of the ICMs are still undifferentiated. In the embryo, OCT4 and NANOG, two genes whose proteins play a critical role in establishing totipotency, are progressively silenced by histone acetylation and histone methylation. The patterns of histone modification differ in the trophectoderm and the ICM corresponding to different timetables for the loss of totipotency and different cellular fates. The importance of these modifications in early embryogenesis is highlighted by the severe abnormal phenotypes caused by experimental mutations in histone-modifying complexes (Jiang et al., 2011; Vastenhouw & Schier, 2012).

4.2. DNA Methylation

DNA methylation, the addition of a methyl group to CpG dinucleotides (sites in the DNA molecule where a cytosine base is followed by a guanine base), acts as a physical barrier to transcription factors and attracts enzymes and proteins that further reduce the transcriptional accessibility of a gene. In general, to establish and maintain methylated DNA, special enzymes called DNA methyltransferases (DNMTs) that facilitate the transfer of a methyl group to DNA are indispensible. Once a sequence of DNA is marked by methylation, another set of proteins is involved in recognizing the methylated cytosines and interacting with chromatin remodelers in order to silence transcription (Bird & Macleod, 2004). Hence, for the most part, hypermethylation is associated with gene silencing and hypomethylation with gene activity (or potential activity).

DNMTs are divided into two groups: de novo methyltransferases recognize something in the DNA that allows them to newly methylate cytosines (this group includes DNMT3a and 3b). These are expressed mainly in early embryonic development and are involved in establishing patterns of methylation involved in cell-type differentiation (Kato et al., 2007). Maintenance methyltransferases add methylation to DNA when one strand is already methylated (this group includes DNMT1). These work throughout the life of the organism to maintain preexisting and de novo methylation patterns during mitotic cellular division, enabling daughter cells of a certain tissue type to inherit the methylation status of their progenitor (Gaudet et al., 2004).

As with histone modification, during preimplantation, DNA methylation levels undergo a series of dynamic changes (Smith et al., 2012). Mature sperm and oocytes are highly methylated until fertilization, indicating little or no transcriptional activity (Smith et al., 2012). In the mouse embryo, shortly after fertilization, the paternal pronucleus is actively demethylated, i.e., enzymes facilitate and accelerate demethylation by a process that is not well understood (Abdalla, Yoshizawa, & Hochi, 2009). The maternal pronucleus, however, appears to be protected from active demethylation by a protein associated with the maternal effect gene STELLA (Minami et al., 2007). Instead, maternal DNA is passively demethylated via a replication dependent loss: epigenetic marks are lost because they are not maintained during several rounds of DNA replication.

Prior to the blastocyst stage, the bulk of embryonic DNA has been demethylated constituting a phase of global hypomethylation corresponding to the establishment of totipotency (Smith et al., 2012). The first cellular differentiation between the trophectoderm and the ICM is accompanied by de novo DNA methylation giving rise to stable silencing of genes involved in the maintenance of pluripotency. The level of methylation in the ICM is higher than that in the trophectoderm, corresponding to different time-tables of cellular differentiation and different cell fates. The differences in methylation between the trophectoderm and the ICM parallel the differences in histone modification, indicating that multiple epigenetic processes work in tandem to regulate gene transcription, pluripotency, and cellular differentiation.

Mouse oocytes and preimplantation embryos lack DNMT1 but express a variant of this protein called DNMT1o, an isoform of DNMT1 that arises from an oocyte-specific alternative splicing of the DNMT1 gene (Mertineit et al., 1998). DNMT1 is classified as an MEG: the mRNA derived from the gene is upregulated in the oocyte, transmitted to the embryo via the oocyte, and plays an important role in embryogenesis (Bressan et al., 2009; Kurihara et al., 2008). DNMT1o-knockout males and females gestated by DNMT1-positive mothers appear grossly normal, but whereas males are fertile, females are infertile, and their embryos die between embryonic day (E)14 and E21 (Howell et al., 2001), making DNMT1 a "bona fide" MEG.

4.3. Imprinting

Imprinting is an epigenetic phenomenon in which specific alleles are expressed in a parent of origin manner. In paternally imprinted genes, the paternal allele is epigenetically modified, preventing its transcription and leading to monoallelic maternal expression; in maternally imprinted genes, the maternal allele is epigenetically modified, preventing its transcription and leading to monoallelic paternal expression. In addition to allele-specific methylation, imprinting is also associated with histone modifications and noncoding RNA (ncRNA), including microRNA (miRNA) (Spahn & Barlow, 2003). Imprinted genes appear to be controlled at differentially methylated regions (DMRs) (Swales & Spears, 2005). A DMR that is differentially methylated in all tissues throughout development is referred to as an imprinting control region (ICR). Such ICRs are hypothesized to be key regulators of imprinting in their particular chromosomal domains (Smallwood & Kelsey, 2012). Another type of DMR exhibits differential patterns of tissue-specific methylation during stages of somatic development.

Approximately 200 imprinted genes have been identified thus far in the mammalian genome, including more than 100 imprinted genes in mice and at least 60 in humans (Shen et al., 2012). A study of the mouse brain suggests that detailed tissue-specific analysis could lead to the discovery of many more imprinted genes. Over ~1300 protein-coding genes and putative ncRNAs have been identified as associated with parent-specific allelic expression in the mouse embryonic and adult brain (Gregg et al., 2010).

Imprinting is a key reason why "monoparental" mouse embryos generated by micromanipulation do not survive. Diploid reconstituted zygotes, containing either two maternal or two paternal pronuclei, display characteristic developmental abnormalities and fail to develop to term (Santos & Dean, 2004). Gynogenetic embryos (diploid maternal) characteristically are growth restricted and fail to derive a functional placenta. In contrast, androgenetic embryos (diploid paternal) while profoundly growth retarded, display a hyperproliferation of extraembryonic tissues.

Imprints in the parental gametes are erased upon every reproductive cycle and reestablished in the immature germ cells of the developing embryo according to their fate as either male or female gametes. Beginning approximately 11.5 days postfertilization (E11.5), the primordial germ cells begin to undergo demethylation to erase the inherited parental imprints (Lucifero, Chaillet, & Trasler, 2004). Primordial germ cell demethylation is complete by ~E13, and subsequent reprogramming of the germ cells occurs when the gender-specific imprinting patterns are once more laid down (Sasaki & Matsui, 2008).

In contrast to the erasure and reestablishment of methylation marks in germ cells, somatic imprinted genes remain methylated throughout embryogenesis. Despite alterations in global methylation levels and chromatin organization during fertilization and early development, patterns of methylation at most somatic imprinted loci in the embryo are specifically conserved during early development without any de novo reprogramming. Cytoplasmic proteins transcribed from maternal genes DNMT1 and STELLA are involved in protecting methylation at imprinted loci during genome-wide preimplantation demethylation, ensuring the correct inheritance of parent-specific epigenetic imprints (Santos & Dean, 2004).

4.4. Noncoding RNAs

Non-coding RNAs (ncRNAs), unlike mRNAs, are not involved in gene transcription. Instead, they function as a vast system of posttranscriptional regulation of DNA expression (Mattick, 2001). Included among ncRNAs are at least 1000 different kinds of micro RNAs (miRNAs)—and the number may be as high as 20,000—short RNA molecules approximately 22 nucleotides long (Osman, 2012). miRNAs are derived from longer primary transcripts (pri-miRNA) by the action of at least two enzymes, Drosha and Dicer. Mature miRNAs associate with a protein complex—the RNA-induced silencing complex—that contains an Argonaute protein at its core, and it is in this context that they carry out their regulatory role. For the most part, RNAs are negative regulators of gene transcription, partially binding to complementary sequences in mRNA resulting in posttranscriptional repression of gene expression.

Maternally inherited miRNAs are abundant in the early embryo (Tang et al., 2007). Dynamic changes in the expression of miRNAs in preimplantation embryos and the increased synthesis of miRNAs after the two-cell stage in mouse embryos suggest that miRNAs have a functional role during this period (Laurent, 2008; Sirard, 2012). Dicer deficiency is lethal during mouse embryogenesis, leading to a lack of detectable stem cells and an acute loss of cell proliferation (Murchison et al., 2007). Similarly, the Argonaute protein Ago2 is required for development through the EGA period. Although the mechanisms of miRNA action are not well defined, it has been hypothesized that one role of miRNA is as a control mechanism in the timely degradation of maternal mRNAs (Schier, 2007). miRNAs may also play roles in cellular differentiation during the blastocyst stage. A cluster of miRNAs from miR-290 to miR-295 were found to be ES cell-specific and may be associated with the maintenance of pluripotency (Shi & Wu, 2009; Yang et al., 2008).

4.5. X-chromosome Inactivation

X-chromosome inactivation (XCI) provides a good example as to how the various epigenetic mechanisms canvassed above can interact. X-chromosome activity changes dynamically in female offspring during PID due to a combination of epigenetic events including DNA methylation, histone modifications, and RNA-mediated silencing. In female embryos with two X chromosomes, one of the two X chromosomes is selected stochastically to be inactivated, a process known as XCI. XCI is triggered by an ncRNA, Xist, which coats the chromosome selected for silencing (Plath, Mlynarczyk-Evans, Nusinow, & Panning, 2002). This is followed by the recruitment of protein complexes involved in multiple epigenetic processes, distinct histone modifications such as H3 K4 demethylation, H3 K9 methylation, H4 deacetylation, and DNA hypermethylation of CpG dinucleotides along X-linked genes (Lee, 2003).

5. CYTOPLASMIC ORGANELLES

5.1. Mitochondria

In addition to providing essential mRNAs, ncRNAs, and proteins, the oocyte also provides the primary source of cellular energy in the form of mitochondria. Mitochondria are intracellular organelles, small membrane-enclosed structures within the cell, in which the end product of the breakdown of glucose in cells is processed to form the primary source of cellular energy, adenosine triphosphate (ATP). While paternal sperm-derived mitochondria are present in the zygote, they are rapidly degraded upon fertilization. Hence, mitochondria are maternally inherited in mammals, and as mitochondria cannot be made de novo, but rather only elaborated from other mitochondria, all of our mitochondria ultimately derive from those in one of our mother's oocytes. In addition to being the primary source of cellular energy, mitochondria also play a central role in a number of critical cellular and metabolic processes (Dumollard, Duchen, & Carroll, 2007) including cellular proliferation; apoptosis or programmed cell death (cellular suicide), a process aimed at destroying a physiologically unwanted cell (Desagher & Martinou, 2000); the regulation and homeostasis of intracellular calcium, which acts as an intracellular signal involved in numerous cellular processes including cellular expression and metabolism; and DNA repair. Mitochondria also play a critical role in oogenesis, fertilization, and early embryogenesis (Dumollard et al., 2007; McBride, Neuspiel, & Wasiak, 2006).

Mitochondria possess their own DNA-mitochondrial DNA (mtDNA)- a circular double-stranded genome. mtDNA exhibits polyploidy, differences in the number of mtDNA copies according to cell type (Clay Montier, Janice, Deng, & Bai, 2009). For example, there are 1075–2794 copies of mtDNA per cell in muscle cells, 1200–10,800 in neurons, and up to 25,000 in liver cells. But by far, the greatest copy number of mitochondria occurs in oocytes, and oocytes from the same female can differ significantly in the number of mtDNA they contain, with human oocytes from the same female containing anywhere from 11,000 to 903,000 mtDNA molecules per oocyte (May-Panloup, Chretien, Malthiery, & Reynier, 2007).

Studies indicate that mitochondrial number, distribution, and structure play essential roles in fertilization and normal embryonic development. mtDNA copy number expands during oocyte maturation (Bentov et al., 2010) and the normality of preovulatory meiotic maturation of the oocyte has been linked to mtDNA copy number. The number of mitochondria and mtDNA in the oocyte at the time of ovulation is critical for both fertilization and ensuing embryo development. Studies indicate that the mean mtDNA copy number in human fertilized oocytes is ~250,000 (Clay Montier et al., 2009), while for unfertilized oocytes, the mean is 164,000, and it has been suggested that a mitochondrial complement of at least ~100,000 copies of mtDNA is required for normal embryonic development (Shoubridge & Wai, 2007).

In all eukaryotic cells (cells containing a membrane enclosed nucleus), energy in the form of ATP is created through two pathways (Pon & Schon, 2007; Scheffler, 2008): In one—anaerobic respiration—glucose is metabolized to pyruvate that can then be converted to lactic acid. The energy created by this mechanism does not involve oxygen (hence the term anaerobic respiration). Anaerobic respiration is inefficient due to the incomplete metabolism of glucose. The other pathway to ATP formation, aerobic respiration, requires oxygen. Aerobic respiration also involves the conversion of glucose to pyruvate, but the pyruvate enters the inner membrane of the mitochondria where it is completely metabolized to carbon dioxide and water. Because the end product of aerobic respiration is the complete breakdown of glucose to water and carbon dioxide, it is ~14 times more efficient than anaerobic respiration.

The role of aerobic respiration in human preimplantation embryo development, although clearly present, is controversial. Mitochondria within human oocytes and preblastocyst-stage embryos appear immature and relatively inactive (Bavister & Squirrell, 2000; Motta et al., 2000). In fact, although oxygen is consumed during preimplantation embryo development (Houghton et al., 1996), estimates of the contribution of mitochondrial respiration to the energetic requirement of mammalian embryo development suggest that as little as 10% of glucose is metabolized through aerobic respiration in the early stages of development, rising to 85% in the blastocyst (Bavister & Squirrell, 2000). However, since aerobic respiration is 14 times more efficient than anaerobic respiration, the 10% of glucose passing through aerobic respiration probably still produces more energy in the form of ATP than the 90% of total glucose metabolized without mitochondria. Overall, the data suggest that both aerobic and anaerobic respiration pathways are active during both oocyte maturation and embryo PID, but aerobic respiration is upregulated during blastocyst development and implantation (Wilding, Coppola, Dale, & Di Matteo, 2009).

Calcium (Ca^{2+}) acts as an ubiquitous intracellular signal that controls various cellular processes including proliferation, transcription, metabolism, and fertilization (Islam, 2012). By taking up and releasing Ca^{2+} and thereby acting as a Ca^{2+} buffer, mitochondria play a key role in its regulation. They can activate or deactivate plasma membrane channels that are regulated by changes in Ca^{2+} concentrations or restrict Ca^{2+} signals to specific cellular domains. Upon fertilization, sperm entry in the cytoplasm triggers repetitive Ca^{2+} waves that traverse the egg. This sperm-triggered Ca^{2+} oscillation is crucial for the initiation of embryonic developmental events including the breakdown of the nuclear membranes of the pronuclei, mitosis, and cytokinesis (the process in which the cytoplasm is divided during cellular division) (May-Panloup et al., 2007; Wilding et al., 2009). During PID, specific and diverse Ca^{2+} signals occur both intra- and intercellularly, and these Ca^{2+} pulses and waves are involved in everything from body axis formation in the blastula to gastrulation to organogenesis (Cao & Chen, 2009; Shoubridge & Wai, 2007).

5.2. Nucleoli

Ribosomes are small organelles involved in translating the nucleotide sequence of mRNA into an amino acid sequence to produce proteins. A cell typically contains anywhere from 1000 to one million ribosomes. A ribosome is composed of ~60% ribosomal RNA (rRNA) and 40% protein. The synthesis of rRNA and ribosome subunit assembly takes place within a structure in the nucleus known as the *nucleolus* (Olson, 2004). Nucleoli are formed around specific genetic loci called nucleolar organizing regions, composed of tandem repeats of rRNA genes found on several different chromosomes (Raska, Shaw, & Cmarko, 2006). Since the nucleolus of spermatozoa is eliminated during spermatogenesis (Schultz & Leblond, 1990), the embryonic nucleolus is inherited in the cytoplasm.

After the second meiotic arrest, oocyte gene transcription of rRNA stops along with all other gene transcription. After fertilization, the maternally derived, transcriptionally inactive nucleolus appears in both male and female pronuclei and in the embryonic nuclei until the four-cell or eight-cell stage; that is, it is present until EGA (Zatsepina, Baly, Chebrout, & Debey, 2003). Although the nucleolus at this stage appears to be inactive and shows a highly compacted structure, studies with mice have revealed that it is essential for early embryonic development (Ogushi et al., 2008; Ogushi & Saitou, 2010). Embryos originating from enucleolated oocytes arrest between the two-cell and four-cell stages (Ogushi et al., 2008). Interestingly, one of the abnormalities noted is related to abnormal chromatin organization in the pronucleus.

6. ASSISTED REPRODUCTIVE TECHNOLOGY

The definition of artificial reproductive technology (ART) varies widely, but the US Center for Disease Control and Prevention (CDC) defines ART as all fertility treatments in which both egg and sperm are handled (Savage, Peek, Hofman, & Cutfield, 2011). Procedures that may be used in the ART process include hormonal stimulation to induce ovulation, in vitro fertilization (IVF), intracytoplasmic sperm injection in which a single sperm is injected directly into the egg in an attempt to achieve fertilization, and the cryogenic freezing of embryos. ART may involve any combination of these, but the most commonly used technique employed in all procedures is IVF, the mixing of eggs with sperm in a specific culture in a Petri dish and the implantation of embryos in the woman's uterus from 1 6 days postfertilization (Glujovsky, Blake, Farquhar, & Bardach, 2012). IVF currently accounts for more than 99% of ART procedures performed worldwide (http://www.ivf-worldwide.com/).

ART has been associated with both pregnancy complications and adverse developmental outcomes. However, studies of adverse outcomes after the use of ART have been questioned, and remain controversial, for at least two reasons. First, adverse outcomes could be due to the underlying causes of subfertility and not ART itself. Second, multiple births have been overrepresented in assisted pregnancies due to the common practice of implanting multiple IVF eggs to increase the likelihood of pregnancy. Since multiple births are themselves associated with an increased risk of negative outcomes, it has been difficult to determine whether pregnancy complications and adverse outcomes are a result of ART or owing to multiple births and their associated complications.

The practice in certain countries of limiting the number of embryos that can be transplanted to just one has resulted in a rise in the number of ART-conceived singletons. Studies of perinatal outcomes of ART singletons appear to show even stronger differences between ART and non-ART singletons compared to ART and non-ART twins (Ceelen, van Weissenbruch, Vermeiden, van Leeuwen, & Delemarre-van de Waal, 2008; Helmerhorst, Perquin, Donker, & Keirse, 2004; McDonald, Murphy, Beyene, & Ohlsson, 2005). ART singletons, when compared with naturally conceived singletons, are at significantly increased risk of placental abnormalities, low birth weight, preterm birth, small for gestational age, perinatal mortality, and congenital malformations (Hansen, Bower, Milne, de Klerk, & Kurinczuk, 2005; Jackson, Gibson, Wu, & Croughan, 2004; McDonald, Han, Mulla, & Beyene, 2010; Rimm et al., 2004). Preterm birth is associated with increased morbidity, mortality, and diminished long-term survival and reproduction (McIntire & Leveno, 2008; Swamy, Ostbye, & Skjaerven, 2008), while low birth weight is associated with chronic diseases expressed later in life such as cardiovascular disease, hypertension, and type 2 diabetes (Barker, 2004).

6.1. IVF and Epigenetics

There is a growing body of evidence that IVF (and/or other ART techniques) can result in epigenetic abnormalities in the preimplantation embryo. Concern for epigenetic effects of IVF has arisen primarily from the observation of an increased incidence of rare genomic imprinting disorders such as Beckwith–Wiedemann syndrome (BWS) and Angelman syndrome (AS) in children born after the use of IVF.

Imprinting disorders can result either from genetic mutations or from nongenetic imprinting defects known as *epimutations* (Moore & Oakey, 2011). The genetic mutations associated with imprinting disorders can consist of (1) large deletions or duplications of chromosomal regions that contain imprinted genes; (2) DNA mutations in genes that are usually imprinted or in their imprinting control centers; or (3) uniparental disomy (UPD), two copies of a chromosome from the same parent. Epimutations involve an epigenetic alteration without any change in the DNA sequence and can arise as a result of errors in imprint establishment, erasure, or maintenance. If primary epimutations occur shortly after fertilization, they can be propagated to multiple tissues.

AS is a debilitating neurodevelopmental disorder that affects approximately one in 15,000 children and is characterized by motor dysfunction, severe mental retardation, speech impairment, frequent seizures, hyperactivity, and a high prevalence of autism (Williams et al., 2006). Recent studies indicate that a failure to inherit a normal maternal copy of the paternally imprinted UBE3A gene accounts for 85–90% of all AS cases. Because of a mutation in the maternal UBE3A allele or paternal UPD, offspring lack an active copy of the maternal UBE3A gene (Greer et al., 2010). Normally, the paternal UBE3A gene is paternally imprinted—hence only the maternal allele is expressed—in specific brain regions including the hippocampus, cerebellum, and regions of the neocortex, but not in nonnervous system tissues (Albrecht et al., 1997). Approximately 3% of patients with AS have an imprinting defect or epimutation as evidenced by the presence of two "normal" copies of the UBE3A gene that are both epigenetically silenced (Greer et al., 2010).

IVF has been associated with AS as a consequence of reports of five IVF-conceived patients with epimutation AS (Cox et al., 2002; Ludwig et al., 2005; Orstavik, Eiklid, van der Hagen, Spetalen, & Kierulf, 2003).

The link between AS and IVF is based on the rarity of AS (1:15,000), the rarity of primary epimutations as a mechanism of AS (~3%) and the relatively infrequent use of IVF as a method of conception (2–3%) (Amor & Halliday, 2008).

BWS is an overgrowth syndrome that affects one in 13,700 children and is characterized by macroglossia, abnormally large abdominal organs, hypoglycemia in infancy, kidney abnormalities, and cancerous and noncancerous tumors (Weksberg, Shuman, & Bruce Beckwith, 2009). The majority of BWS patients have an epimutation affecting the maternal allele of one of the two DMRs at chromosome 11p15. In one region, the maternal H19 imprinted domain acquires a paternal epigenotype. The resulting gain of H19 imprinting center methylation results in silencing of H19 expression and activation of insulin-like growth factor 2 (IGF2) receptor gene expression. The second imprinting defect occurs at the KCNQ10T1 domain and results from loss of methylation on the maternal KCNQ10T1 imprinting center with ensuing biallelic expression of KCNQ10T1 and silencing of KCNQ1 and CDKN1C. The remaining BWS patients have paternal UPD of chromosome 11p or a DNA mutation of the CDKN1C gene. Current estimates are that IVF children are approximately 514 times more likely to develop BWS than non-IVF children although the total numbers remain small due to the rarity of the disorder (Amor & Halliday, 2008; Hiura et al., 2012).

Numerous animal studies have indicated an association between IVF and imprinting abnormalities. For example, it has been known for a number of years that cows and sheep produced through IVF display an increased frequency of large offspring syndrome (LOS), characterized by numerous abnormalities including immunological defects, increased fetal/neonatal death, increased birth weight, organomegaly, and skeletal and placental defects (Behboodi et al., 1995; Young, Sinclair, & Wilmut, 1998). The phenotypes observed in LOS are similar to those observed in BWS and significantly, epigenetic abnormalities of the same loci involved in BWS are observed in calves and sheep with LOS: KCNQ10T1 is hypomethylated with a corresponding increase in KCNQ10T1 expression and decrease in CDKN1C expression (Hori et al., 2010); and LOS sheep also exhibit loss of imprinting for the IGF2 gene (Young et al., 2001). The similarities in IVF-induced epigenetic errors between humans and animal models, where subfertility is not a confounding issue, suggest that manipulation of the early embryo can lead to epigenetic perturbations with potential long-term consequences for offspring (Paolini-Giacobino, 2007; Velker et al., 2012). There is an additional concern that IVF could result in subtle abnormalities that present later in life. For example, in mice, several studies (Calle et al., 2012; Ecker et al., 2004; Fernandez-Gonzalez et al., 2004; Watkins et al., 2007) report long-term consequences of IVF including increased incidence of obesity, elevated systolic blood pressure and heart disease, anxiety, and memory deficits.

6.2. The Oviductal Environment

Although the Fallopian tubes have long been considered a mere conduit for gametes and embryos, numerous studies have demonstrated that the oviduct is involved in a number of important processes that contribute to an optimal environment for fertilization and early embryonic development (Avilés, Gutiérrez-Adán, & Coy, 2010; Lee, Cheong, Chow, Lee, & Yeung, 2009; Tse et al., 2008). The oviductal secretion is a complex fluid formed by secreted components from epithelial cells and blood plasma, and includes growth factors, cytokines (small protein messengers involved in the immune system) and cytokine receptors, hormones and hormone receptors, proteases (enzymes involved in the breakdown of proteins) and protease inhibitors, antioxidants (substances that protect cells against the effects of free radicals produced by ATP production in mitochondria), and chaperones and heat shock proteins that are expressed in response to rises in temperature and other environmental stressors. Preimplantation embryos interact with oviductal epithelial cells to regulate the production of oviductal proteins (Tauber, Wettich, Nohlen, & Zaneveld, 1985), including the production of specific embryotrophic factors that stimulate embryonic growth (Tse et al., 2008).

As noted earlier, there are two main pathways in ATP generation that are necessary for embryonic cellular metabolism: aerobic and anaerobic respiration. Anaerobic respiration predominates in early PID, where pyruvate and lactate are the embryo's main sources of energy, and glucose uptake is minimal. The capacity to metabolize glucose increases significantly during the transition from the morula to blastocyst stage and by the blastocyst stage, glucose has become the preferred nutrient. Significantly, the nutrients available within the human female reproductive tract mirror the changing nutrient preferences of the developing embryo (Lane & Gardner, 2007). The oviductal fluid is characterized by relatively high concentrations of pyruvate and lactate and a relatively low concentration of glucose. In contrast, uterine fluid is characterized by relatively low levels of pyruvate and lactate and a higher concentration of glucose.

Available commercial IVF culture media attempt to reproduce the preimplantation environment, at least to the extent of providing the embryo with nutrients and essential marcomolecules known to be present in the oviduct (Xella et al., 2010). The majority of commercial culture systems are sequential, i.e., different cultures are used at different stages of preblastocyst development in an attempt to mimic the dynamic in vivo environment of early embryo oviductal development (Nelissen et al., 2012). There is a general consensus, however, that all culture systems are "suboptimal" in the sense that the optimal environment for early embryonic development is the oviduct and there is no way to reproduce that environment in a Petri dish. For example, Market-Velker, Fernandes, and Mann (2010) compared five human commercial media systems in a mouse model. IVF mouse embryos produced in all five culture systems displayed a varying, but compromised ability to maintain genomic imprinting in comparison with in vivo-derived mouse embryos.

There is much more to the oviductal environment than oviductal fluid. Physiologically, the preimplantation embryo develops in a hypoxic (low oxygen) environment (oxygen concentration: 5–7%), whereas in vitro embryos are cultured with normal atmospheric oxygen tension (oxygen concentration: 20%) (Chason, Csokmay, Segars, DeCherney, & Randall Armant, 2011). Embryo culture at atmospheric oxygen tension has been associated with increased production of reactive oxidative species (ROS), which are byproducts of aerobic respiration that can damage cell function by modifying the structure of lipids, proteins, and DNA causing strand breaks and inactivation of enzymes (Guerin, El Mouatassim, & Menezo, 2001; Kitagawa, Suzuki, Yoneda, & Watanabe, 2004).

At the same time, ROS serve as key signaling molecules by acting as second messengers through the regulation of key transcription factors, and their disruption can have detrimental developmental consequences (Dennery, 2007). Studies evaluating embryonic development under physiological oxygen concentrations have noted an increase in blastocyst development and embryo cell number across multiple species, although the mechanism is unclear (Bedaiwy et al., 2010; Chason et al., 2011; Kitagawa et al., 2004; Thompson, Simpson, Pugh, Donnelly, & Tervit, 1990; Wale & Gardner, 2010). Furthermore, physiological oxygen tension appears to preserve pluripotency in cultured human ES cell lines, while loss of pluripotency and spontaneous cellular differentiation is more frequent in embryonic cells cultured under atmospheric conditions (Forristal, Wright, Hanley, Oreffo, & Houghton, 2010).

7. CONCLUSIONS

In conclusion, I would like to emphasize four points. First, biological inheritance is not simply a matter of the inheritance of genes. PID demonstrates, in a vivid way, that the genome does not control development in the manner traditionally conceived. This is because a critical period of embryonic development, the pre-EGA, occurs when the embryo effectively lacks a functioning genome. Cytoplasmic inheritance directs the earliest developmental processes and some of that inheritance (e.g., mitochondria and mtDNA) persists throughout the life course. Furthermore, what occurs in the fertilized ovum, although unique in many ways, highlights the importance of processes that occur in every cell. Gene transcription depends upon some combination of enzymatic and epigenetic processes. It is, therefore, necessarily both an epigenetic and epistatic process. Furthermore, transcription is a step in an extraordinarily complex, multilayered, interactive regulatory system that leads from the assembly of pre-mRNA to the assembly of a protein.

Second, PID demonstrates the centrality of epigenetic processes in human development. Embryogenesis is characterized by extensive epigenetic modifications of the oocyte, the pronuclei, and the preimplantation embryo. These epigenetic modifications enable among many other things (known and unknown), the restructuring of paternal and maternal DNA to form the pronuclei, the establishment of pluripotency through the erasure of preexisting epigenetic marks, the maintenance and reestablishment (in the germ cells) of imprinting, XCI, and cellular differentiation.

Third, the oviduct is not a mere conduit for the transmission of the embryo to the uterus. It is an environment designed to promote the early stages of embryonic development. It is also an environment with which the embryo continually communicates during the interactive process of development. Alterations in this environment can affect PID with potential long-term health consequences. The oviduct is, of course, not an environment distinct from the environment of the mother's body, nor is the mother's body an environment distinct from all the environments she occupies. There is growing evidence that all these environments can impact

PID with potentially lifelong consequences (Ashworth, Toma, & Hunter, 2009; Igosheva et al., 2010; Junien, 2006; Kwong, Wild, Roberts, Willis, & Fleming, 2000).

Finally, it is worth noting that ART can be seen as a testament to the ability of the preimplantation embryo to adapt to the changes in its environment, i.e., its phenotypic plasticity. Fertilization to the preimplantation embryo represents a developmental window (one of many), but despite a common assumption that developmental windows entail "fragility," the embryo appears, for the most part, to be able to accommodate itself to a very alien environment. Epigenetic alterations seen in ART embryos may represent adaptive epigenetic responses. Phenotypic plasticity is often discussed in the context of the maternal environment: changes in that environment can transmit information to offspring about the environment they will inhabit. If the cues from the maternal environment (pre and postnatal) are good predictors of the environment in which offspring will find themselves, then the offspring's phenotypic adjustments are adaptive (Qvarnstrom & Price, 2001). What is in some way unique about IVF is that to the extent that the preimplantation IVF embryo receives cues from the environment, it is not a maternal environment but rather an artificially constructed one. Nonetheless, its responses may be such as to enable it to survive, in relative health, until implantation. Yet just as with maternal effects, a high degree of phenotypic plasticity may also imply that sometimes the cues from the environment are maladaptive.

REFERENCES

Abdalla, H., Yoshizawa, Y., & Hochi, S. (2009). Active demethylation of paternal genome in mammalian zygotes. *Journal of Reproduction and Development, 55*(4), 356–360.

Albrecht, U., Sutcliffe, J. S., Cattanach, B. M., Beechey, C. V., Armstrong, D., Eichele, G., et al. (1997). Imprinted expression of the murine Angelman syndrome gene, Ube3a, in hippocampal and Purkinje neurons. *Nature Genetics, 17*(1), 75–78.

Amor, D. J., & Halliday, J. (2008). A review of known imprinting syndromes and their association with assisted reproduction technologies. *Human Reproduction, 23*(12), 2826–2834.

Amundson, R. (2005). *The changing role of the embryo in evolutionary thought: Roots of evo-devo.* Cambridge; New York: Cambridge University Press.

Ashworth, C. J., Toma, L. M., & Hunter, M. G. (2009). Nutritional effects on oocyte and embryo development in mammals: implications for reproductive efficiency and environmental sustainability. *Philosophical Transactions of the Royal Society of London B Biological Sciences, 364*(1534), 3351–3361.

Avilés, M., Gutiérrez-Adán, A., & Coy, P. (2010). Oviductal secretions: will they be key factors for the future ARTs? *Molecular Human Reproduction, 16*(12), 896–906.

Barker, D. J. (2004). The developmental origins of adult disease. *Journal of the American College of Nutrition, 23*(Suppl. 6). 588S–95S.

Bavister, B. D., & Jayne, M. S. (2000). Mitochondrial distribution and function in oocytes and early embryos. *Human Reproduction, 15*(Suppl. 2), S189–S198.

Bedaiwy, M. A., Mahfouz, R. Z., Goldberg, J. M., Sharma, R., Falcone, T., Abdel Hafez, M. F., et al. (2010). Relationship of reactive oxygen species levels in day 3 culture media to the outcome of in vitro fertilization/intracytoplasmic sperm injection cycles. *Fertility and Sterility, 94*(6), 2037–2042.

Behboodi, E., Anderson, G. B., BonDurant, R. H., Cargill, S. L., Kreuscher, B. R., Medrano, J. F., et al. (1995). Birth of large calves that developed from in vitro-derived bovine embryos. *Theriogenology, 44*(2), 227–232.

Bell, C. E., Calder, M. D., & Watson, A. J. (2008). Genomic RNA profiling and the programme controlling preimplantation mammalian development. *Molecular Human Reproduction, 14*(12), 691–701.

Bentov, Y., Esfandiari, N., Burstein, E., & Casper, R. F. (2010). The use of mitochondrial nutrients to improve the outcome of infertility treatment in older patients. *Fertility and Sterility, 93*(1), 272–275.

Bird, A., & Macleod, D. (2004). Reading the DNA methylation signal. *Cold Spring Harbor Symposia on Quantitative Biology, 69*, 113–118.

Bollati, V., & Baccarelli, A. (2010). Environmental epigenetics. *Heredity, 105*(1), 105–112.

Bressan, F. F., De Bem, T. H.C., Perecin, F., Lopes, F. L., Ambrosio, C. E., Meirelles, F. V., et al. (2009). Unearthing the roles of imprinted genes in the placenta. *Placenta, 30*(10), 823–834.

Calle, A., Fernandez-Gonzalez, R., Ramos-Ibeas, P., Laguna-Barraza, R., Perez-Cerezales, S., Bermejo-Alvarez, P., et al. (2012). Long-term and transgenerational effects of in vitro culture on mouse embryos. *Theriogenology, 77*(4), 785–793.

Cao, X., & Chen, Y. (2009). Mitochondria and calcium signaling in embryonic development. *Seminars in Cell and Developmental Biology, 20*(3), 337–345.

Ceelen, M., van Weissenbruch, M. M., Vermeiden, J. P., van Leeuwen, F. E., & Delemarre-van de Waal, H. A. (2008). Growth and development of children born after in vitro fertilization. *Fertility and Sterility, 90*(5), 1662–1673.

Chason, R. J., Csokmay, J., Segars, J. H., DeCherney, A. H., & Randall Armant, D. (2011). Environmental and epigenetic effects upon preimplantation embryo metabolism and development. *Trends in Endocrinology and Metabolism, 22*(10), 412–420.

Chen, M., & Manley, J. L. (2009). Mechanisms of alternative splicing regulation: insights from molecular and genomics approaches. *Nature Reviews Molecular Cell Biology, 10*(11), 741–754.

Christians, E., Davis, A. A., Thomas, S. D., & Benjamin, I. J. (2000). Maternal effect of Hsf1 on reproductive success. *Nature, 407*(6805), 693–694.

Clay Montier, L. L., Deng, Janice J., & Bai, Y. (2009). Number matters: control of mammalian mitochondrial DNA copy number. *Journal of Genetics and Genomics, 36*(3), 125–131.

Cox, G. F., Burger, J., Lip, V., Mau, U. A., Sperling, K., Wu, B. L., et al. (2002). Intracytoplasmic sperm injection may increase the risk of imprinting defects. *American Journal of Human Genetics, 71*(1), 162–164.

De Vries, W. N., Evsikov, A. V., Haac, B. E., Fancher, K. S., Holbrook, A. E., Kemler, R., et al. (2004). Maternal beta-catenin and E-cadherin in mouse development. *Development, 131*(18), 4435–4445.

Dennery, P. A. (2007). Effects of oxidative stress on embryonic development. *Birth Defects Research Part C: Embryo Today: Reviews, 81*(3), 155–162.

Desagher, S., & Jean-Claude, M. (2000). Mitochondria as the central control point of apoptosis. *Trends in Cell Biology, 10*(9), 369–377.

Dumollard, R., Duchen, M., & Carroll, J. (2007). The role of mitochondrial function in the oocyte and embryo. *Current Topics in Developmental Biology, 77*, 21–49.

Ecker, D. J., Stein, P., Xu, Z., Williams, C. J., Kopf, G. S., Bilker, W. B., et al. (2004). Long-term effects of culture of preimplantation mouse embryos on behavior. *Proceedings of the National Academy of Sciences of the United States of America, 101*(6), 1595–1600.

Edelman, G. M., & Gally, J. A. (2001). Degeneracy and complexity in biological systems. *Proceedings of the National Academy of Sciences*, *98*(24), 13763–13768.

Evsikov, A.V., Graber, J. H., Michael Brockman, J., Hampl, A., Holbrook, A. E., Singh, P., et al. (2006). Cracking the egg: molecular dynamics and evolutionary aspects of the transition from the fully grown oocyte to embryo. *Genes and Development*, *20*(19), 2713–2727.

Fernandez-Gonzalez, R., Moreira, P., Bilbao, A., Jimenez, A., Perez-Crespo, M., Ramirez, M. A., et al. (2004). Long-term effect of in vitro culture of mouse embryos with serum on mRNA expression of imprinting genes, development, and behavior. *Proceedings of the National Academy of Sciences of the United States of America*, *101*(16), 5880–5885.

Forristal, C. E., Wright, K. L., Hanley, N. A., Oreffo, R. O., & Houghton, F. D. (2010). Hypoxia inducible factors regulate pluripotency and proliferation in human embryonic stem cells cultured at reduced oxygen tensions. *Reproduction*, *139*(1), 85–97.

Frohnhofer, H. G., & Nusslein-Volhard, C. (1986). Organization of anterior pattern in the Drosophila embryo by the maternal gene bicoid. *Nature*, *324*(6093), 120–125.

Gaudet, F., M Rideout, W., Meissner, A., Dausman, J., Leonhardt, H., & Jaenisch, R. (2004). Dnmt1 expression in pre- and postimplantation embryogenesis and the maintenance of IAP silencing. *Molecular and Cellular Biology*, *24*, 1640.

Gilbert, S. F. (1998). Bearing crosses: a historiography of genetics and embryology. *American Journal of Medical Genetics*, *76*(2), 168–182.

Glujovsky, D., Blake, D., Farquhar, C., & Bardach, A. (2012). Cleavage stage versus blastocyst stage embryo transfer in assisted reproductive technology. *Cochrane Database of Systematic Reviews*, *11*(7).

Greer, P. L., Hanayama, R., Bloodgood, B. L., Mardinly, A. R., Lipton, D. M., Flavell, S. W., et al. (2010). The Angelman syndrome protein Ube3A regulates synapse development by ubiquitinating arc. *Cell*, *140*(5), 704–716.

Gregg, C., Zhang, J., Weissbourd, B., Luo, S., Schroth, G. P., Haig, D., et al. (2010). High-resolution analysis of parent-of-origin allelic expression in the mouse brain. *Science*, *329*(5992), 643–648.

Gu, Z., Steinmetz, L. M., Gu, X., Scharfe, C., Davis, R. W., & Li, W.-H. (2003). Role of duplicate genes in genetic robustness against null mutations. *Nature*, *421*(6918), 63–66.

Guerin, P., El Mouatassim, S., & Menezo, Y. (2001). Oxidative stress and protection against reactive oxygen species in the pre-implantation embryo and its surroundings. *Human Reproduction Update*, *7*(2), 175–189.

Hansen, M., Bower, C., Milne, E., de Klerk, N., & Kurinczuk, J. J. (2005). Assisted reproductive technologies and the risk of birth defects—a systematic review. *Human Reproduction*, *20*(2), 328–338.

Helmerhorst, F. M., Perquin, D. A., Donker, D., & Keirse, M. J. (2004). Perinatal outcome of singletons and twins after assisted conception: a systematic review of controlled studies. *BMJ*, *328*(7434), 23.

Hiura, H., Okae, H., Miyauchi, N., Sato, F., Sato, A., Van De Pette, M., et al. (2012). Characterization of DNA methylation errors in patients with imprinting disorders conceived by assisted reproduction technologies. *Human Reproduction*, *27*(8), 2541–2548.

Hong, S.-H., Rampalli, S., Lee, J. B., McNicol, J., Collins, T., Draper, J. S., et al. (2011). Cell fate potential of human pluripotent stem cells is encoded by histone modifications. *Cell Stem Cell*, *9*(1), 24–36.

Hori, N., Nagai, M., Hirayama, M., Hirai, T., Matsuda, K., Hayashi, M., et al. (2010). Aberrant CpG methylation of the imprinting control region KvDMR1 detected in assisted reproductive technology-produced calves and pathogenesis of large offspring syndrome. *Animal Reproduction Science*, *122*(3–4), 303–312.

Houghton, F. D., Thompson, J. G., Kennedy, C. J., & Leese, H. J. (1996). Oxygen consumption and energy metabolism of the early mouse embryo. *Molecular Reproduction and Development*, *44*(4), 476–485.

Howell, C.Y., Bestor, T. H., Ding, F., Latham, K. E., Mertineit, C., Trasler, J. M., et al. (2001). Genomic imprinting disrupted by a maternal effect mutation in the Dnmt1 gene. *Cell, 104*(6), 829–838. (cited) http://www.ivf-worldwide.com/.

Igosheva, N., Abramov, A. Y., Poston, L., Eckert, J. J., Fleming, T. P., Duchen, M. R., et al. (2010). Maternal diet-induced obesity alters mitochondrial activity and redox status in mouse oocytes and zygotes. *PLoS One, 5*(4).

Islam, S. M. (2012). *Calcium signalling.* (Vol. 740). Dordrecht: Springer.

Jackson, R. A., Gibson, K. A., Wu, Y. W., & Croughan, M. S. (2004). Perinatal outcomes in singletons following in vitro fertilization: a meta-analysis. *Obstetrics and Gynecology, 103*(3), 551–563.

Jammes, H., Junien, C., & Chavatte-Palmer, P. (2011). Epigenetic control of development and expression of quantitative traits. *Reproduction Fertility and Development, 23*(1), 64–74.

Jiang, H., Shukla, A., Wang, X., Chen, W.-yi, Bernstein, B. E., & Roeder, R. G. (2011). Role for Dpy-30 in ES cell-fate specification by regulation of H3K4 methylation within bivalent domains. *Cell, 144*(4), 513–525.

Junien, C. (2006). Impact of diets and nutrients/drugs on early epigenetic programming. *Journal of Inherited Metabolic Disease, 29*(2), 359–365.

Kato, Y., Kaneda, M., Hata, K., Kumaki, K., Hisano, M., & Kohara, Y. (2007). Role of the Dnmt3 family in de novo methylation of imprinted and repetitive sequences during male germ cell development in the mouse. *Human Molecular Genetics, 16*, 2272–2280.

Kitagawa, Y., Suzuki, K., Yoneda, A., & Watanabe, T. (2004). Effects of oxygen concentration and antioxidants on the in vitro developmental ability, production of reactive oxygen species (ROS), and DNA fragmentation in porcine embryos. *Theriogenology, 62*(7), 1186–1197.

Kouzarides, T. (2007). Chromatin modifications and their function. *Cell, 128*(4), 693–705.

Kurihara, Y., Kawamura, Y., Uchijima, Y., Amamo, T., Kobayashi, H., Asano, T., et al. (2008). Maintenance of genomic methylation patterns during preimplantation development requires the somatic form of DNA methyltransferase 1. *Developmental Biology, 313*(1), 335–346.

Kwong, W. Y., Wild, A. E., Roberts, P., Willis, A. C., & Fleming, T. P. (2000). Maternal undernutrition during the preimplantation period of rat development causes blastocyst abnormalities and programming of postnatal hypertension. *Development, 127*(19), 4195–4202.

Lane, M., & Gardner, D. K. (2007). Embryo culture medium: which is the best? *Best Practice & Research Clinical Obstetrics and Gynaecology, 21*(1), 83–100.

Laurent, L. C. (2008). MicroRNAs in embryonic stem cells and early embryonic development. *Journal of Cellular and Molecular Medicine, 12*(6A), 2181–2188.

Lee, J. T. (2003). Functional intergenic transcription: a case study of the X-inactivation centre. *Philosophical Transactions of the Royal Society of London B Biological Sciences, 358*(1436), 1417–1423.

Lee, Y. L., Cheong, A. W., Chow, W. N., Lee, K. F., & Yeung, W. S. (2009). Regulation of complement-3 protein expression in human and mouse oviducts. *Molecular Reproduction and Development, 76*(3), 301–308.

Li, L., Baibakov, B., & Dean, J. (2008). A subcortical maternal complex essential for preimplantation mouse embryogenesis. *Developmental Cell, 15*(3), 416–425.

Lucifero, D., Chaillet, J. R., & Trasler, J. M. (2004). Potential significance of genomic imprinting defects for reproduction and assisted reproductive technology. *Human Reproduction Update, 10*(1), 3–18.

Ludwig, M., Katalinic, A., Groß, S., Sutcliffe, A., Varon, R., & Horsthemke, B. (2005). Increased prevalence of imprinting defects in patients with Angelman syndrome born to subfertile couples. *Journal of Medical Genetics, 42*(4), 289–291.

Mager, J., Schultz, R. M., Brunk, B. P., & Bartolomei, M. S. (2006). Identification of candidate maternal-effect genes through comparison of multiple microarray data sets. *Mammalian Genome, 17*(9), 941–949.

Market-Velker, B. A., Fernandes, A. D., & Mann, M. R. (2010). Side-by-side comparison of five commercial media systems in a mouse model: suboptimal in vitro culture interferes with imprint maintenance. *Biology of Reproduction, 83*(6), 938–950.

Marlow, F. L. (2010). In D. Kessler (Ed.), *Maternal control of development in vertebrates: My mother made me do it.* : Morgan & Claypool Life Science.

Mattick, J. S. (2001). Non-coding RNAs: the architects of eukaryotic complexity. *EMBO Reports, 2*(11), 986–991.

May-Panloup, P., Chretien, M.-F., Malthiery, Y., & Reynier, P. (2007). Mitochondrial DNA in the Oocyte and the Developing Embryo. In C. S. J. Justin (Ed.), *Current topics in developmental biology*: Academic Press.

Ma, J., Zeng, F., Schultz, R. M., & Tseng, H. (2006). Basonuclin: a novel mammalian maternal-effect gene. *Development, 133*(10), 2053–2062.

McBride, H. M., Neuspiel, M., & Wasiak, S. (2006). Mitochondria: more than just a powerhouse. *Current Biology, 16*(14). R551–R60.

McDonald, S. D., Han, Z., Mulla, S., & Beyene, J. (2010). Overweight and obesity in mothers and risk of preterm birth and low birth weight infants: systematic review and meta-analyses. *BMJ, 20*(341).

McDonald, S. D., Murphy, K., Beyene, J., & Ohlsson, A. (2005). "Perinatal outcomes of singleton pregnancies achieved by in vitro fertilization: a systematic review and meta-analysis". *Journal of Obstetrics and Gynaecology Canada, 27*(5), 449–459.

McIntire, D. D., & Leveno, K. J. (2008). Neonatal mortality and morbidity rates in late preterm births compared with births at term. *Obstetrics and Gynecology, 111*(1), 35–41.

Mertineit, C., Yoder, J. A., Taketo, T., Laird, D. W., Trasler, J. M., & Bestor, T. H. (1998). Sex-specific exons control DNA methyltransferase in mammalian germ cells. *Development, 125*(5), 889–897.

Meshorer, E., Yellajoshula, D., George, E., Scambler, P. J., Brown, D. T., & Misteli, T. (2006). Hyperdynamic plasticity of chromatin proteins in pluripotent embryonic stem cells. *Developmental Cell, 10*(1), 105–116.

Minami, N., Suzuki, T., & Tsukamoto, S. (2007). Zygotic gene activation and maternal factors in mammals. *Journal of Reproduction and Development, 53*(4), 707–715.

Moore, G., & Oakey, R. (2011). The role of imprinted genes in humans. *Genome Biology, 12*(3), 106.

Motta, P. M., Nottola, S. A., Makabe, S., & Heyn, R. (2000). Mitochondrial morphology in human fetal and adult female germ cells. *Human Reproduction, 15*(Suppl. 2), S129–S147.

Murchison, E. P., Stein, P., Xuan, Z., Pan, H., Zhang, M. Q., Schultz, R. M., et al. (2007). Critical roles for Dicer in the female germline. *Genes and Development, 21*(6), 682–693.

Nelissen, E. C., Van Montfoort, A. P., Coonen, E., Derhaag, J. G., Geraedts, J. P., Smits, L. J., et al. (2012). Further evidence that culture media affect perinatal outcome: findings after transfer of fresh and cryopreserved embryos. *Human Reproduction, 27*(7), 1966–1976.

Nusslein-Volhard, C., Lohs-Schardin, M., Sander, K., & Cremer, C. (1980). A dorso-ventral shift of embryonic primordia in a new maternal-effect mutant of Drosophila. *Nature, 283*(5746), 474–476.

Ogushi, S., & Saitou, M. (2010). The nucleolus in the mouse oocyte is required for the early step of both female and male pronucleus organization. *Journal of Reproduction and Development, 56*(5), 495–501.

Ogushi, S., Palmieri, C., Fulka, H., Saitou, M., Miyano, T., & Fulka, J., Jr. (2008). The maternal nucleolus is essential for early embryonic development in mammals. *Science, 319*(5863), 613–616.

Olson, M. O. J. (2004). *The nucleolus.* New York: Landes Bioscience/Kluwer Academic.

Orstavik, K. H., Eiklid, K., van der Hagen, C. B., Spetalen, S., & Kierulf, K. (2003). Another case of imprinting defect in a girl with Angelman syndrome who was conceived by intracytoplasmic semen injection. *American Journal of Human Genetics, 72*, 218.

Osman, A. (2012). MicroRNAs in health and disease–basic science and clinical applications. *Clinica y Laboratoria, 58*(5–6), 393–402.

Palini, S., De Stefani, S., Scala, V., Dusi, L., & Bulletti, C. (2011). Epigenetic regulatory mechanisms during preimplantation embryo development. *Annals of the New York Academy of Sciences, 1221*(1), 54–60.

Paolini-Giacobino, A. (2007). Epigenetics in reproductive medicine. *Pediatric Research, 61* (5, Part 2)10.1203/pdr.0b013e318039d978. 51R–7R.

Peterson, C. L., & Laniel, M. A. (2004). Histones and histone modifications. *Current Biology, 14*. R546–R51.

Piko, L., & Clegg, K. B. (1982). Quantitative changes in total RNA, total poly(A), and ribosomes in early mouse embryos. *Developmental Biology, 89*(2), 362–378.

Plath, K., Mlynarczyk-Evans, S., Nusinow, D. A., & Panning, B. (2002). Xist RNA and the mechanism of X chromosome inactivation. *Annual Review of Genetics, 36*, 233–278.

Pon, L. A., & Schon, E. A. (2007). *Mitochondria.* : Amsterdam Academic Press.

Qvarnstrom, A., & Price, T. D. (2001). Maternal effects, paternal effects and sexual selection. *Trends in Ecology and Evolution, 16*(2), 95–100.

Rada-Iglesias, A., Bajpai, R., Swigut, T., Brugmann, S. A., Flynn, R. A., & Wysocka, J. (2011). A unique chromatin signature uncovers early developmental enhancers in humans. *Nature, 470*(7333), 279–283.

Raska, I., Shaw, P. J., & Cmarko, D. (2006). New insights into nucleolar architecture and activity. *International Review of Cytology, 255*, 177–235.

Revil, T., Gaffney, D., Dias, C., Majewski, J., & Jerome-Majewska, L. A. (2010). Alternative splicing is frequent during early embryonic development in mouse. *BMC Genomics, 11*(1), 1–17.

Rimm, A. A., Katayama, A. C., Diaz, M., & Katayama, K. P. (2004). A meta-analysis of controlled studies comparing major malformation rates in IVF and ICSI infants with naturally conceived children. *Journal of Assisted Reproduction and Genetics, 21*(12), 437–443.

Santos, F., & Dean, W. (2004). Epigenetic reprogramming during early development in mammals. *Reproduction, 127*(6), 643–651.

Sasaki, H., & Matsui, Y. (2008). Epigenetic events in mammalian germ-cell development: reprogramming and beyond. *Nature Reviews Genetics, 9*(2), 129–140.

Savage, T., Peek, J., Hofman, P. L., & Cutfield, W. S. (2011). Childhood outcomes of assisted reproductive technology. *Human Reproduction, 26*(9), 2392–2400.

Seydoux, G., & Braun, R. E. (2006). Pathway to totipotency: lessons from germ cells. *Cell, 127*(5), 891–904.

Scheffler, I. E. (2008). *Mitochondria.* Hoboken, N.J: Wiley-Liss.

Schier, A. F. (2007). The maternal–zygotic transition: death and birth of RNAs. *Science, 316*(5823), 406–407.

Schultz, M. C., & Leblond, C. P. (1990). Nucleolar structure and synthetic activity during meiotic prophase and spermiogenesis in the rat. *American Journal of Anatomy, 189*(1), 1–10.

Schupbach, T., & Wieschaus, E. (1986). Germline autonomy of maternal-effect mutations altering the embryonic body pattern of Drosophila. *Developmental Biology, 113*(2), 443–448.

Shen, C.-J., Cheng, W. T. K., Wu, S.-C., Chen, H.-L., Tsai, T.-C., Yang, S.-H., et al. (2012). Differential differences in methylation status of putative imprinted genes among cloned swine genomes. *PLoS One, 7*(2), e32812.

Shi, L., & Wu, J. (2009). Epigenetic regulation in mammalian preimplantation embryo development. *Reproductive Biology and Endocrinology, 7*(1), 59.

Shoubridge, E. A., & Wai, T. (2007). Mitochondrial DNA and the mammalian oocyte. *Current Topics in Developmental Biology, 77*, 87–111.

Sirard, M. A. (2012). Factors affecting oocyte and embryo transcriptomes. *Reproduction in Domestic Animals, 4*, 148–155.

Smallwood, S. A., & Kelsey, G. (2012). De novo DNA methylation: a germ cell perspective. *Trends in Genetics*, *28*(1), 33–42.

Smith, Z. D., Chan, M. M., Mikkelsen, T. S., Gu, H., Gnirke, A., Regev, A., et al. (2012). A unique regulatory phase of DNA methylation in the early mammalian embryo. *Nature*, *484*(7394), 339–344.

Spahn, L., & Barlow, D. P. (2003). An ICE pattern crystallizes. *Nature Genetics*, *35*(1), 11–12.

Swales, A. K. E., & Spears, N. (2005). Genomic imprinting and reproduction. *Reproduction*, *130*(4), 389–399.

Swamy, G. K., Ostbye, T., & Skjaerven, R. (2008). Association of preterm birth with long-term survival, reproduction, and next-generation preterm birth. *JAMA*, *299*(12), 1429–1436.

Tadros, W., & Lipshitz, H. D. (2009). The maternal-to-zygotic transition: a play in two acts. *Development*, *136*(18), 3033–3042.

Tang, F., Kaneda, M., O'Carroll, D., Hajkova, P., Barton, S. C., Sun, Y. A., et al. (2007). Maternal microRNAs are essential for mouse zygotic development. *Genes and Development*, *21*(6), 644–648.

Tauber, P. F., Wettich, W., Nohlen, M., & Zaneveld, L. J. (1985). Diffusable proteins of the mucosa of the human cervix, uterus, and fallopian tubes: distribution and variations during the menstrual cycle. *American Journal of Obstetrics and Gynecology*, *151*(8), 1115 1125.

Thompson, J. G., Simpson, A. C., Pugh, P. A., Donnelly, P. E., & Tervit, H. R. (1990). Effect of oxygen concentration on in-vitro development of preimplantation sheep and cattle embryos. *Journal of Reproduction and Fertility*, *89*(2), 573–578.

Tong, Z. B., Gold, L., Pfeifer, K. E., Dorward, H., Lee, E., Bondy, C. A., et al. (2000). Mater, a maternal effect gene required for early embryonic development in mice. *Nature Genetics*, *26*(3), 267–268.

Tsafriri, A., & Pomerantz, S. H. (1986). Oocyte maturation inhibitor. *Clinics in Endocrinology and Metabolism*, *15*(1), 157–170.

Tse, P. K., Lee, Y. L., Chow, W. N., Luk, J. M., Lee, K. F., & Yeung, W. S. (2008). Preimplantation embryos cooperate with oviductal cells to produce embryotrophic inactivated complement-3b. *Endocrinology*, *149*(3), 1268–1276.

Vastenhouw, N. L., & Schier, A. F. (2012). Bivalent histone modifications in early embryogenesis. *Current Opinion in Cell Biology*, *24*(3), 374–386.

Velker, B. A., Denomme, M. M., & Mann, M. R. (2012). Embryo culture and epigenetics. In G. D. Smith, J. Swain & T. B. Pool (Eds.), *Embryo culture: Methods and protocols*.

Wale, P. L., & Gardner, D. K. (2010). Time-lapse analysis of mouse embryo development in oxygen gradients. *Reproductive BioMedicine Online*, *21*(3), 402–410.

Wang, Q. T., Piotrowska, K., Ciemerych, M. A., Milenkovic, L., Scott, M. P., Davis, R. W., et al. (2004). A genome-wide study of gene activity reveals developmental signaling pathways in the preimplantation mouse embryo. *Developmental Cell*, *6*(1), 133–144.

Wang, E. T., Sandberg, R., Luo, S., Khrebtukova, I., Zhang, L., Mayr, C., et al. (2008). Alternative isoform regulation in human tissue transcriptomes. *Nature*, *456*(7221), 470–476.

Wang, Z., Zang, C., Rosenfeld, J., Schones, D., Barski, A., Cuddapah, S., et al. (2008). Combinatorial patterns of histone acetylations and methylations in the human genome. *Nature Genetics*, *40*, 897–903.

Watkins, A. J., Platt, D., Papenbrock, T., Wilkins, A., Eckert, J. J., Kwong, W. Y., et al. (2007). Mouse embryo culture induces changes in postnatal phenotype including raised systolic blood pressure. *Proceedings of the National Academy of Sciences of the United States of America*, *104*(13), 5449–5454.

Weksberg, R., Shuman, C., & Bruce Beckwith, J. (2009). Beckwith–Wiedemann syndrome. *European Journal of Human Genetics*.

Wilding, M., Coppola, G., Dale, B., & Di Matteo, L. (2009). Mitochondria and human pre-implantation embryo development. *Reproduction*, *137*(4), 619–624.

Wilkinson, L. S., Davies, W., & Isles, A. R. (2007). Genomic imprinting effects on brain development and function. *Nature Reviews Neuroscience, 8*(11), 832–843.

Williams, C. A., Beaudet, A. L., Clayton-Smith, J., Knoll, J. H., Kyllerman, M., Laan, L. A., et al. (2006). Angelman syndrome 2005: updated consensus for diagnostic criteria. *American Journal of Medical Genetics Part A, 140*(5), 413–418.

Wong, C. C., Loewke, K. E., Bossert, N. L., Behr, B., De Jonge, C. J., Baer, T. M., et al. (2010). Non-invasive imaging of human embryos before embryonic genome activation predicts development to the blastocyst stage. *Nature Biotechnology, 28*(10), 1115–1121.

Xella, S., Marsella, T., Tagliasacchi, D., Giulini, S., La Marca, A., Tirelli, A., et al. (2010). Embryo quality and implantation rate in two different culture media: ISM1 versus Universal IVF Medium. *Fertility and Sterility, 93*(6), 1859–1863.

Yang, Y., Bai, W., Zhang, L., Yin, G., Wang, X., Wang, J., et al. (2008). Determination of microRNAs in mouse preimplantation embryos by microarray. *Developmental Dynamics, 237*(9), 2315–2327.

Young, L. E., Fernandes, K., McEvoy, T. G., Butterwith, S. C., Gutierrez, C. G., Carolan, C., et al. (2001). Epigenetic change in IGF2R is associated with fetal overgrowth after sheep embryo culture. *Nature Genetics, 27*(2), 153–154.

Young, L. E., Sinclair, K. D., & Wilmut, I. (1998). Large offspring syndrome in cattle and sheep. *Reviews of Reproduction, 3*(3), 155–163.

Yurttas, P., Morency, E., & Coonrod, S. A. (2010). Use of proteomics to identify highly abundant maternal factors that drive the egg-to-embryo transition. *Reproduction, 139*(5), 809–823.

Zatsepina, O., Baly, C., Chebrout, M., & Debey, P. (2003). The step-wise assembly of a functional nucleolus in preimplantation mouse embryos involves the cajal (coiled) body. *Developmental Biology, 253*(1), 66–83.

Zhang, P., Ni, X., Guo, Y., Guo, X., Wang, Y., Zhou, Z., et al. (2009). Proteomic-based identification of maternal proteins in mature mouse oocytes. *BMC Genomics, 10*(1), 348.

CHAPTER NINE

Evolutionary Psychology: A House Built on Sand

Peter T. Saunders*,†

*Department of Mathematics, King's College London, London, UK
†Institute of Science in Society, London, UK
E-mail: peter.saunders@kcl.ac.uk

Contents

Abstract

While Darwinism has contributed much to our understanding of the living world, it has not given us an adequate account of why organisms are the way they are and how they came to be that way. For that we will need all of science, not just a single algorithm. The crucial contribution of Darwinism to biology is that it explains how we can have functional physical traits without a creator. This is less important in psychology because no one is surprised when people behave in ways that work to their advantage. Evolutionary psychology nevertheless follows the Darwinian model. It assumes from the outset that the brain is largely modular and that human nature is made up of a very large number of functionally specialized psychological mechanisms that have been constructed over time by natural selection. How much confidence one should have in its conclusions depends very much on how far one accepts its premises.

1. INTRODUCTION

Darwinian ideas have been applied in psychology for a long time, but it is only comparatively recently that their supporters have claimed to be revolutionizing the subject. No longer must the social sciences live in the shadow of physics and chemistry. They too can be hard sciences, with the theory of natural selection playing the role of Newton's laws or Einstein's equations. As the philosopher Daniel Dennett (1995) writes, "Darwin's dangerous idea is reductionism incarnate, promising to unite and explain just about everything in one magnificent vision". He is not, it should be added, being ironic.

Darwin's theory of natural selection was certainly one of the most influential steps forward in science. The idea of evolution had been discussed since the early eighteenth century, but it was only with the publication of the *Origin of Species* that it gained widespread acceptance. Even then, it was only when Darwinism was combined with Mendelian genetics to form the so-called synthetic theory, or neo-Darwinism, that it became a major research area.

However, while Darwinism and neo-Darwinism have done a great deal to further our understanding of the living world, they have not provided us with an adequate explanation of how it came to be. Even many neo-Darwinists acknowledge that their theory has made little progress in explaining what they call macroevolution, "the evolution of higher taxa and the production of evolutionary novelties such as new structures" (Mayr, 1991). Yet this is precisely what most people would expect a theory of evolution is for. It is interesting to learn why there have been changes in the colors and patterns of the wings of many species of butterfly (e.g., Joron et al., 2011) but what we really want to know is how there came to be butterflies in the first place.

Besides, even if neo-Darwinism were far more successful than it has been in explaining the world of biology, this would not necessarily make it a suitable foundation for psychology. The importance of natural selection in biology is that it allows us to understand how organisms can be organized and functional without a creator. Because Darwin was able to explain how this could have happened and so refute the "argument from design" he succeeded in convincing his contemporaries about evolution.

In contrast, when we see someone behaving in a way that is to their advantage, while that too might be an adaptation due to natural selection (although this is much easier said by us than done by nature) there is always an obvious competing hypothesis. It might be nothing more than the result of intelligence, experience and culture.

What is more, while it is generally possible to agree that an organism has such and such a heritable physiological trait (but see Gould, 1977; on the human chin), it is far less obvious that a psychological trait is inherited or indeed that it is trait at all. We may argue about how the medusoid got its tentacles but we can all agree that medusoids have tentacles and that this is an inherited trait. It is not at all obvious that the tendency of women to prefer older and richer men or men to prefer younger women is innate or even the sort of thing that could possibly be innate.

The aim of evolutionary biology is to understand why we are the way we are and how we came to be this way. Evolutionary psychology is also about telling us what we are and in particular how much of our behavior is innate, and therefore largely immutable, and how much is not. This is why evolutionary psychology and its predecessor sociobiology can be so controversial. A claim to have shown why something is the case is also a claim that it actually is the case. We can hardly accept the Darwinian explanation of why women tend to prefer older men without also agreeing that this preference is innate and consequently something that we have to take into account in how we organize society.

2. EVOLUTIONARY PSYCHOLOGY

Most of this chapter is not directly about evolutionary psychology. The primary aim is to discuss the shortcomings of the neo-Darwinism on which it is based and whose purported success is claimed to give it strength and credibility. All the same, it will be useful to begin with a description of the theory and an example of how it is applied. Because evolutionary psychologists, like other neo-Darwinists, often object that their critics have set up a straw man, we follow here the suggestion of Tooby, Cosmides, and Barrett (2003) and draw both description and example from one of the founding texts of the subject, *The Adapted Mind* (Barkow, Cosmides and Tooby, 1992).

In the introduction to the volume (Cosmides, Tooby, & Barkow, 1992), we are told that evolutionary psychology "is simply psychology that is informed by the additional knowledge that evolutionary biology has to offer". This sounds modest enough, but the editors immediately add (p. 3): "It unites modern evolutionary biology with the cognitive revolution in a way that has the potential to draw together all of the disparate branches of psychology into a single organized system of knowledge."

They then write: (p. 5)

"The central premise of The Adapted Mind *is that there is a universal human nature, but that this universality exists primarily at the level of evolved psychological mechanisms, not of expressed cultural behaviours. On this view, cultural variability is not a challenge to claims of universality, but rather data that can give one insight into the structure of the psychological*

mechanisms that helped generate it. A second premise is that these evolved psychological mechanisms are adaptations, constructed by natural selection over evolutionary time. A third assumption made by most of the contributors is that the evolved structure of the human mind is adapted to the way of life of our Pleistocene hunter-gatherers, and not necessarily to our modern circumstances."

The assumption that the mind contains many domain-specific specialized mechanisms rather than generalized "learning" or "capacity-for-culture" mechanisms (Symons, 1992) is crucial to the model. Without it, selection could not operate on individual behaviors (or mechanisms, which generally seems to amount to much the same thing) and the project would collapse. Those psychologists and others who do not see the mind as largely modular should be aware that this is an essential part of evolutionary psychology and should be cautious about accepting conclusions that depend on an assumption they do not share. Note also that evolutionary psychologists, like other neo-Darwinists, see all traits as adaptations and natural selection as their creator (cf. Gould & Lewontin, 1979).

For an example of how evolutionary psychologists apply their theory, we turn again to *The Adapted Mind*. According to Buss (1992), empirical studies show that in all cultures, females consider status and resources more important in a potential mate than do males, while males consider youth and physical attractiveness more important than do females. The Darwinian explanation is, naturally, that these preferences are the products of selection. Both increase fitness: women seek males who will be able to support them and their children, while men seek females who are fertile and whose babies are likely to thrive.

This sounds simple and straightforward enough, but only because so much has been left out. Above all, we are being asked to imagine that there is a gene or genes with remarkably direct and specific effects on behavior. Yet there is no argument to support the claim that such genes exist or to suggest how they might act. Still, let us suppose for the sake of argument that they do. That still is not enough, because, as Buss acknowledges, mate selection is often not a matter only for the couple concerned. In many societies, marriages are arranged, and there is even a common theme in folklore of parents trying, and usually failing, to prevent their daughter from marrying a handsome but penniless young man. He proposes, therefore, that if parents turn out to be important, then "the evolutionary account would focus on mechanisms that have evolved in parents to socialize their children in predictable ways".

Mechanisms that cause parents merely to "socialize" their children would not be enough. They would actually have to specify the mating preferences that parents should inculcate into their children. Otherwise the preferences would be cultural, whereas the claim is that they are the result of adaptation and therefore genetic. For a trait to be "genetic" it is not enough that genes are involved somewhere along the chain of causation. In many countries, men button their shirts with the left side over the right and women with the right side over the left. There is a sense, therefore, in which we can say that how a person buttons a shirt ultimately depends on whether they have a Y or an X chromosome, but that does not make it a genetically determined trait and no one would set out to find the shirt-buttoning gene.

It is surely more plausible simply to suppose that because women know that they are likely to bear and nurse children and so require support at some points in their lives, they will seek to ensure that they will have access to the necessary resources when they need them, and societies will develop customs that promote this. There is no need to postulate hypothetical genes with remarkable properties to explain something we would expect to happen anyway, given ordinary human intelligence.

In 1900, about 45% of American men married women 5 or more years younger than themselves and only 3% of American women married men 5 or more years younger than themselves. In 2000, the figures were 27% and 8%, respectively (Coles & Francesconi, 2007). As more women than men now graduate from university and go into well-paid jobs, this converging trend seems likely to continue. We would certainly be ill advised to plan on the assumption that it will not, whatever evolutionary psychology may tell us.

Evolutionary psychologists hold that their theory is an advance on sociobiology because it is concerned with the evolution of mechanisms, not behaviors. This does indeed make it sound more scientific. If, however, nothing is said about the mechanisms apart from the statement that they exist, and if the arguments are about the selective advantages of the behaviors, rather than of the mechanisms that underlie both them and presumably other behaviors, then we are really no further ahead.

3. NEO-DARWINISM

We now return to the main topic of this chapter, the neo-Darwinist theory of evolution. Again, we begin by defining the theory we are going to discuss, although this turns out to be not as easy as you might expect. The problem is that there is no generally accepted or canonical definition and hardly any neo-Darwinist, even those who are philosophers rather than biologists, seems willing to offer one.

One of the very few who ever stated clearly what he understood the theory to be was Maynard Smith (1969). He wrote that Darwinism explains evolution in terms of three properties: heredity, multiplication and variation. Offspring usually resemble their parents, but sometimes they do not. In general, organisms produce more offspring than are required to replace the parents. This leads to competition for resources, and if some of the variations increase the fitness, i.e., if they make some organisms more likely than others to survive and leave offspring, then the proportion of individuals with those variations will increase from generation to generation. Eventually they will replace the original forms. In this way, natural selection can bring about adaptive evolution.

This is Darwinism; neo-Darwinism is obtained by adding the theory of Mendelian genetics, according to which characteristics are passed from one generation to another through discrete entities called genes. Variations are caused by mutations, i.e., changes in the genes. Because it arose through the bringing together of two earlier theories, neo-Darwinism is also known as the synthetic theory of evolution.

As Maynard Smith realized, however, the above is not a theory. It is a set of properties that neo-Darwinism assumes organisms possess, together with an inference that can be drawn from them. He therefore continued, "The theory of neo-Darwinism states that these properties are necessary and sufficient to account for the evolution of life on this planet to date."

The last bit is crucial, even though it is often glossed over. Neo-Darwinism is neither the theory that the natural selection of random genetic mutations can lead to adaptive evolution nor is it the theory that this process really does occur in nature. Hardly anyone could disagree with either of these statements, but neither of them is a theory of evolution. Neo-Darwinism is the theory that the design and history of the whole of the living world can be explained by the natural selection of random genetic mutations. This is a very bold claim indeed, which is presumably why most neo-Darwinists hesitate to make it explicitly. In fact, many will deny it is what they believe. It is, however, the theory they actually use. When you read their work, you find that the explanations they offer are indeed in terms of genes and selective advantages.

A century ago, Romanes (1900, p. 21) wrote the following about A.R. Wallace, the codiscoverer with Darwin of the theory of evolution by natural selection:

"Mr Wallace does not expressly maintain the abstract impossibility of laws and causes other than those of utility and natural selection… Nevertheless, as he nowhere recognises any other law or cause…, he practically concludes that, on inductive or empirical grounds, there is no such other law or cause to be entertained."

Whatever neo-Darwinists may say when they are defending their theory, in practice it is as described by Maynard Smith, or, more dramatically by Dennett (1995, p. 59):

"Here then, is Darwin's dangerous idea: the algorithmic level is the level that best accounts for the speed of the antelope, the wing of the eagle, the shape of the orchid, the diversity of species and all the other occasions for wonder in the world of nature." [emphasis in the original]

4. NEO-DARWINISM, NATURAL SELECTION AND EVOLUTION

Neo-Darwinists tend to use the terms "neo-Darwinism", "evolution", and "natural selection" as though they were synonyms. In fact, they are not, and using them interchangeably leads to confusion.

In the first place, the expression "theory of evolution" can mean two quite different, although related things. On the one hand, it can be the theory that evolution has taken place, that the organisms we see today are the products of a process of descent with modification from earlier forms and ultimately from inorganic matter. This is often referred to as "the fact of evolution". Alternatively, it can mean one particular theory about how this process occurred, viz. neo-Darwinism.

Neo-Darwinists generally conflate the two meanings. This allows them to ignore criticisms of their theory on the grounds that anyone who doubts their account of evolution can be dismissed as some sort of creationist. At the same time, it gives the creationists a much easier target than they deserve because weaknesses in the neo-Darwinist account of the evolutionary process can be used to cast doubt on the fact of evolution.

As for natural selection, while it is obviously important in evolution, it needs raw material in the form of variation. Nature can hardly select variations that have not appeared. Surely, therefore, the study of evolution cannot be only the study of selection; it must include the origin and nature of the variations from which the selections are made. Physics, chemistry, physiology and other subjects as well must be important in the study of evolution. All of science, not just one simple algorithm, must be brought to bear on the problem.

There is, however, a way in which selection could conceivably be the whole story. Suppose the variations were random, not just in the sense that they do not occur preferentially in the directions in which they are needed, but in the stronger sense that there is almost nothing we can say about them, except possibly that they must be small. Then if we assumed that just about any small change was possible and, what is more, that it was not significantly more or less likely to occur than any other, it would follow that the origin of the variations had nothing to contribute to our understanding of the process. Selection would then be the only factor with a significant effect.

But even if they are due to random mutations in genes, the variations will not be strongly random. They occur in accordance with the laws of physics and chemistry, and in higher organisms, they follow the patterns of developmental biology. Some variations are far more readily produced than others, and while small ones are far more common, large coordinated changes are by no means ruled out. It is not easy to determine which variations are the most likely to occur—it is certainly a lot harder than making up *Just So* stories about selective advantages—but as we shall see, the problem is by no means intractable.

Neo-Darwinists (e.g., Mayr, 1982, 1984; Pinker, 2012; but see Saunders and Ho, 1984) confuse the two meanings of random. Those who define it at all claim that they mean it in the weak sense but their theory depends on the variations being strongly random.

Darwinists sometimes ask us to imagine a large room full of monkeys who are sitting at typewriters and hitting keys at random. A human is watching them, and when, purely through chance, a monkey produces *Hamlet* or the *Origin of Species*, he sends it off to be published. In that scenario, the human can be said to be the real author of the work, and selection was the creative process. The monkeys really make no significant contribution because their typing was random in the strong sense.

A better metaphor would be a room full of humans. They produce real words, real sentences, and real chapters. All the same, very little of their output is worth publishing. There still has to be an editor to select the small proportion that can be used and to suggest improvements and correct mistakes. The editor is important, but he is not the creator of the work.

5. THE GENE

While the gene is at the center of the neo-Darwinist story of evolution, there are two major problems with the concept that are seldom if ever addressed. In the first place, it is very difficult to pin down exactly what a gene is. And given that the role of a gene, whatever it is, is to code for the production of a protein, and that as even neo-Darwinists now acknowledge, genes generally act in complex webs of interactions rather than individually (e.g., Tooby et al., 2003), how can we justify the claim that the connection between gene and trait is sufficiently direct to allow selection for behavior to act on the gene in the way the theory demands?

By about 1961, a century or so after Mendel's experiments in the Brno monastery, scientists believed they finally understood what genes are and how they work. Mendel's hypothetical units had become real physical entities, segments of molecules of DNA. These directed the production of proteins through the now famous triplet code and so determined all the many traits that constitute an organism, whether bacterium or human being.

Unfortunately, the research did not stop there. No sooner had molecular biologists created this simple and elegant picture of how heredity works than they set about demolishing it. The more they discovered about the genome, the less tenable the simplistic concept of the gene became. The idea of a gene as a piece of DNA, acting on its own to produce a gene product and hence a phenotypic trait, is hopelessly out of date.

The genome and the processes that go on within it are so complicated that molecular biologists now define a gene in terms of what it does rather than by what it is, rather as Mendel did. In particular, as a well known (and comparatively old) textbook reminds us, it is neither to be thought of as a single contiguous stretch of DNA nor even just as several stretches on the same chromosome (Darnell, Lodish, & Baltimore, 1986). The genome too is not as it is was imagined but fluid, subject to changes in the course of normal development and in response to feedback from the environment (see Ho, 2003). Another field that depends on the obsolete view of the gene is genetic engineering, which we are assured has the potential to produce all sorts of desirable traits by transferring single stretches of DNA (see Ho, 1998).

There is no question that changes in the genome can cause changes in behavior. Everyone knows that introducing a new chemical into the body or changing the concentration of one that is already there can affect behavior in ways that are both significant and largely predictable. But these generally involve more than one behavioral trait (or mechanism, if you prefer). We are all familiar with the characteristic effects of alcohol or some hormones, and in neither case are they restricted to one "behavior". Alcohol does not specifically reduce our ability to drive or cause us only to sing rude songs; neither would a gene whose product had the same effect on the brain as alcohol.

Can we really speak of "a gene for" the sorts of traits that evolutionary psychologists insist their theory can explain? You can get an idea of how difficult it is from an attempt made by Dawkins (1982, p. 23) (and quoted enthusiastically by Dennett, 1995, p. 116). "Reading," Dawkins says, "is a learned skill of prodigious complexity, but this provides no reason in itself for skepticism about the possible existence of a gene for reading. All we would need in order to establish the existence of a gene for reading is to discover a gene for not reading, say a gene which induced a brain lesion causing specific dyslexia." He argues that such a gene is not implausible and that it is reasonable to call it a gene for not reading because that might well be its only noticeable effect.

He then continues, "… it follows from the ordinary conventions of genetic terminology that the wild-type gene at the same locus, the gene that the rest of the population has in double dose, would properly be called a gene 'for reading'. If you object to that, you must also object to our speaking of a gene for tallness in Mendel's peas, because the logic of the terminology is identical in the two cases."

The logic is indeed the same, but what concerns us here is not whether there is a way in which the internal logic of theoretical genetics permits us to ascribe a meaning to the expression "a gene for reading". It is whether the concept has anything to do with evolution, and, in particular, whether it can help us to understand how we have come to have the ability to learn to read.

Clearly it does not. Indeed, in a similar discussion of eggshell removal by black-headed gulls (p. 24), Dawkins acknowledges, "It most definitely does not follow that this particular locus 'for' eggshell removal was one of the ones on which natural selection worked during the evolution of the adaptation. On the contrary, it seems much more probable that a complex behavioral pattern like eggshell removal must have been built up by selection on a large number of loci, each having a small effect in interaction with the others." And a bit later on he adds, with commendable candor (p. 26), "It is too bad if geneticists usually are forced to concentrate on loci that are convenient rather than evolutionarily important." In other words, most of the evidence from genetics that is supposed to underpin the neo-Darwinist theory of evolution is actually irrelevant to it.

From time to time we read of the discovery of a gene for some behavioral trait or other, language for instance. It is worth bearing in mind that, as in these examples borrowed from Dawkins, the function was usually inferred from observations of individuals who *lacked* the gene in question and that this makes it a gene *for* the trait only in a very limited sense, although of course hardly anyone acknowledges this in their report. For more on genes and dyslexia (and cognition in general) see Fisher (2006).

Every biology student is taught the simple model of Mendelian genetics in which a trait is determined by a single gene with two alleles. If no genotype gives a selective advantage to individuals that possess it, the proportions of the two alleles in the population will reach equilibrium. If, however, there is a selective advantage associated with one allele, its proportion will increase from generation to generation and ultimately the other will disappear altogether.

This provides the basis for evolution by natural selection. Imagine a population in which at some locus (i.e., for one gene) there is only one allele. Everyone has the same version of the trait that the gene determines. From time to time random mutations produce different alleles. Most will make no difference at all, others will be deleterious. Once in a while, purely by chance, a new allele will appear that gives some advantage to those individuals that possess it. Even if the advantage is very small, the mutant allele will eventually replace the original one and the entire population will have it.

This, we are told, is how evolution occurs. Mutations occur at random and those alleles that give a small advantage replace the earlier forms of the gene. This happens over and over again, and slowly a trait such as an eye or a preference for older men or younger women is built.

A neo-Darwinist reading this account might object that it is a caricature. No one, he would insist, believes in "one gene/one character" any more. There are a few single gene traits, usually involving some heritable disease or a feature such as blood groups and eye color where it is easy to see how one protein can make all the difference. Most traits, however, are determined through interactions among many genes.

What neo-Darwinists do not acknowledge is how little we know about how selection operates when genes interact in complex ways with many genes influencing a trait and the same gene influencing a number of traits. Most of the arguments that are used depend on either a single gene or, at the most, a few genes that interact in a very simple way or have to be considered together because they are linked, i.e., close together on a chromosome and therefore tending to be passed on together to the next generation.

It may well be the case that no neo-Darwinist really believes in one gene/one character. But it is what they assume when they are constructing their stories about natural selection. The arguments they use depend on it whether they realize it or not. And as Bernard Shaw wrote in *Man and Superman*, "What a man believes may be ascertained, not from his creed, but from the assumptions on which he habitually acts."

The story that it is on account of natural selection that women tend to prefer older men sounds simple and straightforward, until you remember how much has been left out. We are being asked to believe either that a single mutation could lead to the production of a protein that somehow acts in the brain to make women be attracted to older men or, even less plausibly, that there was a long sequence of mutations each of which made women just a bit more likely to be attracted to older men. What is more, the mutations would have to act in very specific ways. It would not be enough for them to make women more intelligent. That might indeed make them more capable of realizing that marrying older men was in their best interests, but it would not fit the neo-Darwinist story.

6. FALSIFIABILITY

A number of years ago, there was a debate about whether neo-Darwinism is a scientific theory, by which was meant whether it satisfies Popper's criterion of demarcation. Popper (1934) had argued that for a theory to be considered scientific it must make at least one prediction that could be falsified and, if it were, would cause us to abandon the theory. General Relativity passed the test because it would have been rejected if the famous expedition of 1919 had not found that light rays were bent as they passed near the sun. Marxism failed because while it made falsifiable predictions, when these turned out to be false, Marxists merely added ad hoc hypotheses to make the facts fit into the framework of their theory.

For a number of reasons, Popper's criterion is not now generally accepted as drawing a boundary between science and nonscience, but it still provides a useful distinction between different theories: those that can be tested, even if not conclusively, and those that cannot. The more a theory is open to testing, the more confidence we can have in its results.

Darwinism falls into the latter class. It does lead to predictions that are falsifiable and indeed are sometimes falsified. When that happens, however, no one argues that the theory should be abandoned. Instead, what is abandoned is a particular hypothesis about adaptation, and another one is soon put forward in its place (cf. Gould & Lewontin, 1979).

For example, it used to be argued that natural selection has made women inclined to be faithful because they can only be impregnated once every year or so, and so gain no advantage from having more than one mate. Promiscuous men, on the other hand, can impregnate many women and so pass on more of their genes to the next generation (Bateman, 1948).

When Darwinists eventually noticed that women too are often unfaithful, they did not abandon their theory. Instead, the story became that promiscuity in women allows them to pass on more of their genes if they are impregnated by alpha males and then find inferior males to help them care for the offspring. Alternatively, both sexes may be either faithful or not, using more complex strategies, where "strategy" means a genetically based program (Gangestad & Simpson, 2000). How genes are supposed to achieve this is not explained.

If we are too easily satisfied with adaptive accounts, we will fail to look for deeper explanations of a trait, or indeed to question whether the trait actually exists at all. For example, considerable effort was expended for a while in analyzing weaning as a competition between the infant, who wants to continue being fed to increase its chances of survival, and the mother, who wants to maximize her lifetime reproductive success by having more offspring. Eventually Gomendio, Cassinello, Smith, and Bateson (1995) demonstrated that it was not a matter of Darwinian parent–offspring conflict (Trivers, 1974) at all: most young mammals soon lose the ability to digest milk and so have nothing to gain from continuing to suckle. Weaning turns out to be more cooperative than competitive.

7. THE ALTERNATIVE

The waves of the sea, the little ripples on the shore, the sweeping curve of the sandy bay between the headlands, the outline of the hills, the shape of the clouds, all these are so many riddles of form, so many problems of morphology, and all of them the physicist can more or less easily read and adequately solve: solving them by reference to their antecedent phenomena,

*in the material system of mechanical forces to which they belong, and
to which we interpret them as being due. They have also, doubtless, their
immanent teleological significance; but it is on another plane of thought from
the physicist's that we contemplate their intrinsic harmony and perfection
and 'see that they are good'.*

*Nor is it otherwise with the material forms of living things. Cell and tissue,
shell and bone, leaf and flower, are so many portions of matter, and it is
in obedience to the laws of physics that their particles have been moved,
moulded and conformed.*

D'Arcy Thompson (1917, p. 10)

Neo-Darwinists are of course aware that there is a long and complex process of development between egg or seed and organism. But they confidently ignore it when they are studying evolution. For them, developmental biology is the study of construction, not of architecture and design. Physics and chemistry are relevant to how an individual organism comes into being but not to why it has the features that it does. That is to be explained by natural selection. What is more, there is no plausible scientific alternative in sight. Anyone who doubts the neo-Darwinist account of evolution must be some sort of creationist.

There is indeed a scientific alternative, but it is not just another theory that, like neo-Darwinism, purports to give a complete explanation of evolution. On reflection, of course, we should ask why anyone would ever have expected that there could ever be such a thing in the first place. Can there really be a single theory, capable of being stated in a sentence or two, that can tell us about the whole of the living world and how it came to be, that can "explain just about everything in one magnificent vision"? That sounds more like a religion than science.

The alternative is really more an approach than a theory. It amounts to nothing more than looking very carefully at whatever it is whose evolution we are trying to elucidate and using all of science to assist us in our task. We will learn about organisms and their evolution by studying biology, about behavior and its evolution by studying psychology and anthropology, and so on. (see e.g., Ho, 2011; Ho and Saunders, 1984)

To see the difference, consider how we might set out to explain the complicated shape of the medusoid *Cordylophora*, shown on the left of Fig. 9.1. A Darwinist would naturally begin with the a priori assumption that the medusoid had evolved from a simpler ancestor through a sequence of intermediates, each slightly more ramified and slightly fitter than the one

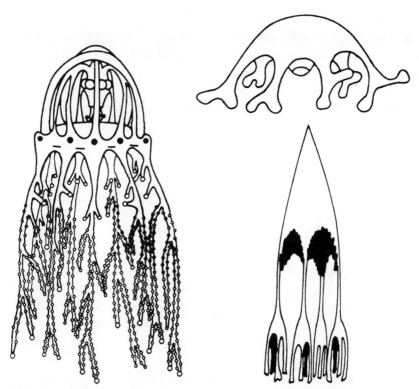

Figure 9.1 *Cordylophora, Caldomena* and a drop of fusel oil in paraffin. *After D'Arcy Thompson (1917).*

before. The problem would then be to work out a hypothetical sequence, at which point the problem would be considered solved.

The shape on the lower right, however, is not a medusoid; in fact it is not an organism at all. It is a drop of fusel oil that has been allowed to fall into paraffin. Now a drop of fusel oil does not have ancestors, it does not have a genome, and it is not subject to natural selection. It has that shape because the undirected physical forces between two fluids of appropriate densities, viscosities and surface tensions, and with the right relative velocity, can produce it. This can happen as a drop of one liquid falls into another, but it can also happen as a small, round, juvenile organism grows in the sea. So this particular form can appear relatively easily in nature, although you might not have thought so just from looking at it. And that is largely why we observe it.

The medusoid's peculiar shape is useful to it, and no doubt if it were a severe handicap, it would have been eliminated by natural selection. But this explains why the shape has *persisted*. It tells us neither how it arose in the first place nor why the medusoid has this particular form instead of some other. Selection was indeed involved, but studying selection will not tell us why medusoids look like they do. Unless, that is, you prefer to believe that natural selection has acted on a long sequence of random variations only to produce one of the relatively small number of forms that are also observed in inert drops of fluid.

Even if we are convinced that we have identified the function that some trait serves and how it contributes to the organism's fitness, it does not mean we have explained how the organism came to have that trait (cf. Gould & Lewontin, 1979). Within neo-Darwinism, traits (including psychological mechanisms, Buss, Haselton, Shackleford, Bleske, & Wakefield, 1998) whose evolution cannot be explained by their present contribution to fitness are called exaptations, the term reflecting the assumption that the explanation must lie in their contribution to fitness, if not in the present then surely in the past. That fitness may have had very little to do with it is not considered.

A medusoid is a comparatively simple organism, and what happens in more complicated ones is harder to determine; but then real science tends to be hard. On the other hand, in more complicated organisms, there is another factor to be taken into account, the nature of the developmental process. An organism is not assembled from prefabricated parts like an automobile on an assembly line. Instead, an embryo develops as a whole, which is why organisms fit together so well. If an engineer decides to make a piston slightly longer and heavier, he has to modify many other components to match it. If a mutation were to alter the shape of a bone, the developmental process can ensure that the other bones and tissues that are connected to it adjust automatically. The trabeculae would also naturally align themselves to cope with the changed stresses. There would be no need to wait for separate corresponding mutations for each of these.

This adaptability makes large changes possible. For example, many babies whose mothers took thalidomide during pregnancy were born with seriously deformed limbs. Yet the limbs were properly organized and the babies grew into adults who could lead normal lives. To be sure, these particular changes were disadvantageous to the individuals in which they occurred, but then many variations are. The point is that large coordinated changes can and do happen. There is, incidentally, a rare genetic disorder that produces much the same symptoms as thalidomide, so it is not a question of a possible difference in effect between an externally supplied chemical and a natural gene product.

8. THE EPIGENETIC LANDSCAPE

Variations in organisms are essentially variations in their developmental processes, so if we are looking for general principles in evolution, we should begin by looking for common features of development.

About 70 years ago, the British embryologist C. H. Waddington (1940) pointed out that developing organisms share some important properties. Organisms of the same species have characteristic developmental pathways, "chreods", that they follow as they develop. These are stable in the sense that if the embryo is perturbed during development, it generally returns not to the state it was in when perturbed, but to the same pathway further down. The end result will be much the same as if the embryo had been left alone. Waddington called this property homeorhesis, to stress that it is the flow down the pathway that is stable, not any particular point along it. Development can typically proceed only to restricted number of alternative end states rather than a whole spectrum. The effect of this "canalization" is that we observe distinguishable species rather than a continuum of forms.

Waddington illustrated this by what he called the epigenetic landscape, using the word epigenetic in its original sense of development. Most biologists now define epigenetics as the study of heritable changes in gene expression not due to changes in the DNA, and this can be yet another source of confusion. The title of a paper may lead you to suppose that it is about the importance of development, whereas it turns out to be all about genes.

Waddington imagined the developmental system as a mountainous terrain. The valleys represent possible developmental pathways. The precise shape of the landscape depends on a network of guy-ropes beneath it; the ropes stand for the effects of the genes and the complexity of the network is to remind us that while genes certainly affect development, they do so in very complicated ways (Figs. 9.2 and 9.3).

A ball rolling down a valley represents the developing organism, and you can see the properties Waddington had identified clearly illustrated. There are alternate pathways, and these lead to a discrete set of end points rather than a continuum of outcomes. The pathways are stable against both perturbations, represented by the ball being deflected up the side of a valley

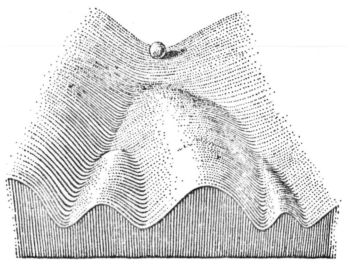

Figure 9.2 Part of an epigentic landscape. The path followed by the ball, as it rolls down toward the spectator, corresponds to the developmental history of a particular part of the egg. There is first an alternative, toward the right or the left. Along the former path, a second alternative is offered; along the path to the left, the main channel continues leftward, but there is an alternative path that, however, can only be reached over a threshold. *(The strategy of the genes, Waddington, © 1957 George Allen and Unwin. Reproduced by permission of George Allen and Unwin).*

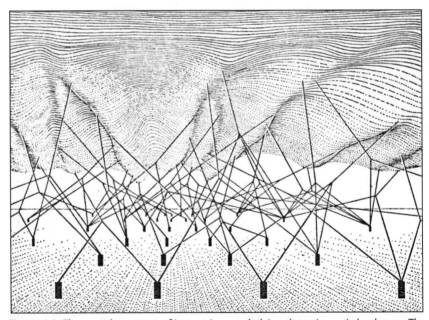

Figure 9.3 The complex system of interactions underlying the epigenetic landscape. The pegs in the ground represent genes; the strings leading from them are the chemical tendencies that the genes produce. The modeling of the epigenetic landscape, which slopes down from above one's head toward the distance, is controlled by the pull of these numerous guy-ropes that are ultimately anchored to the genes. *(The strategy of the genes, Waddington, © 1957 George Allen and Unwin. Reproduced by permission of George Allen and Unwin).*

(and returning to the valley floor further down, like a bobsled on a run), and mutations, represented by changes in the ropes that permanently alter the shape of the landscape.

If the ball is disturbed so much that it is forced out of the valley, it will probably reach a dead end, but it is possible that it will pass over a watershed and then continue down a different valley, just as stable as the original one.

Mutations are most likely to have significant effects only if they disturb the landscape near where one valley divides into two. At such a point, even a small alteration in the topography can be enough to send the ball down a different path, i.e., a small genetic change can bring about a large change in the phenotype, and without necessarily affecting the action of any other genes. The exceptions to this are likely to be toward the end of the process when all the major features have been laid down and the valleys have flattened; these are of course the most commonly studied.

We would not expect major changes to come about through the accumulation of small ones. If pulling a little on a rope shifts the landscape in the desired direction at first, eventually the tensions from the other ropes to which it is linked will prevent it from continuing further in that direction. Moreover, the complexity of the network will mean that other parts of the landscape will be affected, and in ways that were not intended, and are therefore more likely than not to be deleterious.

It is easy enough to think of real examples of the properties of the epigenetic landscape. For example, there are limits to what can be accomplished even by artificial selection. The great plant breeder Luther Burbank (1931) famously remarked that he could get plums of many different sizes but not as small as a pea or as big as a grapefruit. Both pleiotropy (one gene affecting more than one character) and genetic linkage (the fact that genes that are physically close together on a chromosome tend to be passed on together) will oppose strong directional selection.

The existence of alternative pathways is illustrated by the well-known phenomenon of phenocopying: environmental perturbations bringing about the same changes in organisms as mutations. Treating the larvae of normal *Drosophila* with ether can produce adults that closely resemble the *bithorax* mutant, which has four wings instead of the normal two (Ho, Tucker, Keeley, & Saunders, 1983). That a mutation can produce the same effects as thalidomide is another example. It would be very hard to explain how a mutation and a perturbation could have such very similar complex effects if the alternative trajectory did not already exist.

These properties are certainly important, so we may ask how organisms acquired them. For a neo-Darwinist, this must have happened by natural selection acting on small selective advantages over many generations.

There is, however, an alternative explanation. Whatever else organisms are, they are highly complex nonlinear dynamical systems. What is more, they are in some sense stable, as they have to be, more or less by definition. For if they were not, we would not consider them to be organisms. And almost any nonlinear dynamical system, if it is stable at all, will have these properties (Saunders, 1993, in press).

It should not be entirely surprising even to the nonmathematician that nonlinear dynamical systems should have the properties illustrated by the epigenetic landscape. For example, everyone learns at school that a linear algebraic equation has one solution, but a quadratic has two. What is more, if the coefficients are real numbers, the root of the linear equation is also real, whereas the roots of the quadratic can be complex. Thus even in these simplest of all mathematical systems, nonlinearity can make a significant difference.

We expect nonlinear systems to have multiple steady states and also, because of the sort of systems we are talking about, alternative trajectories, i.e., chreods. Not all of these will be stable, but those that are will have the homeorhesis that Waddington described. The system will also exhibit canalization because there will be distinct trajectories rather than a continuum of possible pathways. There did not necessarily have to be such things as organisms, but once we know there are, we should expect them to have these properties.

9. PUNCTUATED EQUILIBRIA

If the living world came into being through a long process of descent with gradual modification, this ought to be reflected in the fossil record. We should find long sequences of fossils beginning with the earliest forms and slowly changing into those we observe today. In fact we do not. The fossil record is full of large gaps, which opponents of evolution from Cuvier to the present day have always claimed are evidence against the fact of evolution.

Evolutionists have traditionally responded that the gaps are merely artifacts. Not all organisms leave fossils and not all the fossils that are formed are found. This defense has become less and less convincing as paleontologists have become more confident that if the intermediates really existed they would have found them by now. Besides, if evolution had been gradual, we would expect the fossil record to at least show steady progress with some bits missing. Instead, the record shows a succession of long periods in which very little seems to have happened and with no intermediate forms between them. This pattern is known as punctuated equilibria (Eldredge & Gould, 1972).

Darwinists, however, still insist that evolution is gradual. In the *Origin of Species*, we read "*natura non facit saltum*" and over a century later Dawkins (1985) writes, "Gradualism is of the essence. In the context of the fight against creationism, gradualism is more or less synonymous with evolution itself. If you throw out gradualism you throw out the very thing that makes evolution more plausible than creation."

Why are evolutionists so committed to gradualism when that is not what the evidence shows? Why do they offer the creationists such an easy target by allowing the argument about evolution to focus on the question of the gaps? Would it not be better to make the case for evolution on other, stronger grounds, such as carbon dating and the observation that the deeper we dig the simpler the organisms we find, and leave the question of whether evolution has been gradual to be decided later?

Not for neo-Darwinists, because for them it is an article of faith that evolution occurs by the natural selection of random variations, random in the strong sense that there is little or nothing that can be said about them. If the variations are strongly random, they must necessarily be small because the chance that a large strongly random variation could be anything but disastrous for the organism in which it occurs must be infinitesimal. Hence the scorn with which neo-Darwinists dismissed Goldschmidt's (1940) macromutations, or "hopeful monsters". And that is why evolution must be gradual, whatever the evidence may say. If, however, we focus less on the genes and think instead about how organisms develop and how the process can be affected by mutations or other perturbations, then we realize that whether mutations are strongly random, weakly random, or for that matter not random at all, *phenotypic* variations are not strongly random and the problem disappears.

In fact, neo-Darwinists have found a way of explaining away the gaps. As their theory is unfalsifiable, it would be astonishing if they could not. The major changes, they tell us, must have occurred by small groups becoming isolated, evolving rapidly, and then returning to outcompete the main population. There were indeed intermediates but we do not find them partly because the group that was doing all the evolving was very small and partly because they were not in the same place as the population whose history we are trying to trace. In effect, all the evolution that really mattered took place just when no one was looking.

But the gaps in the fossil record and punctuated equilibria are not awkward facts that require a special and inherently unverifiable explanation. On the contrary, they are precisely what we should expect. This is because one of the predictions of the epigenetic landscape, i.e., of the recognition that organisms are complex nonlinear dynamical systems, is that large changes are unlikely to come about through successions of small ones. The stability of the system means that most mutations will have only small effects, if any at all, and these will not accumulate.

If however, a mutation, or an environmental change, or a combination of the two takes a reaction over a threshold and so an organism onto a new developmental pathway, then there can be a substantial change. Macroevolution does not generally occur by a long sequence of microevolutionary steps; it is an essentially different process. And on that picture, punctuated equilibria is precisely what we would expect, rather than something that has to be explained by a more or less plausible but intrinsically unverifiable story.

The medusoids mentioned in the previous section provide an example. If we release a drop of fusel oil close to the surface of the paraffin, we do not observe the ramified form shown in the figure. It appears only if the drop enters the paraffin at above a certain speed, and then it is in more or less its final form. There is no sequence of intermediate states, with first one "tentacle", then two, and so on.

Over a long period, we would expect changes in parameters such as the size of the organism or the elasticity of the membrane that surrounds it. These may or may not have been gradual, but even if they were, the change from no tentacles to many would have been abrupt, as it is with the drop of fusel oil. The reason we do not observe the intermediates in the fossil record is simply that there never were any.

10. THE PROBLEM OF GOOD

By its very nature, Darwinism invokes selfishness and competition as the most important driving forces in nature. The full title of Darwin's book includes the phrase *the Struggle for Life* and of course the most famous twentieth century work on the subject was Dawkins' (1976) *The Selfish Gene*. This has always been a problem for Darwinists, because it implies a view of humanity that is not only dismal but also does not correspond to what we see about us. From time to time, there have been attempts to introduce the idea of group selection, in which there can be adaptations that favor the group at the expense of the individual, but this idea has never been widely accepted because it does not fit the Darwinian model (cf. Dawkins, 1976; Pinker, 2012; Wilson & Wilson, 2007, 2008; Wynne-Edwards, 1962).

It is possible to account for apparently altruistic behavior through devices such as kin selection (Hamilton, 1964) and reciprocal altruism or the iter-ated prisoner's dilemma (see Axelrod, 1984), but the explanation based on selfishness is the one that comes to the mind first. It follows that accounts with selfishness at the core will tend to predominate. This is true both in the scientific literature and in the popular press, and because of the subject matter of the social sciences, what appears in the popular press is important. Such accounts serve as a justification for selfish behavior because while phi-losophers insist, correctly, that we must not confuse "is" with "ought", the two are frequently conflated in people's minds. And it is harder to criticize someone for acting selfishly if we believe that what he is doing is "only natural". Moreover, should we ever observe what appears to be unself-ish behavior, we are to interpret it as merely another form of selfishness. "Scratch an 'altruist,' and watch a 'hypocrite' bleed" (Ghiselin, 1974, p. 247).

Throughout the ages, many Christians have worried about the so-called Problem of Evil. If the universe was created by a God who is both omnipotent and good, how can there be evil in the world? The eighteenth century natural theologian William Paley, best known as the author of the famous parable about the watch and the watchmaker, offered a solution to this paradox (Paley, 1819, p. 436). As a good Newtonian, he held that the world must run according to laws. These generally work well, but it might sometimes happen that cases of "apparent evil" (Paley seems to have been reluctant to accuse his Creator of allowing genuine evil) would arise "out of the thwartings and crossings of laws whose effects are for the most part beneficent."

Neo-Darwinists, in contrast, worry about the Problem of Good. If the living world was created by natural selection, which is both very powerful and inherently selfish, how can organisms ever be good? Their solution is that while selfishness and greed are the general rule, it can sometimes hap-pen that cases of apparent good may arise out of the subtle workings of laws whose effects are for the most part selfish.

Those who do not believe that the universe was created by the Judeo-Christian God need not be unduly concerned about the Problem of Evil.

Those who do not believe that the living world was created by natural selection alone need not be unduly concerned about the Problem of Good.

11. CONCLUSIONS

While Darwinism has contributed much to our understanding of the living world, it does not explain why organisms are the way they are and how they came to be that way. Nor can it. For that, we will have to draw on all of science, not just the natural selection of small random variations.

The most important accomplishment of Darwinism is that it provides an explanation for adaptation. That was how Darwin convinced the world of the fact of evolution and it remains important today. But adaptation is nowhere near as difficult a problem for the psychologist as it is for the biologist. That humans should usually behave in ways that are to their advantage seems so obvious as not to require any special explanation. At the same time, while the link between gene and physiological trait is nowhere near as direct as neo-Darwinists assume, that between gene and psychological trait is even less so. It is also much harder to be sure what counts as a psychological "trait" at all.

Neo-Darwinism can appear very impressive until you look at it closely and notice how many things neo-Darwinists ignore. When challenged, however, they generally respond that they are indeed aware of these factors. They know that it is not a matter of one gene/one character. They know that genes are not contiguous stretches of single DNA molecules. They know that genes generally act in large webs of interactions. They insist that when they speak of random variations they only mean that the variations are not preferentially in the directions in which they are required. They accept that the process of development means that a lot happens between genotype and phenotype, and so on. Yet when you look carefully at their work, you find it is firmly based on the assumptions they insist they reject.

No model can include every possible factor, so we inevitably have to ignore some and simplify others. When we do this, however, we should be prepared to give some justification for what we have done and some estimate of the error that is likely to have been introduced. Of course the error may not be just quantitative. Leaving out an important factor can entirely change the behavior of a system.

That the Earth is round makes no difference at all when we are laying out a garden or a city. We do, however, have to take it into account when we are working on a larger scale. For example, the southeastern part of the Canadian province of Saskatchewan is divided into rural municipalities most of which appear at first glance to be squares. In fact, many of them are not; the grid had to be distorted so it would fit between lines of longitude on a round Earth. The amount of distortion is, however, quite small.

On the other hand, if Columbus had planned his voyage to the Orient on the assumption that the Earth is nearly, but not quite, flat he would never have tried to sail around it. That we can circumnavigate the Earth is a property of its being round. We would not discover it by working with a flat Earth model and adding a small correction; the roundness has to be included in the analysis from the beginning.

In the same way, the fact that punctuated equilibria should be a natural feature of evolution, not something that has to be explained by a hypothetical and inherently unobservable mechanism, is something we discover only if we consider development to be important in evolution. If we think of it as making at the most minor adjustments to the broad picture provided by neo-Darwinism, we will miss such phenomena completely, just as it would never occur to a flat-earther to try to reach Asia from Europe by sailing west.

Neo-Darwinists are clearly confident that their results are basically correct and that barring some totally unexpected developments in biology, in the future there will be only small modifications. But they give little or no justification for this belief. That would not matter so much in what we might call exploratory research, where the aim is to try out an idea and see what sorts of consequences it might entail. When, however, they claim to have the explanation for the speed of the antelope and all those other things that Dennett (1995) mentions, and when they also claim to be unifying all the "disparate branches of psychology", then it is incumbent on them to demonstrate that were they to include what they are omitting, their conclusions would still stand up. If they cannot do this, then others need not take these conclusions so seriously, certainly not where they conflict with what they discover in their own research.

Cosmides et al. (1992) urge psychologists to allow for what they call conceptual or vertical integration: different disciplines should be mutually consistent. Psychology, they insist, must be consistent with evolutionary biology, by which of course they mean neo-Darwinism. Its results must fit into the adaptationist framework.

That is why the Problem of Good (or, as sociobiologists and evolutionary psychologists prefer to call it, the problem of altruism) has been such an important issue for them and why Hamilton's (1964) theory of kin selection was considered such a remarkable achievement. If we do not believe that human behavior can be seen as a collection of "behaviors" each of which has evolved by natural selection acting on individuals, then there is no problem. It is significant that J.B.S. Haldane (among others) had thought of kin selection long before Hamilton, but no one had considered the question worth pursuing until neo-Darwinism had become the dominant paradigm and altruism was seen as a potentially serious problem within it.

Psychologists should be concerned instead that their theories are consistent with what is being discovered about the workings of the brain (e.g., Ramachandran, 2011), about child development (e.g., Karmiloff-Smith, 2000), about theoretical linguistics (e.g., Clark & Lappin, 2011) and in many other fields where the mechanisms of the mind (in a much broader sense of the term than is envisaged by evolutionary psychologists) are being investigated.

They should also bear in mind that evolutionary psychology assumes from the outset that the brain is essentially modular and that behavioral traits or the mechanisms that generate them are adaptations created by natural selection. The only question is what they are adaptations for, or perhaps what they were adaptations for in the Stone Age. For those who do not accept the premises, the conclusions are correspondingly less convincing.

REFERENCES

Axelrod, R. (1984). *The evolution of co-operation*. New York, NY: Basic Books.
Barkow, J. H., Cosmides, L., & Tooby, J. (Eds.), (1992). *The adapted mind*. Oxford, England: Oxford University Press.
Bateman, A. J. (1948). Intra-sexual selection in *Drosophila*. *Heredity*, 2(3), 349–368. http://dx.doi.org/10.1038/hdy.1948.21.
Burbank, L. (1931). *The harvest of the years*. New York, NY: Houghton Mifflin Co.
Buss, D. M. (1992). Mate preference mechanisms: consequences for partner choice and intra-sexual competition. In Barkow et al., (pp. 249–266).
Buss, D. M., Haselton, M. G., Shackleford, T. K., Bleske, A. L., & Wakefield, J. C. (1998). Adaptations, exaptations and spandrels. *American Psychologist*, 53, 533–548.
Clark, A., & Lappin, S. (2011). *Linguistic Nativism and the poverty of the stimulus*. Chichester, England: Wiley.
Coles, M. G., & Francesconi, M. (2007). *On the emergence of toyboys*. Frankfurt, Germany: Institute for the Study of Labour (IZA).
Cosmides, L., Tooby, J., & Barkow, J. H. (1992). Evolutionary psychology and conceptual integration. In Barkow et al., (pp. 3–15).
Darnell, J. E., Lodish, F., & Baltimore, D. (1986). *Molecular cell biology*. New York, NY: W. H. Freeman.
Dawkins, R. (1976). *The selfish gene*. Oxford, England: Oxford University Press.
Dawkins, R. (1982). *The extended phenotype*. Oxford, England: WH Freeman.
Dawkins, R. (1985). What was all the fuss about? (Review of *Time Frames* by N. Eldredge), *Nature*, *316*, 683–684.
Dennett, D. C. (1995). *Darwin's dangerous idea*. New York, NY: Simon and Schuster.

Eldredge, N., & Gould, S. J. (1972). Punctuated equilibria: an alternative to phyletic gradualism. In T. J. M. Schopf (Ed.), *Models in paleobiology* (pp. 82–115). San Francisco, CA: Freeman-Cooper.

Fisher, S. F. (2006). Tangled webs: tracing the links between genes and cognition. *Cognition, 101*, 270–297.

Gangestad, S. W., & Simpson, J. A. (2000). The evolution of human mating: trade-offs and strategic pluralism. *Behavioral and Brain Sciences, 23*, 573–587.

Ghiselin, M. (1974). *The economy of nature and the evolution of sex.* Berkeley CA: University of California Press, Berkeley.

Goldschmidt, R. B. (1940). *The material basis of evolution.* Seattle, WA: University of Washington Press.

Gomendio, M., Cassinello, J., Smith, M. W., & Bateson, P. P. G. (1995). Maternal state affects intestinal changes of rat pups at weaning. *Behavioral Ecology and Sociobiology, 37*, 71–80.

Gould, S. J. (1977). *Ontogeny and phylogeny.* Cambridge, MA: Belknap Press.

Gould, S. J., & Lewontin, R. C. (1979). The spandrels of San Marco and the Panglossian paradigm: a critique of the adaptationist programme. *Proceedings of the Royal Society of London, B205*, 581–598.

Hamilton, W. (1964). The genetical evolution of social behaviour. *Journal of Theoretical Biology, 7*, 1–16.

Ho, M. W. (1998). *Genetic engineering, dream or nightmare?* Bath, England: Gateway Books.

Ho, M. W. (2003). *Living with the fluid genome.* London, England: Institute of Science in Society.

Ho, M. W. (2011). Development and evolution revisited. In K. E. Hood, C. T. Halpern, G. Greenberg & R. M. Lerner (Eds.), *Handbook of developmental science, behavior and genetics* (pp. 61–109). Hoboken, NJ: Wiley.

Ho, M. W., & Saunders, P. T. (Eds.), (1984). *Beyond neo-Darwinism: An introduction to the new evolutionary paradigm.* London, England: Academic Press.

Ho, M. W., Tucker, C., Keeley, D., & Saunders, P. T. (1983). Effects of successive generations of ether treatment on penetrance and expression of the bithorax phenocopy in *Drosophila melanogaster. Journal of Experimental Zoology, 225*, 357–368.

Joron, M., Frezal, L., Jones, R. T., Chamberlain, N., Lee, S. F., Haag, C. R., et al. (2011). Chromosomal rearrangements maintain a polymorphic supergene controlling butterfly mimicry. *Nature, 477*, 203–206.

Karmiloff-Smith, A. (2000). Why babies' brains are not Swiss Army knives. In H. Rose & S. P. R. Rose (Eds.), *Alas poor Darwin: Arguments against evolutionary psychology* (pp. 144–156). London, England: Jonathan Cape.

Maynard Smith, J. (1969). The status of neo-Darwinism. In C. H. Waddington (Ed.), *Towards a theoretical biology 2: Sketches* (pp. 82–89). Edinburgh, Scotland: Edinburgh University Press.

Mayr, E. (1982). *The growth of biological thought.* Cambridge, MA: Belknap Press.

Mayr, E. (1984, November 2). The triumph of the evolutionary synthesis. (Review of Ho and Saunders, 1984), *Times Literary Supplement*, 1261–1262.

Mayr, E. (1991). *One long argument.* Cambridge, MA: Harvard University Press.

Paley, W. (1819). *Natural theology; or, evidences of the existence and attributes of the deity, collected from the appearances of nature* (16th ed.). London, England: F.C. and J. Rivington.

Pinker, S. (2012, June 18). The false allure of group selection. *Edge.* http://edge.org/conversation/the-false-allure-of-group-selection.

Popper, K. R. (1934). Logik der Forschung Vienna, Austria: J. Springer. (English translation (1959): *The logic of scientific discovery*, London, England: Hutchinson.

Ramachandran, V. S. (2011). *The Tell-Tale brain.* New York: W. W. Norton.

Romanes, G. J. (1900). The Darwinism of Darwin and of the post-Darwinian schools. (new edition). *Darwin, and after Darwin.* (Vol. 2). London, England: Longmans, Green & Co.

Saunders, P. T. (1993). The organism as a dynamical system. In F. Varela & W. Stein (Eds.), *Thinking about biology: SFI studies in the sciences of complexity Lecture Notes* (Vol. 3, pp. 41–63). Reading, MA: Addison Wesley.

Saunders, P. T. Dynamical systems, the epigenetic landscape and punctuated equilibria. In P. C. M. Molenaar, K. M. Newell, & R. M. Lerner (Eds.), *Handbook of developmental systems theory and methodology*. New York, NY: Guilford Press, in press.

Saunders, P. T., & Ho, M. W. (1984, November 23). Neo-Darwinism (Reply to Ernst Mayr). *Times Literary Supplement*, 1342.

Symons, D. (1992). On the use and misuse of Darwinism in the study of human behavior. In Barkow et al., (pp. 137–159).

Thompson D'A, W. (1917). In *On growth and form*. Cambridge University Press. Cambridge. (abridged edition, ed. J. T. Bonner, 1961).

Tooby, J., Cosmides, L., & Barrett, H. C. (2003). The second law of thermodynamics is the first law of psychology: Evolutionary developmental psychology and the theory of tandem, coordinated inheritances: comment on Lickliter and Honeycutt. *Psychological Bulletin, 129*, 858–865.

Trivers, R. (1974). Parent–offspring conflict. *American Zoologist, 14*, 249–264.

Waddington, C. H. (1940). *Organisers and genes*. Cambridge, England: Cambridge University Press.

Waddington, C. H. (1957). *The strategy of the genes*. London, England: George Allen and Unwin.

Wilson, D. S., & Wilson, E. O. (2007). Rethinking the theoretical foundation of sociobiology. *Quarterly Review of Biology, 82*, 327–347.

Wilson, D. S., & Wilson, E. O. (2008). Evolution "for the good of the group". *American Scientist, 96*, 380–389.

Wynne-Edwards, V. C. (1962). *Animal dispersion in relation to social behaviour*. Edinburgh, Scotland: Oliver & Boyd.

CHAPTER TEN

A Contemporary View of Genes and Behavior: Complex Systems and Interactions

Douglas Wahlsten
Department of Psychology, University of North Carolina Greensboro, Greensboro, NC, USA
E-mail: dlwahlst@uncg.edu

Contents

Abstract

Several large-scale searches for genes that influence complex human traits, such as intelligence and personality, in the normal range of variation have failed to identify even one gene that makes a significant difference. All previously published claims for genetic influences of this kind now appear to have been false positives. For more serious psychiatric and medical disorders such as schizophrenia and autism, several genes have been found where a rare mutation contributes to abnormal behavior, but in many instances they are de novo mutations not obtained from a parent. Despite the many disappointments in the search for genes influencing human behavior, the field of molecular genetics has made remarkable progress to the extent that several broadly

Advances in Child Development and Behavior, Volume 44   © 2013 Elsevier Inc.
ISSN 0065-2407, http://dx.doi.org/10.1016/B978-0-12-397947-6.00010-6   All rights reserved.

applicable principles can now be affirmed. These principles show how development is regulated by networks of interacting genes that function in an environmental context. They invalidate several key assumptions of statistical genetic analysis that are made when estimating heritability. There is now a need to reform the teaching of genetics to our students and to restrict the funding of further searches for elusive genes that account for so little variance in normal behaviors.

1. INTRODUCTION

From the beginnings of modern genetics, there have been two broad approaches to the genetic analysis of behavioral and other phenotypes: 1) statistical genetics; and 2) single gene analysis. The identification of single-gene inheritance and the study of how genes are involved in development has become the domain of molecular genetics (Wahlsten, 1999), where virtually all genes possessed by humans have now been identified, and great advances are being made in learning about how genes function in the development of complex organisms (Davidson, 2006; Davidson & Erwin, 2006). Statistical genetic analysis (Fisher, 1918; Lush, 1945), on the other hand, has been used to study the resemblance of relatives in a breeding population when no specific gene relevant to a phenotype is known. The analysis relies on polygenic models that assume (a) large numbers of genetic variations, each with a very small effect, combine to generate phenotypic differences; and (b) genetic and environmental effects are independent and algebraically additive.

One of the promises of the Human Genome Project and subsequent large-scale searches for genes pertinent to behavioral variation has been that, given enough genetic markers distributed widely across all chromosomes and large samples of human subjects, the specific genes responsible for variation in complex traits such as intelligence and personality can be identified. The new genetic information could then supposedly be used to predict a person's future phenotypic value and possibly devise drug and other treatments to enhance desirable features and suppress the undesirable ones. Human behavior genetics could thereby spawn an applied science.

2. A FUTILE SEARCH FOR SPECIFIC GENE INFLUENCES ON COMPLEX HUMAN TRAITS

Several large-scale studies of specific genes related to interesting behavioral phenotypes have been reported recently, and the fate of statistical genetic analysis has become clear (Wahlsten, 2012). The genes contributing to variation in complex human traits in the normal range of individual variation cannot be identified because the effect of any one gene is just too small. Lacking knowledge of which specific genes are most pertinent, none of the marvelous tools of modern molecular genetics can be applied to complex psychological traits for any worthy purpose.

The evidence that statistical genetic analysis has reached a dead end comes from two kinds of studies. One approach is to review previous claims of single-gene effects and then study fresh samples of thousands of people to see if the published claims can be confirmed. Remarkably, for personality traits and intelligence, researchers have failed to confirm virtually all published claims (Flint, 2011; Posthuma & de Geus, 2006; Wahlsten, 2011b). In a recent examination of previously published claims for 12 specific gene effects on IQ in three fresh samples totaling 9771 subjects, only one gene appeared to have a significant influence in just one of the three samples (Chabris et al., 2012); not one of the published claims could be consistently replicated. It now is clear that virtually the entire world literature of claims for single-gene effects on IQ consists of false positives.

The other kind of study (GWAS, genome-wide association study) scans the entire genome using hundreds of thousands of genetic markers that maybe related to psychological traits. As reviewed recently (Wahlsten, 2012), recent studies have failed to identify even one gene that is significantly and consistently related to IQ. One very large study (Davies et al., 2011) assessed 599,011 single nucleotide polymorphisms (SNPs) in DNA from 3782 adults in six samples from Scotland, England, and Norway. The authors concluded: "Analyses of individual SNPs and genes did not result in any replicable genome-wide significant association." It now appears that any specific gene that somehow is important for population variation in intelligence must account for less than 0.1% of the phenotypic variance, a trivially small amount. Any gene with a larger effect should be detectable with current methods.

3. NEW SOURCES OF GENETIC INFORMATION

While the search for genes of particular interest to psychology has found little or nothing worthy of note, vast quantities of information about the genes themselves and how they work are being accumulated. The linear sequence of nucleotide bases in one person's DNA has been known since 2001, and the identity of more than 20,000 genes coding for proteins has been well established. In October of 2012, the results of the next generation of molecular studies, the massive ENCODE project (Encyclopedia of functional DNA elements), announced preliminary findings about the 98% of the genome that does not code for protein and made data available via the Internet (Bernstein et al., 2012). The project involved more than 442 professional scientists from 32 institutes worldwide who conducted more than 1600 complex experiments on 147 types of human cells (Birney, 2012) at a cost of more than \$308 million, resulting in a burgeoning flood of new data with no end in sight that has been compared with a "runaway train" (Maher, 2012). None of the ENCODE studies collected data on pheno-types of interest to psychology. Any benefit for psychological science will be indirect.

Several popular websites that provide access to recent genetic informa-tion on humans are listed in Table 10.1. The National Center for Biological Information (NCBI) site on its human Mapview page shows each chro-mosome, and a click on a chromosome gives detailed statistics, such as the

Table 10.1 Websites providing genetic information about humans and other species

Website URL	Sponsor	Kinds of information
www.genecards.org	Weizmann Institute of Science	Extensive data on all known human genes. Search by gene name, symbol, or phenotype. Many links to other databases. See Table 10.2.
omim.org	Online Mendelian Inheritance in Man	List of all genetic variants relevant to a phenotype reported in the literature, some from very old reports and not verified. Many links to other sites. Many references.
www.ncbi.nlm.nih. gov/projects/ mapview	Natl. Center for Biol. Information	Extensive data on genes arranged by human chromosome.
www.ensembl.org	Ensembl genome browser	Extensive data on genome of many species. Queries must be written in Perl language. Requires considerable exper-tise.
biogps.org	BioGPS is a team of three scientists supported by NIH	Convenient source of information on gene expression in many kinds tissue plus other kinds of data.

number of genes located there and a sampling of named genes. The Online Mendelian Inheritance in Man site has historical data, extensive references to published literature, and links to many other sites of interest. GeneCards gives quick access to a prodigious amount of information on a specific gene that can be searched by phenotype or disease. Clicking on "disease genes" provides a list of 3931 genes where a mutation has been associated with some medical disorder, and another click on any of those gene symbols gives a display of numerous things known about the specific gene (Table 10.2). The BioGPS site gives a convenient display of the magnitude of expression of the RNA from a specific gene in several dozen types of cells in the body.

Those who are not geneticists should exercise caution when using the websites. Considerable expertise is needed to access and interpret many kinds of data. For example, the Ensembl site stores vast amounts of data, but the user must write a query in the Perl language in order to generate a meaningful report. There is so much data available that finding something of interest amongst the forest of facts can be a major challenge. Many websites collect data with automatic scans of the Internet for certain keywords, and there is little quality control. If a search of a site for "intelligence" yields dozens of genes, this does not insure that each of them has a verified role

Table 10.2 Useful features of GeneCards entry for a gene

Aliases	All previous names for this gene; needed for search of older literature
Summaries	Brief description of what the gene does at the molecular level
Genomic views	Genomic location of the gene on a small diagram of a chromosome
Function	Short paragraph summarizing molecular and physiological function; animal models: mouse knockouts that yield abnormal phenotypes
Pathways	Metabolic and functional pathways; pharmacodynamics; interaction networks
Drugs	Compounds having a direct effect on the gene product or function
Expression	mRNA expression levels in different organs
Genomic variants	Variations in the gene structure that maybe related to protein structure; gives some suggestions about function; provides population frequency of variants
Disorders	Disorders or diseases that are associated with the name of the gene in the published literature; not a valid indication of any causal link

in mental development. In some instances, it simply tells the user that some published article used both the key word "intelligence" and the name of a gene. Thus, the sites provide a good place to begin a review of what is known about specific genes influencing intelligence and other traits, but they cannot be relied on to give authoritative answers.

4. BROAD PRINCIPLES AFFIRMED

Despite disappointments about recent attempts to analyze complex traits genetically, there has been remarkable progress in understanding how genes function in a more general sense. There is sufficient accumulated evidence to affirm several broadly applicable principles that regard the gene as a part of a system of interacting chemical constituents and environmental features. The principles have important implications for attempts by psychologists to model and understand the role of genetic variation in individual differences in complex psychological traits.

4.1. No Gene Codes for a Specific Phenotype

Since the pioneering work of Johannsen (1911), it has been widely understood that a genotype is inherited from the parents, while the measurable phenotype develops during the lifetime of the individual and many phenotypes can occur for the same genotype. Virtually every phenotype of interest to psychology is influenced by both genotype and environment. This is true even for seemingly "hard wired" aspects of the brain as well as more labile behavioral states. A vivid example is provided by the corpus callosum (CC) that connects the two cerebral hemispheres in placental mammals. In certain highly inbred mouse strains where each animal has an identical genotype, some have a normal CC, whereas other littermates conceived at the same time lack the CC entirely and have a deep fissure in the cerebral midline (Wahlsten, Bishop, & Ozaki, 2006). Within a strain, the severe cerebral abnormality is not hereditary (Wahlsten, 1982) and instead arises from exceedingly small differences in timing of crucial events in the embryo.

This principle obtains at every level from the molecule to behavior (Gottlieb, 1992, 2007). For many years, it was held by geneticists that a gene (DNA) codes for a specific kind of protein molecule, often an enzyme, through an intermediate messenger mRNA molecule. Now the one gene–one enzyme view is no longer tenable. Most mammalian genes consist of several *exons* along a region of DNA that are separated by *introns*. Each exon codes for the structure of a corresponding mRNA molecule, and that mRNA sequence in turn codes for the sequence of amino acids that are assembled into a linear polypeptide molecule. The polypeptides are then spliced together into a longer protein molecule that has biological functions. It is now recognized that the polypeptides arising from a single gene can be spliced together in several ways to yield different kinds of proteins (Alberts, Johnson, Lewis, Roberts, & Walter, 2008; Matlin, Clark, & Smith, 2005). Thus, alternative splicing allows for one gene to code for several kinds of proteins. For the α-tropomyosin gene, for example, there are 18 distinct exons, and their mRNAs can be assembled to make at least 10 forms of the protein that are expressed in different kinds of tissue (Lees-Miller, Goodwin, & Helfman, 1990).

Alternative splicing occurs for more than 90% of known human genes (Pan, Shai, Lee, Frey, & Blencowe, 2008), and many known genetic disorders result from splice abnormalities (Matlin et al., 2005). Which specific protein is produced in a specific cell can be influenced by the surroundings of that cell and even psychological processes such as stress (Shaked, Zimmerman, & Soreq, 2008; Singh, Tapia-Santos, Bebee, & Chandler, 2009). Thus, even at the molecular level, the DNA of a gene does not code for any specific molecular entity other than RNA segments. What protein actually is generated depends on the local environmental context, which in turn can be strongly influenced by events in the environment outside the organism. Rich and sometimes confusing details about alternative splice variants for any specific gene are provided by GeneCards, NCBI, and Ensembl.

4.2. Names of Genes Often Do Not Describe Normal Function

The common conventions for naming genes sometimes contribute to misunderstandings about what genes do. Since the time of Morgan who studied heredity in fruit flies (Morgan, 1914), a gene has been named for its most salient phenotypic effect observed by the scientist who discovered it (e.g., white eye, Japanese waltzer, etc.). In mice, for example, two copies of the *obese* gene typically result in fat mice in the typical lab environment with free feeding, and two copies of the *diabetes* gene also typically lead to high blood sugar and high body weight. Nevertheless, the genes do not code for diabetes or obesity. Restriction of caloric intake in the *diabetes* mice can prevent the obesity and diabetes in these animals (Lee & Bressler, 1981), whereas caloric restriction in *obese* mutants of course lessens body weight but does not eliminate the diabetes. With better knowledge of how appetite is regulated by hormones, the names were changed. A protein named leptin was discovered that is released from white fat cells and stimulates the leptin receptor in the hypothalamus, which in turn reduces hunger (Zhang et al., 1994). It turns out that the *obese* gene codes for the structure of leptin, and a mutation in that gene makes leptin ineffective in signaling satiety, so the *obese* gene is now designated as the *obese* allele of the leptin gene (lep^{ob}), and the *diabetes* gene is now the *diabetes* allele of the leptin receptor gene ($lepr^{db}$). Renaming the gene for the protein, however, does not overcome the problem of misunderstanding. The name leptin itself is derived from the Greek word *leptos* for thin. Invoking an obscure Greek surrogate name covertly implies that the gene codes for body size, but we know that body size itself is a phenotypic outcome of a complex physiological and behavioral system. Physiologists who understand the complex web of interactants are not confused by names for the parts, but consumers of genetic information can easily be misled by mere names into thinking the gene codes for the phenotype.

It is now a common practice to name human genes for their properties at the molecular level, but when function is not well understood, the gene is sometimes named for the rare disease that results when it is defective. For example, the full name of the human CFTR gene is cystic fibrosis transmembrane conductance regulator. The gene does not exist, however, to cause trouble for some people and give them cystic fibrosis; a rare mutation in the gene whose function is unknown causes the disease state. Another example is the ACLS or acrocallosal syndrome gene believed to be in the p, 13 band of chromosome 7, for which the protein involved or the function of the normal form of the gene is not yet known. Some "disease genes" are named for a medical disorder, while others are named for their biochemical action or affinity. Once the molecular function of a gene currently named for a disease is discovered, the official gene name most likely will be changed.

4.3. Any One Gene Alters Several Phenotypes

When a genetic mutation alters several different phenotypes, the gene is said to exhibit pleiotropy. For example, albinism is a kind of genetic disorder where the person cannot synthesize any melanin pigment in the hair, skin, or eyes because the enzyme tyrosinase (gene TYR) is abnormal, so that the hair appears white and eyes are red (Summers, 2009). Pigmentation is a phenotype, and it is not the only phenotype altered in albinism; vision is usually very poor, the eyes are often crossed, and the risk of skin cancers is elevated (Perez-Carpinell, Capilla, Illueca, & Morales, 1992).

Most severe mutations of a specific gene usually alter many phenotypes, so that there is no one-to-one correspondence of a gene and a phenotype (Fig. 10.1a). The DNA in a gene codes for the sequence of amino acids in a protein. What that protein typically does in a living organism depends on the organs in which it is expressed. Effects can be quite diverse when expression of a gene is shut down in some cell lines but not others (Youngson, Chong, & Whitelaw, 2011).

Another example is the serotonin transporter gene (SLC6A4) that is located in band q11.2 of chromosome 17. It is certainly involved in regulation of human behavior, and drugs that alter anxiety (e.g., citalopram, fluoxetine) often act on the serotonin transporter protein that enables the neurotransmitter serotonin to be removed from the synaptic cleft by re-uptake into the presynaptic side where it can be metabolized. The transporter is expressed

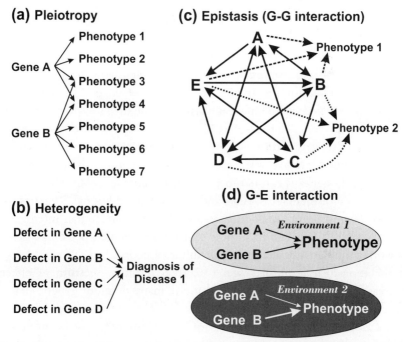

Figure 10.1 (a) A specific gene influences many phenotypes, and most phenotypes are in turn affected by many different genes. No gene simply codes for a particular measurable phenotype, even though the name of a gene sometimes suggests this. (b) Many kinds of genetic defect can result in a very similar phenotypic change in thinking or behavior that warrants the same psychiatric diagnosis. An individual receiving the diagnosis usually suffers from just one of several possible genetic defects. (c) Genes function in networks of interactants. A phenotype maybe influenced by several genes, many of which interact with each other. Some interactions can be unidirectional, while others maybe bidirectional. (d) The influence of a gene on a phenotype depends on the organism's environment, such that the strength of the effect, here represented by arrow thickness, can change more for one gene than another when the environment is changed. The strength of the influence thereby is determined jointly by two factors, and their influences cannot be separated statistically.

in several kinds of tissue, however, including the small intestine that is not generally thought to regulate anxiety. Thus, serotonin mediates several kinds of physiological functions, some of which are also likely to be affected by a drug that is intended to alter anxiety, thereby generating "side effects."

A few genes have highly localized tissue expression and are associated closely with a single phenotypic domain. The opsin genes OPN1LW, OPN1MW, and OPN1SW code for opsin molecules that are most sensitive to long, medium, and short wavelength lights, respectively, in the cones of the retina. Although the cellular localization of the opsins is very limited, the phenotypic behavioral consequences of color blindness are extensive (Sacks, 1997).

4.4. Many Kinds of Genetic Defect can Yield the Same Behavioral Abnormality

Recent large-scale genetic investigations of psychiatric and psychological disorders such as schizophrenia (Sebat, Levy, & McCarthy, 2009; Stefansson et al., 2008) and autism spectrum disorder (Levy et al., 2011; Sanders et al., 2011) have revealed that people can earn the same diagnosis for very different genetic reasons (heterogeneity; Fig. 10.1b). Likewise, a severe mutation in any one of dozens of different genes can impair mental development and reduce intelligence test scores. A recent study of mental disability confirmed the role of 23 genes identified in previous research as being relevant to at least one case of mental disability, and it found an additional 50 genes that had not been reported previously in the literature on human intelligence (Najmabadi et al., 2011). In most instances, the afflicted person possessed only one of the genetic defects, while different people had different defects. At the molecular level, each gene exerts a distinct spectrum of effects, whereas at the level of behavior and mind, many different genetic defects appear to have similar consequences. This is expected when a complex trait results from the combined effects of many different causal influences (Wahlsten, 2011b).

4.5. A Defect in the Genetic Material Possessed by an Individual May Not be Part of Heredity

It has been recognized for many years that some alterations of the chromosomes occur de novo; the child shows a genetic defect that is not present in either parent. A common example is Down syndrome, where an extra copy of chromosome 21 is obtained from one parent, while both parents have normal chromosomes. The person who has the *trisomy*, three copies of chromosome 21, is sterile and never transmits the condition to a child.

Thus, Down syndrome is a genetic defect but is not hereditary. When the defect is quite harmful, it is not likely to be transmitted to the next generation and is generally rare in the population. In fact, in many instances the person suffering from a mental disorder carries a new mutation in an important gene and did not inherit it from either parent (previous Section 4.4). For several complex psychiatric and psychological abnormalities, de novo mutations comprise a large fraction of total cases. The prevalence of de novo mutations for medical disease challenges current views of heredity and what it means to say a disorder is hereditary. Clearly, a defect can be genetic without being hereditary in the sense of transmission across generations. Table 10.3 presents some of the terms in common use and offers concise definitions.

Table 10.3 Genetic terms in common use with concise definitions

Gene	Region of DNA that codes for sequence of amino acids in a specific protein. Regions involved in regulating the gene's expression may also be considered part of the gene, although this definition is not universally accepted.
Genome	The entire sequence of all DNA molecules for a species or individual
Gene locus	The position along the DNA molecule where a specific part of gene occurs.
Genetic	Very general term pertaining to the chemical composition of a gene and its role in development and function.
Genetic polymorphism	Difference in one or more nucleotide bases of a gene that exists in a breeding population of individuals. Different forms of the gene are termed *alleles*.
Genetic marker	A DNA polymorphism that occurs outside the protein coding portion of a specific gene and maybe associated with phenotypic variation.
Genotype	Set of all genes possessed by an individual; may include regulatory regions. Sometimes denotes just the pair of alleles at one specific locus.
Epigenetic	Involves material in chromosomes but is not part of the DNA sequence itself; changes in gene expression transmitted across several cell divisions from early life to the adult; maybe transgenerational to offspring.
Environment	Aggregate of influences originating outside the organism that impinge on it.
Phenotype	Measureable characteristic of the organism, other than DNA, RNA, or epigenetic marks. Could be pigmentation, blood sugar, brain size, IQ, etc.

Continued

Table 10.3 Genetic terms in common use with concise definitions—cont'd

Heredity	Aggregate of things transmitted from parent to offspring in a breeding population. Includes chromosomes as well as maternally transmitted mitochondria but can also include family environment and epigenetic marks.
Hereditary	Tendency for a phenotype to show parent–offspring resemblance.
Heritability	Percent of offspring phenotypic variance in a population associated in a linear manner with individual differences in heredity in the parents; average influence over many relevant genes and many kinds of interactions as well as environmental correlates.
Heritable	Vague term that sometimes means a specific genetic variant is transmitted from parent to offspring and other times means heritability in a population is greater than zero.

4.6. The Biological Effects of a Specific Gene Depend on the Properties of Numerous Other Genes with Which It Interacts

The proteins that are synthesized on the basis of genetic information are parts of complex networks involving many genes and proteins. Accordingly, the protein product of one gene can alter the level of expression of a different gene, and the products of two different genes can contribute to the same chemical process in a cell (Fig. 10.1c). These kinds of biochemical interactions are so prevalent that the term "interactome" is now used to denote the large-scale pattern of interactions among thousands of genes and their products (Li et al., 2004). Internet databases such as GeneCards display some of the more important connections known for each specific gene.

Interaction also has a statistical meaning when there are different variants (alleles) of a gene in a population. If the consequences of a mutation in one gene depend on the individual's genotype at another gene, then there is statistical or epistatic interaction at the population level. Comprehensive assays of thousands of gene–gene pairings in single-celled organisms, such as yeast, have demonstrated the widespread nature and importance of such interactions (Ito et al., 2001). Networks of interactions have also been documented in multicellular animals (Li et al., 2004). Interactions among genes most likely occur at the level of the population because of interactions that take place at the molecular level.

Demonstrating epistasis in mammals is more challenging because of the large samples that are needed. It is feasible in mice by breeding animals afflicted with two different targeted mutations or placing the same mutation on two different genetic backgrounds (Wolfer, Crusio, & Lipp, 2002). Specific instances are known that affect phenotypic outcomes, but no large study of hundreds of pairwise interactions has yet been done in mammals. The lack of controlled breeding in humans severely limits the information available about gene–gene interactions at the population level, but there is nothing known about humans that would exempt us from the phenomenon. The prevalence of gene–gene interactions in smaller organisms strongly suggests that epistasis is widespread in living things.

4.7. Response to a Changed Environment Depends on Genotype and Effects of a Genetic Mutation Depend on Environment

The actions of a gene are not the results of some endogenous, autonomous script contained in the gene itself. The transcription of information in the DNA into mRNA is controlled by intricate mechanisms that help to coordinate the activities of diverse genes during development (Davidson, 2006) and daily life. These mechanisms in many cases are very sensitive to the organism's environment. For example, it has been shown that a sudden increase (heat shock) in the temperature of the medium in which nematode worms are living produces an abrupt increase in the expression levels of more than 28 genes (GuhaThakurta et al., 2002). Likewise, the transition from the light to the dark phase of the day–night cycle alters expression levels of more than 200 genes (Hughes et al., 2007; Yan, Wang, Liu, & Shao, 2008). Sensory experience, social relationships and different kinds of learning exert their influences on behavior in many species through changes in gene expression (Robinson, Fernald, & Clayton, 2008).

One of the most puzzling things about DNA in the recent past was the great abundance (about 98%) of "junk" DNA sequences in mammals that do not code for protein. Some of these are the introns between protein encoding exons within the bounds of a gene, while other regions are outside the limits of a fairly well-defined gene. Recent large-scale analysis of the human genome (ENCODE) has revealed, however, that at least 80% of the DNA has a clear connection with function, and much of it is involved in regulation of gene action (Bernstein et al., 2012; Gerstein et al., 2012). Whereas portions of DNA that control transcription of a gene are promoters that are adjacent to it, other regulatory regions are much further away

(Sanyal, Lajoie, Jain, & Dekker, 2012), and this mosaic of control regions challenges simple definitions of what is a gene. It now appears that many genetic mutations that cause disease do not alter the protein coding portion of the DNA but instead affect the regulation of gene action (Boyle et al., 2012; Maher, 2012; Schaub, Boyle, Kundaje, Batzoglou, & Snyder, 2012). A seemingly normal gene product produced at the wrong time and place can be just as damaging, and maybe even more so, than a change in the protein product itself. The widespread distribution of controlling elements across the genome provides abundant opportunities for alteration of gene expression by chemical reactants as well as the external environment.

There are many specific examples in diverse species showing how a phenotypic response to environmental change can be greater in one genotype at a locus than another (Fig. 10.1d). A recent example is provided by the foraging gene (*for*) in fruit flies where there are two well-studied alleles, rover and sitter, that differ greatly in the extent of foraging activity when the flies are well fed. Under starvation conditions, however, the sitter flies greatly increase their exploratory activity and behaviors of the two genotypes become similar (Burns et al., 2012).

Studies of standard lines or strains of lab animals that differ at many genes have provided dramatic examples of interaction with environment. An experiment was done with lines of mice bred selectively for high or low amounts of fighting with an intruder mouse (Hood & Cairns, 1989). The original selection study had been conducted with males who were housed singly for several weeks before the fighting bouts. When untested males were instead housed in groups, however, none of them, not even those from the high line, showed much fighting during the test.

Examples of interactions involving specific genes are also abundant in humans (Caspi & Moffitt, 2006). For example, in the recessive genetic disorder phenylketonuria, those with two copies of the defective gene coding for the enzyme phenlyalanine hydroxylase are much more sensitive than normal individuals to the amount of the amino acid phenylalanine in the diet (Scriver & Waters, 1999).

A widely discussed instance of interaction involves a polymorphism in the promoter region of the serotonin transporter gene (5-HTT, now SLC6A4); there are three common genotypes termed s/s, l/l, or s/l involving short (s) or long (l) alleles. Caspi et al. (2003) and Caspi and Moffitt (2006) studied a birth cohort of more than 1000 New Zealanders, 17% of whom had reported a depressive episode by age 26. These researchers also had gathered information on their subjects' experiences of stressful life events. In general, the more stressful events that were experienced between ages 16 and 21, the higher the risk of depression, but the relation was not strong for individuals homozygous for the long allele, whereas s/s homozygotes and l/s heterozygotes showed a substantial increase in depressive symptoms after several stressors. Thus, an environmental influence was differentially effective depending on genotype at a single important gene. This study also suggests that some genetic effects will be apparent only when relevant environmental risk factors are also taken into account. There has been some controversy about the finding, with one metaanalysis of several published studies failing to support it (Risch et al., 2009) and a more comprehensive analysis substantiating it (Karg, Burmeister, Shedden, & Sen, 2011).

Because there are no pure strains of humans and their environments cannot be rigorously controlled, it is not possible to demonstrate the importance of gene–environment interactions in any general sense. Furthermore, many realistic kinds of interaction are very difficult to detect because of the low statistical power (Wahlsten, 1990, 1991, 2011a). Nevertheless, the preponderance of evidence on this topic from experimental animals and single-gene variations in humans strongly suggests that interactions play an important and pervasive role in human behavioral development.

4.8. Environmentally Induced Changes in an Individual May Become Part of Heredity

Several situations are recognized where events during early development alter regulatory elements along the DNA molecule and the change lasts into adult life (Szyf, 2009). Such changes must be transmitted from one cell to another when division occurs by mitosis. This is a kind of cell-to-cell transmission that we do not term hereditary because it does not extend to the next generation of individuals. One mechanism underlying such change is binding of methyl groups that can silence a gene (Fagiolini, Jensen, & Champagne, 2009). Other kinds of chemical alterations of chromosomal material can mediate similar changes. Those *epigenetic* effects were initially documented in transformations of normal cells into cancerous tumors. Regions where the changes occur comprise the *epigenome* (Szyf, 2009). Subsequently, it was found that epigenetic transformation is important in the differentiation of the brain into various kinds of cells (Davies et al., 2012; Feng, Fouse, & Fan, 2007). Furthermore, epigenetic changes can be occasioned by social experience and nutritional insufficiency (Szyf, 2009). Sites along the DNA where methyl groups are known to bind are numerous and widespread (ENCODE), yielding a pattern of DNA binding termed the *methylome*, and this pattern implies that epigenetic changes are pertinent to a wide range of developmental phenomena.

Several reports of transmission of epigenetic changes across generations have caused us to rethink the nature of heredity (Skinner, Anway, Savenkova, Gore, & Crews, 2008; Wolstenholme et al., 2012; Youngson & Whitelaw, 2008). Transgenerational epigenetic effects appear to be part of heredity, but they do not involve changes in the DNA sequence itself. They must involve continuity of the epigenetic modification across meiotic cell division that creates gametes. Examples in plants are well known, but few of these are bona fide instances where the epigenetic change was induced by altered environment (Paszkowski & Grossniklaus, 2011). Likewise in mammals, transmission of environmental modifications across two or more generations is not well established. There are specific instances that seem to qualify, but crucial control conditions have not always been included in the experimental designs, and the preponderance of evidence suggests that "transgenerational inheritance via the gametes is likely to be rare" (Daxinger & Whitelaw, 2012).

By definition, gene or genotype consists of nucleotide base sequence in a DNA molecule, while heredity can extend well beyond this narrow domain. It is clear that environmental conditions that have an impact on a parent, especially the mother, can cause epigenetic changes in the offspring. Thus, epigenetic effects can clearly give rise to parent–offspring correlation. Whether they can extend beyond the first generation of offspring remains an open question. The issue is a serious one for the human behavior genetics, because parent–offspring and sibling correlations are the crucial facts justifying imputation of genetic influence. Substantiation of hereditary influences across two or more generations of humans is almost completely absent from the published literature, and in any event, is very difficult to interpret because of major environmental changes over several decades.

5. IMPLICATIONS FOR STATISTICAL ANALYSIS OF HEREDITY

Phenotypes, including behavioral phenotypes, are generated by interacting systems of genes that are also responsive to diverse aspects of the organism's environment. Even seemingly simple phenotypes such as pigmentation are quite complex and dependent on numerous genes. The number of genes involved in what psychologists regard as truly complex traits is very large indeed. The role of more than 1000 distinct genes would need to be taken into account in any reasonably complete picture of how a behavioral phenotype emerges and is regulated.

Great progress has been made in understanding the molecules and molecular interactions that are involved in normal brain development and neural function in adults. Many neurotransmitters, neurohormones, and complex systems of interactants have been thoroughly explored, and drugs have been devised to alter specific parts of these systems. Gene knockout methods have been employed in experimental animals, especially mice, to investigate which parts of a system are critically important for function. The roles of hundreds of specific genes in regulating normal behavioral domains have been studied and are reasonably well understood. The molecular genetic discoveries from the Human Genome Project and ENCODE are of considerable use in research that investigates function at the molecular level.

Among those hundreds or thousands of genes that are important for a psychological domain such as anxiety or intellectual ability, a few probably involve genetic polymorphisms in a population that contribute to individual phenotypic differences, whereas many other genes are effectively the same in almost all members of a population. For the psychology of individual differences, it has been critically important to identify some of the relevant genetic polymorphisms. For phenotypes in the normal range of human psychological variation, all attempts to locate such genes have failed, despite the almost heroic efforts of many well-funded investigators. It is clear that any such genetic polymorphisms involve variants that have very small phenotypic effects, each accounting for less than 0.1% of phenotypic variance. Hence, the culprit genes remain hidden. Nothing of their nature or functions is known whatsoever.

The models used for the statistical genetic analysis of human behavior assume effects of genes and environments are additive and independent, and they also assume the effects of the various genes are also independent. It is now very clear from modern molecular genetic research that these assumptions are not valid. Different genes do interact with each other on a large scale as expressed in the interactome. The activities of many genes are regulated by features of both the internal and external environments, and responses to environmental change are dependent on genotype. At one time, critics of quantitative genetic analysis argued that the models in common use were unverifiable and, as such, lack scientific credibility. Now we can take the critique one step further and argue that the models are just plain wrong (Wahlsten, 1994); they lack validity and violate principles that are firmly based on a large body of research in molecular genetics. Gene–gene and gene–environment interactions are ubiquitous and important. Any statistical model that posits they are absent is just a figment of a wishful and blissfully ignorant imagination. Many theorists in psychological science debate the statistical genetic issues earnestly, but to the modern genetic scientist swimming happily in a rich sea of brightly colored facts those theorists appear more like ancient creatures struggling laboriously through a lifeless mire of irrefutable conjectures.

Recent observations call into question the interpretation of simple correlations among relatives in studies of human behavior. For one thing, it is now apparent that parent–offspring correlation is not a valid indicator of strength of genetic influence. Epigenetic and other transgenerational environmental effects can inflate the correlation for nongenetic reasons, especially when these data involve only first degree relatives, while de novo mutations can deflate it for real genetic reasons. If a population parameter such as heritability based on correlations among relatives cannot be estimated in a manner that is free from implausible assumptions, bias and artifact, it has no valid implications for educational policies or medical practices. The calculation of heritability should mainly be explained to today's students when teaching them about the history of the field.

6. CONCLUSIONS

The teaching of genetics to students in Psychology needs to be revised to reflect our current understanding of what genes are and how they work. A vast amount is known about the role of genes in brain development and function. This knowledge is crucial for understanding the origins of many bona fide genetic disorders as well as the mechanisms of actions of many drugs and other signals in the environment. We should explain to students why quantitative genetic analysis of phenotypic variation in the normal range has not been successful and is not likely to yield important insights into causes of human behaviors.

Funding of large-scale studies that search for genes affecting intelligence and personality should be curtailed. Recent failures to detect genes with small or moderate effects have occurred despite good methodology with very large samples and huge numbers of genetic markers. Some scientists now call for new studies with even larger samples. This would be an extravagant waste of public wealth. If a study of vast proportions costing many millions of dollars were to find a few genes with small effects, nothing of practical or even theoretical use could be done with such knowledge, although it might provide a little morale boost for the exponents of quantitative genetics in psychology.

REFERENCES

Alberts, B., Johnson, A., Lewis, J., Roberts, K., & Walter, P. (2008). *Molecular biology of the cell* (5th ed.). New York: Garland Science.

Bernstein, B. E., Birney, E., Dunham, I., Green, E. D., Gunter, C., & Snyder, M. (2012). An integrated encyclopedia of DNA elements in the human genome. *Nature, 489,* 57–74.

Birney, E. (2012). The making of ENCODE: lessons for big-data projects. *Nature, 489,* 49–51.

Boyle, A. P., Hong, E. L., Hariharan, M., Cheng, Y., Schaub, M. A., Kasowski, M., et al. (2012). Annotation of functional variation in personal genomes using RegulomeDB. *Genome Research, 22,* 1790–1797.

Burns, J. G., Svetec, N., Rowe, L., Mery, F., Dolan, M. J., Boyce, W. T., et al. (2012). Gene-environment interplay in *Drosophila melanogaster*: chronic food deprivation in early life affects adult exploratory and fitness traits. *Proceedings of the National Academy of Sciences of the United States of America, 109*(Suppl. 2), 17239–17244.

Caspi, A., & Moffitt, T. E. (2006). Gene-environment interactions in psychiatry: joining forces with neuroscience. *Nature Reviews Neuroscience, 7,* 583–590.

Caspi, A., Sugden, K., Moffitt, T. E., Taylor, A., Craig, I. W., Harrington, H., et al. (2003). Influence of life stress on depression: moderation by a polymorphism in the 5-HTT gene. *Science, 301,* 386–389.

Chabris, C. F., Hebert, B. M., Benjamin, D. J., Beauchamp, J., Cesarini, D., van der Loos, M., et al. (2012). Most reported genetic associations with general intelligence are probably false positives. *Psychological Science.*

Davidson, E. H. (2006). *The regulatory genome.* Amsterdam: Academic/Elsevier.

Davidson, E. H., & Erwin, D. H. (2006). Gene regulatory networks and the evolution of animal body plans. *Science, 311,* 796–800.

Davies, G., Tenesa, A., Payton, A., Yang, J., Harris, S. E., Liewald, D., et al. (2011). Genome-wide association studies establish that human intelligence is highly heritable and polygenic. *Molecular Psychiatry, 16,* 996–1005.

Davies, M. N., Volta, M., Pidsley, R., Lunnon, K., Dixit, A., Lovestone, S., et al. (2012). Functional annotation of the human brain methylome identifies tissue-specific epigenetic variation across brain and blood. *Genome Biology, 13,* R43.

Daxinger, L., & Whitelaw, E. (2012). Understanding transgenerational epigenetic inheritance via the gametes in mammals. *Nature Reviews Genetics, 13,* 153–162.

Fagiolini, M., Jensen, C. L., & Champagne, F. A. (2009). Epigenetic influences on brain development and plasticity. *Current Opinion in Neurobiology, 19,* 207–212.

Feng, J., Fouse, S., & Fan, G. (2007). Epigenetic regulation of neural gene expression and neuronal function. *Pediatric Research, 61,* 58R–63R.

Fisher, R. A. (1918). The correlation between relatives on the supposition of Mendelian inheritance. *Transactions of the Royal Society of Edinburgh, 52,* 399–433.

Flint, J. (2011). Personality genetics. In *Challenging genetic determinism* (pp. 78–98). Montreal: McGill-Queen's University Press.

Gerstein, M. B., Kundaje, A., Hariharan, M., Landt, S. G., Yan, K. K., Cheng, C., et al. (2012). Architecture of the human regulatory network derived from ENCODE data. *Nature, 489,* 91–100.

Gottlieb, G. (1992). *Individual development and evolution. The genesis of novel behavior.* New York: Oxford University Press.

Gottlieb, G. (2007). Probabilistic epigenesis. *Developmental Science, 10,* 1–11.

GuhaThakurta, D., Palomar, L., Stormo, G. D.,Tedesco, P., Johnson, T. E., Walker, D. W., et al. (2002). Identification of a novel cis-regulatory element involved in the heat shock response in *Caenorhabditis elegans* using microarray gene expression and computational methods. *Genome Research, 12,* 701–712.

Hood, K. E., & Cairns, R. B. (1989). A developmental-genetic analysis of aggressive behavior in mice: 4. Genotype-environment interaction. *Aggressive Behavior, 15,* 361–380.

Hughes, M., Deharo, L., Pulivarthy, S. R., Gu, J., Hayes, K., Panda, S., et al. (2007). High-resolution time course analysis of gene expression from pituitary. *Cold Spring Harbor Symposia on Quantitative Biology, 72,* 381–386.

Ito, T., Chiba, T., Ozawa, R., Yoshida, M., Hattori, M., & Sakaki, Y. (2001). A comprehensive two-hybrid analysis to explore the yeast protein interactome. *Proceedings of the National Academy of Sciences of the United States of America, 98,* 4569–4574.

Johannsen, W. (1911). The genotype conception of heredity. *American Naturalist, 45,* 129–159.

Karg, K., Burmeister, M., Shedden, K., & Sen, S. (2011). The serotonin transporter promoter variant (5-HTTLPR), stress, and depression meta-analysis revisited: evidence of genetic moderation. *Archives of General Psychiatry, 68,* 444–454.

Lee, S. M., & Bressler, R. (1981). Prevention of diabetic nephropathy by diet control in the *db/db* mouse. *Diabetes, 30,* 106–111.

Lees-Miller, J. P., Goodwin, L. O., & Helfman, D. M. (1990). Three novel brain tropomyosin isoforms are expressed from the rat alpha-tropomyosin gene through the use of alternative promoters and alternative RNA processing. *Molecular and Cellular Biology, 10,* 1729–1742.

Levy, D., Ronemus, M., Yamrom, B., Lee, Y. H., Leotta, A., Kendall, J., et al. (2011). Rare de novo and transmitted copy-number variation in autistic spectrum disorders. *Neuron, 70,* 886–897.

Li, S., Armstrong, C. M., Bertin, N., Ge, H., Milstein, S., Boxem, M., et al. (2004). A map of the interactome network of the metazoan *C. elegans. Science, 303,* 540–543.

Lush, J. L. (1945). *Animal breeding plans.* Ames, Iowa: Iowa State College Press.

Maher, B. (2012). ENCODE: the human encyclopaedia. *Nature, 489,* 46–48.

Matlin, A. J., Clark, F., & Smith, C. W. (2005). Understanding alternative splicing: towards a cellular code. *Nature Reviews Molecular Cell Biology, 6,* 386–398.

Morgan, T. H. (January 1914). The mechanism of heredity as indicated by the inheritance of linked characters. *Popular Science Monthly,* 1–16.

Najmabadi, H., Hu, H., Garshasbi, M., Zemojtel, T., Abedini, S. S., Chen, W., et al. (2011). Deep sequencing reveals 50 novel genes for recessive cognitive disorders. *Nature, 478,* 57–63.

Pan, Q., Shai, O., Lee, L. J., Frey, B. J., & Blencowe, B. J. (2008). Deep surveying of alternative splicing complexity in the human transcriptome by high-throughput sequencing. *Nature Genetics, 40,* 1413–1415.

Paszkowski, J., & Grossniklaus, U. (2011). Selected aspects of transgenerational epigenetic inheritance and resetting in plants. *Current Opinion in Plant Biology, 14,* 195–203.

Perez-Carpinell, J., Capilla, P., Illueca, C., & Morales, J. (1992). Vision defects in albinism. *Optometry & Vision Science, 69,* 623–628.

Posthuma, D., & de Geus, E. J.C. (2006). Progress in the molecular-genetic study of intelligence. *Current Directions in Psychological Science, 15,* 151–155.

Risch, N., Herrell, R., Lehner, T., Liang, K. Y., Eaves, L., Hoh, J., et al. (2009). Interaction between the serotonin transporter gene (5-HTTLPR), stressful life events, and risk of depression: a meta-analysis. *JAMA, 301,* 2462–2471.

Robinson, G., Fernald, R., & Clayton, D. (2008). Genes and social behavior. *Science, 322,* 896–900.

Sacks, O. (1997). *The island of the colorblind.* New York: Knopf.

Sanders, S. J., Ercan-Sencicek, A. G., Hus, V., Luo, R., Murtha, M. T., Moreno-De-Luca, D., et al. (2011). Multiple recurrent de novo CNVs, including duplications of the 7q11.23 Williams syndrome region, are strongly associated with autism. *Neuron, 70*, 863–885.

Sanyal, A., Lajoie, B. R., Jain, G., & Dekker, J. (2012). The long-range interaction landscape of gene promoters. *Nature, 489*, 109–113.

Schaub, M. A., Boyle, A. P., Kundaje, A., Batzoglou, S., & Snyder, M. (2012). Linking disease associations with regulatory information in the human genome. *Genome Research, 22*, 1748–1759.

Scriver, C. R., & Waters, P. J. (1999). Monogenic traits are not simple - lessons from phenyl-ketonuria. *Trends in Genetics, 15*, 267–272.

Sebat, J., Levy, D. L., & McCarthy, S. E. (2009). Rare structural variants in schizophrenia: one disorder, multiple mutations; one mutation, multiple disorders. *Trends in Genetics, 25*, 528–535.

Shaked, I., Zimmerman, G., & Soreq, H. (2008). Stress-induced alternative splicing modulations in brain and periphery: acetylcholinesterase as a case study. *Annals of the New York Academy of Sciences, 1148*, 269–281.

Singh, R. K., Tapia-Santos, A., Bebee, T. W., & Chandler, D. S. (2009). Conserved sequences in the final intron of MDM2 are essential for the regulation of alternative splicing of MDM2 in response to stress. *Experimental Cell Research*. 10.1016/j.yexcr.2009.07.017.

Skinner, M. K., Anway, M. D., Savenkova, M. I., Gore, A. C., & Crews, D. (2008). Transgenerational epigenetic programming of the brain transcriptome and anxiety behavior. *PLoS ONE, 3*, e3745.

Stefansson, H., Rujescu, D., Cichon, S., Pietilainen, O. P., Ingason, A., Steinberg, S., et al. (2008). Large recurrent microdeletions associated with schizophrenia. *Nature, 455*, 232–236.

Summers, C. G. (2009). Albinism: classification, clinical characteristics, and recent findings. *Optometry & Vision Science, 86*, 659–662.

Szyf, M. (2009). The early life environment and the epigenome. *Biochimica et Biophysica Acta, 1790*, 878–885.

Wahlsten, D. (1982). Genes with incomplete penetrance and the analysis of brain development. In I. Lieblich (Ed.), *Genetics of the Brain* (pp. 367–391). Amsterdam: Elsevier.

Wahlsten, D. (1990). Insensitivity of the analysis of variance to heredity-environment interaction. *Behavioral and Brain Sciences, 13*, 109–120.

Wahlsten, D. (1991). Sample size to detect a planned contrast and a one degree-of- freedom interaction effect. *Psychological Bulletin, 110*, 587–595.

Wahlsten, D. (1994). The intelligence of heritability. *Canadian Psychology, 35*, 244–258.

Wahlsten, D. (1999). Single-gene influences on brain and behavior. *Annual Review of Psychology, 50*, 599–624.

Wahlsten, D. (2011a). *Mouse behavioral testing*. London, UK: Elsevier.

Wahlsten, D. (2011b). Special challenges of complex behavioural traits: discovery and applications. In L. Maheu & R. A. Macdonald (Eds.), *Challenging genetic determinism. New perspectives on the gene in its multiple environments* (pp. 51–77). Montreal: McGill-Queen's University Press.

Wahlsten, D. (2012). The hunt for gene effects pertinent to behavioral traits and psychiatric disorders: from mouse to human. *Developmental Psychobiology, 54*, 475–492.

Wahlsten, D., Bishop, K. M., & Ozaki, H. S. (2006). Recombinant inbreeding in mice reveals thresholds in embryonic corpus callosum development. *Genes, Brain and Behavior, 5*, 170–188.

Wolfer, D. P., Crusio, W. E., & Lipp, H. P. (2002). Knockout mice: simple solutions to the problems of genetic background and flanking genes. *Trends in Neurosciences, 25*, 336–340.

Wolstenholme, J. T., Edwards, M., Shetty, S. R., Gatewood, J. D., Taylor, J. A., Rissman, E. F., et al. (2012). Gestational exposure to bisphenol a produces transgenerational changes in behaviors and gene expression. *Endocrinology, 153*, 3828–3838.

Yan, J., Wang, H., Liu, Y., & Shao, C. (2008). Analysis of gene regulatory networks in the mammalian circadian rhythm. *PLoS Computational Biology*, *4*, e1000193.

Youngson, N. A., Chong, S., & Whitelaw, E. (2011). Gene silencing is an ancient means of producing multiple phenotypes from the same genotype: common mechanisms and functions in epigenetic processes can be seen throughout all life forms. *BioEssays*, *33*, 95–99.

Youngson, N. A., & Whitelaw, E. (2008). Transgenerational epigenetic effects. *Annual Review of Genomics and Human Genetics*, *9*, 233–257.

Zhang, Y., Proenca, R., Maffei, M., Barone, M., Leopold, L., & Friedman, J. M. (1994). Positional cloning of the mouse obese gene and its human homologue. *Nature*, *372*, 425–432.

CHAPTER ELEVEN

Genetic Causation: A Cross Disciplinary Inquiry

Sheldon Krimsky

Department of Urban and Environmental Policy and Planning, Tufts University, Medford, MA, USA
E-mail: sheldon.krimsky@tufts.edu

Contents

Abstract

The growth of genome-wide and Candidate Gene Association Studies have elevated genetic causality in medicine and behavioral science. This chapter explores the concept of causality as it has been applied in genetic explanations and distinguishes the varieties of methods used to establish genetic causation. The essay ends with a cautionary note on applying genetic causation to compelx human behaviors and neurocognitive abnormalities.

"Science is about causes, period."

E. Turkheimer (Turkheimer, 2011).

"Causality in any particular form does not need to be a feature of all successful scientific explanations."

D. Noble (Noble, 2008).

1. INTRODUCTION

How we ascertain causes and find agreement about causes depends largely on the methods and tools of science—methods that vary among the disciplines. The issue of genetic causation began to gnaw at me when

Advances in Child Development and Behavior, Volume 44
ISSN 0065-2407, http://dx.doi.org/10.1016/B978-0-12-397947-6.00011-8

I started to read published studies that found genetic causes or genetic determinants for: breakfast eating patterns, (Keski-Rahkoren et al., 2004) loneliness, (Boomsma et al., 2006) religiousness, (Koenig et al., 2005) agreeableness, (Garpenstrand et al., 2002) voting behavior, novelty seeking, (Strobel, Wehr, Michel, & Brocke, 1999) altruism, cooperation, (Mertins et al., 2011) credit card debt, (De Neve & Fowler, 2010) antisocial behavior, binge eating and drinking, criminal behavior, (DeLisi, Beaver, Vaughn, & Wright, 2009) attitudes of fairness, attitudes toward infidelity, aggression, (McDermott et al., 2009) individualism vs collectivism, number of sexual partners, political ideology, shyness, utilitarian moral judgments, and social networking (Fowler, Settle, & Christakis, 2011). Genetics is beginning to play a prominent role in subdisciplines of political science, developmental and behavioral psychology, and anthropology. Genetic causation, whether weak or strong, deterministic or probabilistic, is a term introduced through a variety of methods in these disciplines, the results of which have gained little attention in the field of genetics itself. Nevertheless, the papers connecting genes to behavior are published in some of the premier journals in the social and behavioral sciences. Genetic causation is also used in disciplines that seek a genetic etiology of disease. This chapter begins with a discussion of "causality" generally and then focuses on "causality" in genetics. I shall argue that for many claims the concept of "genetic cause" does not stand up to critical scrutiny.

2. CAUSALITY IN SCIENCE

The term "causality" has a long lineage in philosophy and in the history of science. Aristotle introduced four causes: material, formal, efficient, and final cause. The material cause is the composition of an object—thus in biochemistry the material cause of a protein is its composition of amino acids. Its formal cause is the arrangement of the matter in it or the three-dimensional structure of the amino acids making up the protein. The efficient cause of a substance is what changes its motion—forces acting on it like an electric field acting on a molecule. The final cause is the purpose it serves or the end to which it is directed. For example, a protein designed to deliver oxygen through the blood stream would be its final cause. During the growth of modern science with its emphasis on the physics of moving bodies, the efficient and material causes became the primary focus of philosophical discussion. The expression "A is the Cause of B" meant that "A" preceded "B" in time and that "A" was a sufficient condition for the determination of "B". Newtonian causality was based on law-like or nomological statements such as "all metals expand when heated." Some classical views of causality locate a power in the cause such as "A has the power to bring about B." In other views, A and B are causally related when they appear together, but not necessarily before or after—as in an antecedent and consequent events.

Newtonian causality also manifests a functional relationship between A and B. For example, the ball "B" of mass "m" was brought to acceleration "a" by force A (efficient cause) according to the relationship "$F = ma$". Certain fields of natural science connect causality to prediction and claim there is symmetry between causal explanation and prediction. To say that "A" is the cause of "B" is to say that from "A" and law L we can predict the occurrence of "B" by logic or statistical inference.

3. CAUSALITY IN MODERN GENETICS

Modern genetics, both in its scientific and cultural discourse, has introduced causality in various forms. Genes are said to be the templates for the synthesis of proteins. In response to the question, "what caused this protein to be abnormal?", the answer is "the gene that codes for the protein had a mutation." Causality, in this sense, means that the mutated gene is the reason for the appearance of the abnormal protein.

It is also common to hear the term "genes" used as the cause of a trait, such as "She has a gene for blue eyes," that suggests that her genes caused or were determinative of her "blue eyes". In addition, genes are cited as a cause of a disease or developmental abnormality such as the genes for amyotrophic lateral sclerosis or achondroplasia (dwarfism). Genetic causes have also been cited for behavioral characteristics, such as aggression, infidelity, and political choices. Finally, the idea that a gene or a group of genes can predispose someone to an effect is another form of a causal statement. To say that A presupposes someone to an effect B is to declare A necessary (with some probability) but not sufficient for B. Before we can fully understand the different forms of causality used in genetics, it would be useful to provide an overview of the changes that have taken place in the theory of molecular genetics over the past 50 years.

4. GENETIC THEORY IN TRANSITION

During the last half century in which genetic theory developed many hypotheses and revisions to those hypotheses were made regarding the nature of the animal and plant genomes, the relationship between the genes and phenotype (such as disease, behavior, physical traits), gene–environment interactions, the process by which cells decode the information stored on DNA, and the estimated number of genes in the human chromosome. Even the concept of the gene has been revised. The early concept of the genome was likened to a Lego structure, composed of segments of DNA linked together and differentiated by the sequence of four nucleotides (the bases A, G, C, T).

An early physical model of the DNA molecule, which James Watson and Francis Crick had fabricated, consisted of metallic double helix "made of flat plates of galvanized metal with narrow brass tubes for bonds."(Ridley, 2003, p. 70). This static model served as an early representation of the structure of the DNA molecule. Initially, genes were viewed as the segments of DNA, which held coding information about proteins.

While some segments of the DNA in the human genome have a coding function, most of the 3 billion base pairs were considered junk DNA with no coding or regulatory functions—the flotsam and jetsam of evolution. The remaining functional DNA (estimated at about 2% of the genome) was divided into discrete segments, each assigned to the coding for one of the unique 100,000 proteins in the human body. Recently the junk DNA hypothesis has been questioned as scientists have discovered that more and more of the non-coding DNA is transcribed into RNA with uncharacterized functions. Also, new estimates put the number of human genes at between 20,000–30,000, signaling that some DNA segments contain the code for more than one protein.

Francis Crick postulated the Central Dogma of molecular genetics theory in 1958 at a meeting of the Society of Experimental Biology in a talk titled "On Protein Synthesis". According to Crick, genetic information transfers from nucleic acid (DNA or RNA) to nucleic acid, or from nucleic acid to protein, but never from protein to nucleic acid. In other words, proteins do not contain the information for duplicating themselves. The Central Dogma has often been simplified as "DNA makes RNA makes Protein." Early popular conceptions of the genetic mechanism gave the false impression that DNA is a self-actualizing master molecule. In fact, proteins play a critical role in directing the orchestral process of protein synthesis (Proteins). "DNA may be a large complex molecule, but alone it does nothing. It does not have powers of self-replication, nor [does it have power] to create new generations of life"(Richards, 2002). In a popular magazine article, Barry Commoner gave a more realistic view of the role of proteins in all aspects of DNA replication and transcription. "…in the living cell the gene's nucleotide code can be replicated faithfully only because an array of specialized proteins intervenes to prevent most of the errors—which DNA by itself is prone to make and or repair the few remaining ones…genetic information arises not from DNA alone but through its essential collaboration with protein enzymes—a contradiction of the central dogma's precept that inheritance is uniquely governed by the self-replication of the DNA double helix."(Commoner 2002).

Initially, molecular geneticists believed that the function of a gene was to control the production of a single polypeptide. Then it was discovered that genes carried the code for forms of RNA that do not become polypeptides. From the late 1960s to the present, the details of the Central Dogma have been filled in or revised with some variations in how information flows in viruses and retroviruses. An example of replication without nucleic acid is given by prions, newly discovered molecules responsible for mad cow (kuru) disease. Prions can replicate even though they do not contain nucleic acid. They alter normal brain proteins to adapt to the prion's shape, thus in a sense, replicating themselves. If information can be transferred from protein to protein, then this raises in question one of the core ideas in the Central Dogma of molecular biology.

The Lego model of the genome is as simplistic as the Bohr model of the atom. Rather than genes being fixed entities in a static structure waiting to be self-activated, the current conception views the genome as more characteristic of an ecosystem—more fluid, more dynamic, and more interactive than the Lego model implies. "…the assumption that identifiable bits of DNA sequence are even"genes"for particular proteins has turned out not to be generally true. Alternative splicing of fragments of particular sequences, alternative reading frames, and post-transcriptional editing—some of the things that happen between the transcription of DNA and the formatting of a final protein product—are among the processes the discovery of which had led to a radically different view of the genome"(Dupré, 2008).

Within a decade, scientists began to acknowledge that such a view was far too simplistic and the complexity of the genome began to reveal itself. To begin, gene–gene interactions defy a linear model of genetic causality represented by: DNA → RNA → Protein → Disease. Second, DNA sequences may express different products when situated in different parts of the chromosome (the position effect), which means that the role of DNA must be seen in the context of other parts of the cell and organism. Also, the same segment of DNA may be read differently in the same location because of different reading frames (Li et al., 2011). Furthermore, genes may function differently during different periods in a person's life.

By 2001, scientists at the Food and Drug Administration recognized this complexity when reviewing food safety issues arising from genetically modified crops. An agency document included the following statement: "It is also possible with bioengineering that the newly introduced genetic material may be inserted into the chromosome of a food plant in a location that causes the food derived from the plant to have higher levels than normal of toxins, or lower levels of a significant nutrient. In the former case the food may not be safe to eat, or may require special preparation to reduce or eliminate the toxic substance. In the latter case the food may require special labeling, so that consumers would know that they were not receiving the level of nutrients they would ordinarily expect from consuming a comparable food."(Food and Drug Administration, 2001) This model of the plant genome is far afield from the piano metaphor where the added or subtracted key does not interact with the other keys or affect the system as a whole, other than to add a new protein. Also, there are other reasons to revise the idea of DNA → RNA → Protein. According to a study published in 2001, the authors found that information in DNA is not always faithfully transferred to RNA in transcription—RNA bases did not match the corresponding DNA sequence (Mingyao et al., 2011).

DNA once thought to be the fixed components of heredity is now viewed as a much more fluid idea. Genes may change during a person's development. Genes interact with one another as well as with the environment. Genes may be turned on and off resulting from factors outside the DNA, as exemplified by loss-of-function (LoF) variants in the human genome with at least 20 genes having been completely lost (Quintana-Murci, 2012). The epigenome, which regulates how and when genes get expressed, can be inherited through the germline.

5. VARIETIES OF GENETIC CAUSATION

Causality in genetics is determined by different methods used by scientists in different disciplines. For my first example, genetic causality can be inferred from laboratory experiments. In knock-out mice certain genes are removed from the egg before fertilization. Those mice can be compared with the non-knock out variety (the controls). Scientists can examine gene expression, a particular pathway in the gene–protein complex or phenotypic effects—even behavioral change in the animal. Replicability and the use of controls are expected of such experiments for establishing causation.

The argument that causation can be established is as follows. A and B are genetically identical mice except that B has one gene (g_x) deleted. A has characteristic C; B lacks characteristic C. The studies can be replicated and the effects are totally deterministic. When g_x then "C" is observed; when not-g_x then "C" is not observed. Therefore, g_x causes "C"(Dyck et al., 2009; Quintana-Murci, 2012). Even in this highly controlled experiment, the gene in question does not act independently of the proteins, lipids, and cellular machinery such as the ribosomes that go into synthesizing the protein—whether normal or defective. Strictly speaking, the gene is never a sufficient cause of the abnormality. But, as a matter of convention, we can call it the cause if everything else remains constant. This may be traced to John Stuart Mill's Method of Difference in his Methods of Induction.

> *"If an instance in which the phenomenon under investigation occurs, and an instance in which it does not occur, have every circumstance in common save one, that one occurring only in the former; the circumstance in which alone the two instances differ, is the effect, or the cause, or an indispensable part of the cause, of the phenomenon."*
>
> **(Mill, 1859, p. 225).**

A simple model of genetic causation in human disease is illustrated by Phenylketonuria (PKU). This is an autosomal recessive metabolic genetic disorder characterized by a mutation in the gene for the liver enzyme phenylalanine hydroxylase (PAH), rendering it nonfunctional. This enzyme is necessary to metabolize the amino acid phenylalanine (Phe) to the amino acid tyrosine. Without the enzyme, the Phe builds up and can cause mental retardation. Recessive Gene Mutation + phenylalanine → disease. This causal factor is fully predictable based on a genetic marker. One author explained the use of "causality" in such cases. "The Causal nature of Mendelian (recessive or dominantly acting) mutations is fairly easy to establish because there is essentially a strong correspondence between the presence of a mutation and a disease phenotype" (Geschwind, 2011). While it is not probabilistic, the intensity of the effect can vary depending on the amount of the amino acid Phe in the diet. Phe must also be seen as a "necessary" but not sufficient cause. In a world where Phe were ubiquitous in the diet, the cause would be both necessary and sufficient as discussed above in Mill's Methods.

When we get to diseases involving many genes, causality is much more complex not only because the genetic components are acting nonlinearly, but because the disease phenotype is diverse and may be arising from different pathways. This is the case with autism spectrum disorder (ASD). The language of the neurogeneticists is cautious with respect to causality. The methods of analysis include whole genome association studies seeking to find Copy Number Variants that could correlate with ASD. Instead of "causality," scientists speak about "susceptibility genes." "Several dozen ASD susceptibility genes have been identified in the past decade, collectively accounting for 10–20% of ASD cases"(Geschwind, 2011). No specific gene could be found for the majority of ASD; the most common susceptibility genes account for not more than 1–2% of cases.

A third method of establishing some form of causation in the social sciences involves the use of statistical techniques in combination with twin studies. Although in other behavior genetics work, estimates are made of shared and nonshared environmental variance, in this instantiation of behavior genetics research, monozygote and dizygote twins are compared for some variable, and the environment is assumed constant for each twin pair. If the phenotype is more highly correlated with monozygote twins than dizygote twins then it is deemed heritable— or genetically caused. The National Longitudinal Study of Adolescent Health had a sample of 90,000 adolescents and 1,110 twins that continues to be mined because the survey contains genetic factors (Fowler, Dawes, & Christakis, 2009).

There has been considerable debate about the methodology of twin studies, especially the assumption of environmental uniformity among twins and "whether the influences of genes as opposed to environment on a given trait can be neatly partitioned into percentages…."(Charney, 2008a). Among the claims made by using twin methodology are: "Genes predict voter turnout";"Genes found for congeniality"; and "genes influence social networks".

A fourth method for establishing genetic causality, albeit indirect causality, is the method of Candidate Gene Associations (CGA). This involves large datasets of genetic information (polymorphisms). Candidate genes are selected from prior studies where associations have been found. Investigators use statistical techniques to correlate particular alleles with a phenotype and ascertain whether a gene is more frequently seen in participants with the disease than in participants without the disease. No one claims that the candidate gene studies by themselves yield causality. The phenotype could be a disease or a behavior. If there is a strong correlation between the gene variant and the phenotype, the gene is said to predict the behavior in question. Some of the results from these studies have claimed: "gene said to predict voting behavior,"(Fowler & Dawes, 2008) "…we hypothesize that people with more transcriptionally efficient alleles of the MAOA and 5HTT genes are more likely to vote"(Fowler & Dawes, 2008). Other studies assert that genes account for "punishing behavior" in an experimental setting, (McDermott et al., 2009) partisanship and party identification, (Dawes & Fowler, 2009) liberal political ideology, (Settle et al., 2010) credit card debt, (De Neve & Fowler, 2010) antisocial personality, (DeLisi et al., 2009) leadership, (De Neve & Fowler, 2010), and preferences for the voluntary provision of social goods (Mertins et al., 2011).

Familial studies have been used to determine whether a phenotype is manifest in family members at a higher frequency than in the general public—if so, suggesting a genetic cause.

Causality has also been inferred from proxy variables when a direct genetic variable was not found. For example, in the case of the heart drug BiDil, the Food and Drug Administration approved the use of a clinical trial where "race" was a proxy for some as yet undetermined genetic factor (Kahn, 2011).

They used self-reporting of race as the independent variable and correlated it with outcome measures of a drug treatment for congestive heart failure. When the statistics showed a sufficient correlation from self-identified African Americans enrolled in the study, the drug was approved. It was inferred that the drug improved the patients treated more effectively than a placebo and thus was considered to have had a therapeutic effect.

While social scientists have discussed "causal" or "quasi-causal relationships" between genes and complex human behaviors, many medical geneticists have not found genetic links to schizophrenia, diabetes and hypertension.

Genetic causality is complicated by the fact that "gene–environment interactions" have become part of the new genetic paradigm shift. Back in 1968, MacMahon summarized the complexity of the gene–environment interaction making the sorting out of these two factors in a causal analysis an unlikely enterprise. He wrote:

1. It has become clear that there is no disease that is determined entirely by either genetic or environmental factors.
2. There is evidently more overlap in the time of operation of genetic and environmental factors than was previously suspected.
3. Just as the environment may exert its effect through the genetic mechanism of mutation, so may genetic factors operate by changing the environment.
4. The role of gene and environment and the nature of specific factors involved may be quite different in individuals with identical manifestations.

He cites the case of "yellow shanks", a characteristic observed in certain fowl when fed on yellow corn. He noted that a farmer using only yellow corn as feed and owning several strains of fowl would observe the trait appeared in select strains. He believed the trait was genetically determined.

Another farmer feeding some of his flock on yellow and some on white corn would note that the trait only appeared on those fed yellow corn. He concluded that the trait was environmentally determined. Neither environment nor genetics accurately described the cause of the condition. It is more accurate to say, that within a specified range of genetic background, environmental factors determine the occurrence; within a specified range of environment, the trait is genetically determined. Denis Noble notes that "genetic causality" is vexing "not only because the concept of the gene has become problematic,…but also because it is not usually a proximal cause."(Noble, 2008).

6. COMPLEX BEHAVIOR AND GENES

Let's consider the study by James Fowler and Christopher Dawes published in the *Journal of Politics* titled "Two Genes Predict Voter Turnout"(Fowler, Baker, & Dawes, 2008). They reported results of their study that a gene and a gene–environment interaction increased the likelihood of voting. Before we enter into the nature of their methodology and how they infer causality, let's first consider what genes do.

They provide a code (CGAT sequences) that the cell uses to synthesize a protein—from its component amino acids that circulate in the cells. How does a protein get turned into voting behavior? Also, given the multitude of ways that a person's voting behavior can be affected, such as peers, family, education, political affiliation, type of employment (farm worker who is peripatetic versus a banker), if someone were to develop a hypothesis that a gene causes voting behavior, it would have to be indisputable given that common sense suggests that this behavior is embedded in a complex culture that is far more likely to affect behavior than a protein.

The authors of the study were able to access genetic data from a National Adolescent Study of Adolescent Health. They had information on eight genes from twins and full siblings from 2,574 respondents. They looked at variants of a gene called MAOA, which encodes the enzyme monamine oxidase. Variants of this gene produce more or less quantities of serotonin, which affect the brain. They also correlated the MAOA variant with religious affiliation. A difference in transcription rates is supposed to exhibit different levels of bioavailable MAOA in the brain, which is supposed to correspond to different levels of bioavailable serotonin, which can be associated with different behavior. Their analysis includes a gene–environment interaction where religious affiliation and MAOA are supposed to work together. This type of gene–environment interaction requires another quite imaginative causal framework where proteins and religious affiliation co-interact to produce a behavioral outcome.

Their use of the Candidate Gene Association method is based on a case–control design. The gene frequency that is associated with certain behavior in one group is compared to another group, which does not exhibit the behavior. You can either go backwards (explain) or forwards (predict) the results. One group has the requisite candidate gene and you observe the behavior; the other group lacks the candidate gene and you observe the behavior. You can also divide two groups by behavior and then look at the genes to determine whether a correlation exists. This method of inferring causality is highly inductive. It is based purely on statistical correlations and not on deep neurological science where causal connections can be found between serotonin levels and human choice. There is no telling what type of correlations we can find with a .05 significance criterion if we generate enough statistical tests. But for such a result to really hold our attention at a causal level, we need an explanatory framework that is so powerful that it excludes all other more reasonable possibilities between proteins and behavior.

The genetic model used in these studies, and in particular the causal view of genes, is part of the old paradigm and not the new paradigm. The new view, which dramatically affects the idea of genetic causality, includes the epigenome—the proteins wrapped around DNA that provide the switching mechanisms for genes. As one author puts it: "the extent to which a gene can be transcribed is controlled by the epigenome, the complex biochemical regulatory system that turns genes on and off, is environmentally responsive, can influence phenotype via environmentally induced changes to gene transcribability with no changes to the DNA sequence"(Charney & English, 2012). Even at the rudimentary level of protein synthesis, DNA alleles do not tell the story. Therefore, correlating DNA to behavior has more problems than the multiple levels of causation between protein synthesis and human choice. There is no linear, unidirectional, DNA → protein causality.

Second, there is nothing stable about DNA. Its role in an organism changes through time and through environmental interactions; there are jumping genes, mutations and that "one and the same allele in one and the same individual might be completely inactivated in one set of tissues (e.g., the brain), partially inactivated in another, and completely active in a third."(Charney & English, 2012).

Third, one commentator notes that most human traits with genetic components are affected by a vast number of genes—each with a small effect. This framework would substantially discount the idea there is one gene, one cause, one effect. "There is an ever-growing consensus that complex traits, among which certainly be included all politically relevant behaviors, are influenced by hundreds or thousands of proteins encoded in hundreds or thousands of genes of small effect that interact with one another, the environment and the epigenome in complex ways"(Charney & English, 2012).

Let's take a closer look at the twin-studies method. John Alford, Carolyn Funk, and John Hibbing published a paper in 2005 using twin studies in which they found that political orientations are highly heritable and thus strongly determined by genetics (Alford, Funk, & Hibbing, 2005). How far can twin studies take you in asserting genetic causality of behavior? The twin studies typically use sets of monozygote (identical) and dizygote (fraternal) twins. If a behavioral trait correlated more highly with monozygote twins over dizygote twins, then it is presumed to be an indication of higher heritability of that behavior. Alford et al. (2005) reached their conclusion that genes shape political behavior when they found significantly higher correlations for MZ twins vs. same sex DZ twins on the Wilson-Patterson Attitude Inventory Score. The environment is assumed equal or randomly distributed among the twins.

Psychiatric geneticists have reported for decades that twin studies are confounded and that one's conclusions come at considerable risk (Rosenthal, 1979). One of the most important assumptions of such studies is the "equal environment assumption." Identical twins may be treated differently than fraternal twins, which could affect behavioral phenotypes. Some critics of the twin method are in wonderment that it has survived and believe that its survival is no longer a scientific question. "The twin method survives today not because the critics have been successfully 'rebuffed' but rather [because it is] the outcome of a power struggle, not the resolution of a debate among scientists" (Joseph, 2010).

Familial studies may be a source of hypotheses about genetic causation—but findings among family members may equally suggest a hypothesis of environmental causation. As one behavioral geneticist observed "Many behaviors run in families," but family resemblance can be due to nature or nurture (Plomin et al., 2008, p. 70). The question remains of how methodologically to disentangle environmental from genetic factors in causation.

There are so many levels of structure and development and so many possible interactions between the production of the hormone or the structure of a brain region and any behavior that, without irrefutable causal evidence, it would take an act of pure imagination to make a leap that there is a voting gene or a social network gene. Or as Richard Lewontin noted: "It is a sign of the foolishness into which an unreflective reductionism can lead us that we seriously argue from protein similarity to political similarity." In a genocentric framework, there is a tendency to draw the simplest reductionist explanation for a behavior or trait and neglect the complexities of multigenetic, gene-environment, epigenetic interactions or some complex combination involving multiple causation. As Martin Richards noted: "molecular genetics often has the feel of greedy reductionism, trying to explain too much, too fast, under-estimating the complexity and skipping over whole levels of process in the rush to link everything to the foundations of DNA"(Richards, 2002).

Turkheimer does not believe that current methods can disentangle genetic and environmental causation for complex human behavior. I am inclined to agree.

"…individual differences in complex human characteristics do not in general, have causes, neither genetic nor environmental. Complex human behavior emerges out of a hyper-complex developmental network into which individual genes and individual environmental events are inputs. The systemic causal effects of any of these inputs are lost in the developmental complex of the network."

(Turkehimer, 2001).

"Causal explanations of complex difference among humans are therefore not going to be found in individual genes or environments any more than explanations of plate tectonics can be found in the chemical composition of individual rocks."

(Turkehimer, 2001).

There is a rancorous debate among political scientists and between some political scientists and geneticists and behavioral psychologists. Referring to the study by Alford, Funk, and Hibbing (AFH) that political orientation can be genetically transmitted, Evan Charney wrote:

"I could perform the exact same study as AFH using a different questionnaire and claim to have determined what percentage of an individual's belief concerning the doctrine of the Trinity is due to genes and what percentage to environment—or to what extent whether one favors the New York Yankees or the Boston Red Sox, or Mercedes or BMWs, or Lowes or Home Depot is "heritable""

(Charney, 2008b).

Political scientists are introducing causality for political behavior by a hotly contested method (Twin Studies) or the use of statistical correlations without biological pathways by citing gross neurological hypotheses that would be unrealistic to medical geneticists. After 15–20 years of research on the genetic basis of autism spectrum disorders, which is characterized by a combination of abnormalities in language, social cognition, and mental agility, one neurogeneticist concluded: "Several dozen QASD susceptibility genes have been identified in the past decades, collectively account for 10–20% of ASD cases." This is after an army of researchers have done case-controlled studies. Yet, in far more complex and subtle human developmental behavior, political scientists make stronger claims about the genetic basis of political choices. No one assumes that autism is a personal choice, rather a neurological condition that affects or comes along with other organ pathologies. Nearly everyone assumes we have free will to determine whether, when, and for whom we vote—notwithstanding the fact that there are environmental influences. To make a claim about the genetic basis of political choice flies against all the a priori, personal and empirical knowledge we have accumulated over centuries about human choice.

A growing number of geneticists do not hold DNA to have sole primacy in causation of building a molecule, much less a phenotype. "Genes, as we now define them in molecular biological terms, lay a long way from their phenotypic effects, which are exerted through many levels of biological organization and subject to many influences from both those levels and the environment"(Noble, 2008). Some geneticists are more explicit about avoiding causal language in candidate gene studies or genome-wide studies. "I avoid using terms such as"due to"or"caused by"when referring to the statistical relations between an independent variable and a dependent variable, but instead use terms such as"associated with"to avoid deterministic implications"(Stoltenberg, 1997). What is often neglected in the association studies of individual alleles or polymorphisms is the context in which the genetic pathways operate. One commentator notes that a single genetic variant may have a different impact on the health depending on the other genetic variants that exist in the genome, environmental factors, or a combination of both (Drmanac, 2012).

7. CONCLUSION

In some very important respects, nothing happens in a living organism without its DNA and genes playing a role. Almost all illnesses, even those we attribute to viruses or bacteria, have a genetic component to them. Some of the infectious agents will be deactivated by the immune system. In other cases, the foreign agents will overcome the body's defenses resulting in a disease outcome.

Notwithstanding the close linkage of genetics and environment with disease, we may speak solely of an environmental or genetic cause of a disease. In cases where an individual's genetics is not unusual, we may speak of an environmental cause, even though the genetic–environmental interaction is a sufficient cause of the outcome. On the other hand, if the genetic factor (polymorphism or mutation) is rare, even though it is not itself a sufficient cause, we speak of a genetic cause. In these cases, there is usually a well-defined mechanism that supports the causal language.

In Candidate Gene Association studies, where statistical methods are used to identify probabilistic connections between a genetic locus and a phenotype, the use of causal language is problematic for several reasons. First, the statistical association could be a secondary correlate to another factor that is part of a causal network. As an analogy, if we find an association between the stock market and the crime rate, that could be a spurious causal relationship with the real cause being a rise in poverty. Second, as was previously mentioned, the factor that has been identified as "statistically associated with" may not be proximate to the effect. Therefore, there may be many intervening variables that may or may not diminish the efficacy of the genetic factor. Third, there may be numerous nonlinear genetic interactions, each of which may contribute to a small part of the effect. Only one locus in the genetic quilt of interactions can mistakenly be viewed as *the cause*. Autism is a case in point. "The candidate genes most strongly implicated in NDD [neurodevelopmental disabilities] causation encode for proteins in synaptic architecture, neurotransmitter synthesis…No single anomaly predominates. Instead, autism appears to be a family of diseases with common phenotypes, linked to a series of genetic anomalies, each of which is responsible for no more 2–3% of cases" (Landrigan, Lambertini, & Birnbaum, 2012). The authors argue that the causal mechanism for neurodevelopmental diseases like autism defy most of the classical models of causality. They postulate that genes and environmental factors are responsible for an effect, but how much of each and what combination makes up the causal tree is unknown and may be unknowable.

Large DNA databanks have provided the grist for associational studies that seek genetic determinants of human phenotype. If my analysis is correct then the methods, by themselves, at best can produce testable hypotheses that some alleles or mutations may contribute to one of the pathways in the complex system linking genetic, somatic, neurological, and environmental components. Failing to account for the complex quilt of interactions, mistakenly, affords DNA causal efficacy that cannot be supported.

REFERENCES

Alford, J. R., Funk, C. R., & Hibbing, J. H. (2005). Are political orientations genetically transmitted? *American Political Science Review, 99*(2), 153–167.

Boomsma, D. I., Cacioppo, J. T., Slagboom, P. E., et al. (2006). Genetic linkage and association analysis for loneliness in Dutch twin and sibling pairs points to a region on chromosome 12q23-24. *Behavior Genetics, 36*, 137–146.

Charney, E., & English, W. (February 2012). Candidate genes and political behavior. *American Political Science Review*, 7–10. http://dx.doi.org/10.1017/S0003055411000554.

Charney, E. (2008a). Genes and ideologies. *Perspectives on Politics, 6*(2), 299–319. at p. 301.

Charney, E. (2008b). Politics, genetics, and "greedy reductionism". *Perspectives on Politics, 6*(2), 337–343. at p. 341.

Commoner, B. (2002). Unraveling the DNA myth, Harpers Magazine, pp. 30–47. at p.

Dawes, C. T., & Fowler, J. H. (2009). Partisanship, voting and the Dopamine D2 receptor gene. *Journal of Politics, 7*, 1157–1171 (July).

De Neve, J.-E., & Fowler, J. H. (2010). The MAOA gene predicts credit card debt. *Social Science Research Network*. Accessed 10.11.11 http://papers.ssrn.com/sol3/papers.cfm?abstract_id=1457224.

DeLisi, M., Beaver, K. M., Vaughn, M. G., & Wright, J. P. (2009). All in the family. *Criminal Justice and Behavior, 36*(11), 1187–1197.

Drmanac, R. (June 1, 2012). The ultimate genetic test,. *Science, 336*, 1110–1111. at p. 1110.

Dupré, J. (2008). What genes are and why there are not genes for race. In B. A. Koenig, S. S.-J. Lee & S. S. Richardson (Eds.), *Revisiting race in a genomic age* (pp. 39–55). New Brunswick, NJ: Rutgers University Press. at p. 41.

Dyck, B. A., Skoblenick, K. J., Castellano, J. M., Ki, K., Thomas, N., & Mishra, R. K. (August 2009). Behavioral abnormalities in synapsin II knockout mice implicate a causal factor in schizophrenia. *Synapse, 63*(8), 662–672.

Fowler, J. H., & Dawes, C. T. (July 2008). Two genes predict voter turnout. *Journal of Politics, 70*(3), 579–594.

Fowler, J. H., Baker, L. A., & Dawes, C. T. (2008). Genetic variation in political participation. *American Political Science Review, 102*(2), 233–248.

Fowler, J. H., Dawes, C. T., & Christakis, N. A. (February 10, 2009). Model of genetic variation in human social networks. *PNAS, 106*(6), 1720–1724.

Fowler, J., Settle, J. E., & Christakis, N. A. (2011). Correlated genotypes in friendship networks. *Proceedings of the National Academy of Sciences of the United States of America, 108*(5), 1993–1997.

Garpenstrand, H. N., Norton, M., Damberg, G., et al. (2002). A regulatory monoamine oxidase a promoter polymorphism and personality traits. *Neuropsychobiology, 46*(4), 190–193.

Geschwind, D. H. (2011). Genetics of autism spectrum disorders. *Trends in Cognitive Sciences, 15*(9), 409–415. at p. 411.

Joseph, J. (2010). The genetics of political attitudes and behavior: Claims and refutations. *Ethical Human Psychology and Psychiatry, 12*, 200–217, at p. 213.

Kahn, J. (2011). BiDil and racialized medicine. In S. Krimsky & K. Sloan (Eds.), *Race and the genetic revolution* (pp. 129–141). New York: Columbia University Press.

Keski-Rahkoren, A., Viken, R. J., Raprio, J., et al. (2004). Genetic and environmental factors in breakfast eating patterns. *Behavior Genetics, 34*, 503–514.

Koenig, L. B., McGue, M., Kreuger, R. F., et al. (2005). Genetic and environmental influences on religiousness: findings for retrospective and current religiousness ratings. *Journal of Personality, 73*, 471–488.

Landrigan, P., Lambertini, L., & Birnbaum, L. S. (July 7, 2012). A research strategy to discover the environmental causes of autism and neurodevelopmental disabilities. *Environmental Health Perspectives, 120*(7), A258–A260. at p. 258.

Li, M., Wang, I. X., Li, Y., Bruzel, A., Richards, A. L., Toung, J. M., & Cheung, V. G. (2011). Widespread RNA and DNA sequence differences in the human transcriptome. *Science, 333*, 53–58.

MacMahon, B. (1968). Gene-envirionment interactions in human disease. In D. Rosenthal & S. Kety (Eds.), *The Transmission of Schizophrenia: American Psychologist*. Oxford: Pergamon Press.

McDermott, R., Tingley, D., Cowden, T. J., et al. (2009). Monoamine oxidase a gene (MAOA) predicts behavioral aggression following provocation. *Proceedings of the National Academy of Sciences of the United States of America, 106*(7), 2118–2123.

Mertins, V., Schote, A. B., Hoffeld, W., et al. (2011). Genetic susceptibility for individual cooperation preferences: the role of monoamine oxidase a gene (MAOA) in the voluntary provision of public goods. *PLoS One, 6*(6), 16.

Mingyao, L., Wang, I. X., Li, Y., Bruzel, A., Richards, A. L., Toung, J. M., et al. (1 July 2011). Widespread RNA and DNA sequence differences in the human transcriptome. *Science*, 53–58. http://dx.doi.org/10.1126/science.1207018. Published online 19 May 2011.

Mill, J. S. (1859). *System of logic*. New York: Harper & Brothers Publishers.

Noble, D. (2008). Genes and causation. *Philosophical Transactions of the Royal Society A, 366*, 3001–3015. at p. 3012.

Plomin, R., DeFries, J. C., McClearn, G. E., et al. (2008). *Behavioral genetics* (4th ed.). New York: Worth.

Food and Drug Administration. (2001). Premarket notice concerning bioengineered foods: proposed rule. Federal Register 66:4706–4738. Retrieved from http://www.cfsan.fda.gov/~lrd/fr0101.html.

Proteins operate in many ways in the process of turning information from DNA into the synthesis of a polypeptide molecule including their role in the ribosome—the staging area for the assembly of amino acids into proteins.

Quintana-Murci, L. (February 17, 2012). Gene losses in the human genome. *Science, 335*, 806–807.

Richards, M. (2002). How distinctive is genetic information? *Studies of the History, Philosophy, Biology & Biomedical Sciences, 32*(4), 663–687. at p. 668.

Ridley, M. (2003). *Nature via nurture*. New York: Harper Collins.

Rosenthal, D. (1979). Genetic factors in behavioral disorders. In M. Roth & V. Cowie (Eds.), *Psychiatry, genetics and pathography: A tribute to Eliot Slater* (pp. 22–33). London: Oxford University Press.

Settle, J. E$., Dawes, C. T., Christakis, N. A., & Fowler, J. H., et al. (2010). Friendships moderate an association between a Dopamine gene variant and political ideology. *Journal of Politics, 72*, 1189–1198.

Stoltenberg, S. F. (1997). Coming to terms with heritability. *Genetica, 99*, 89–96. at p. 90.

Strobel, A. A., Wehr, A., Michel, A., & Brocke, B. (1999). Association between the dopamine D4 receptor (DRD4) exon III polymorphism and measures of novelty seeking in a German population. *Molecular Psychiatry, 4*(4), 378–384.

Turkheimer, E. (2011). Commentary: variation and causation in the environment and genome. *International Journal of Epidemiology, 40*, 598–601. p. 598.

CHAPTER TWELVE

Pathways by which the Interplay of Organismic and Environmental Factors Lead to Phenotypic Variation within and across Generations

Lawrence V. Harper
Department of Human Ecology, University of California, Davis, CA, USA
E-mail: lharper@ucdavis.edu

Contents

Abstract

The range of responses made to environmental exigencies by animals, including humans, may be impacted by the experiences of their progenitors. In mammals, pathways have been documented ranging from transactions between a mother and her developing fetus in the womb through continuity of parenting practices and cultural inheritance. In addition, phenotypic plasticity may be constrained by factors transmitted by the gametes that are involved in the regulation of gene expression rather than modifications to the genome itself. Possible mediators for this kind of inheritance are examined, and the conditions that might have led to the evolution of such transmission are considered. Anticipatory adjustments to possible environmental exigencies are likely to occur when such conditions recur regularly, but intermittently across generations and endure for substantial periods of time, and when adjusting to them after the fact is likely to be biologically costly, even life-threatening. It appears that physical growth and responses to nutrient availability are domains in which

Advances in Child Development and Behavior, Volume 44   © 2013 Elsevier Inc.
ISSN 0065-2407, http://dx.doi.org/10.1016/B978-0-12-397947-6.00012-X All rights reserved.

anticipatory, epigenetically inherited adjustments occur. In addition, given the fact that humans have oppressed one another repeatedly and for relatively long periods of time, such behavioral tendencies as boldness or innovativeness may be behavioral traits subject to such effects. The implications of these factors for research and policy are discussed.

1. INTRODUCTION

To fully appreciate the factors that shape behavioral development, it will be necessary to consider concepts drawn from a broad spectrum of disciplines beyond psychology—from evolutionary to molecular biology—and how they interrelate. As Rutter (2002) noted, while sometimes useful for testing "causal hypotheses," the conceptual separation of "nature, nurture, and development" more often than not impedes discovery. Thus, he argued: "…it is essential that psychosocial research be a part of biology, and not separate from it" (Rutter, 2002, p. 10). Likewise, from an examination of conceptual issues in evolutionary biology, Laland, Sterelny, Odling-Smee, Hoppitt, and Uller (2011) concluded that in order to understand causation, "how" and "why" questions are not really alternatives but, rather complementary, and conceptions that emphasize reciprocal, as opposed to unidirectional, effects are most likely to prove fruitful. Arguably then, if our conceptions of behavioral development are to provide a solid foundation for both policy and practice, it will be necessary to identify the biological substrates, the environmental conditions impacting them, and the dynamics of their interplay within and across generations.

When attempting to account for individual differences, there has been a tendency to separate the influences of "inheritance," often invoked to account for that which is relatively invariant phenotypically, from "nurture," or "experience," referenced when examining factors that facilitate plasticity or adjustment to prevailing conditions. However, as exemplified by physical differences between human monozygotic twins reared apart (e.g., Tanner, 1990), essentially the same genome can yield substantial phenotypic variation. Thus, identifying the phenotypic characteristics that are likely to be (more) plastic and the pathways that mediate such plasticity are core issues. On one hand, ontogenetic processes must lead to the expression of species-typical phenotypes, and on the other, it is clear that within the constraints of species-typicality, there must be the potential for adaptive adjustments to prevailing conditions (e.g., Johnson & Tricker, 2010; Rando & Verstrepen, 2007)—that highlights the need to identify the relevant conditions for typical development, and for adjustments within those boundaries, along with the processes by which they are accomplished.

While mammalian species differ even with respect to the details of otherwise common processes such as those underlying the inactivation of one X chromosome (e.g., Morison, Ramsay, & Spencer, 2005), animal models can still inform us of where to look in order to locate components of the pathways responsible for human phenotypic variation. For example, studies of the effects of maternal behavior on responsiveness of the hypothalamic–pituitary–adrenal axis in offspring of rats (e.g., Cameron et al., 2005) provided a fruitful model for examining the effects of childhood maltreatment on humans (McGowan et al., 2009).

This chapter will focus on transgenerational epigenetic inheritance, the transmission from one generation to the next of phenotypic adjustments made in response to conditions experienced by one's ancestors. It can be seen as a modification of an individual's range of response to certain conditions, which, to be fully understood, both in terms of the underlying processes and the conditions activating them, requires an integration of concepts ranging from molecular genetics to evolutionary biology. The possibility of this kind of inheritance has been advanced repeatedly over time. However, until recently, probably due to the existence of a number of plausible alternative pathways to account for continuity across generations, and the lack of a conceptual model for the underlying mechanism(s), epigenetic inheritance has not been taken seriously as a potential contributor to individual differences (see Gottlieb, 2002; Jablonca & Lamb, 1995; Weber & Depew, 2003; West-Eberhard, 2003 for reviews and commentaries on these issues).

Current models of developmental processes hold that since a multicellular organism originates from a single cell, the fertilized egg, most if not all of the cells, in the individual share the same genome—as supported by the successful cloning of whole animals from such highly differentiated cells as olfactory receptors (Eggan et al., 2004). Thus, as the "totipotent" zygote begins to divide, in part in response to conditions resulting from proliferation itself (e.g., Stevens & George, 2005), cellular differentiation results from the selective activation of subsets of gene networks and restriction of the activities of alternative networks thereby giving rise to specialized tissues. The DNA sequence in the chromosomes remains the same while "epigenetic" modifications occur to the factors that control which segments of the chromosomal strands will be read out to determine cellular identity (Reik, 2007). In development, genes are now thought to be regulated via networks in which certain products form the core elements of multifaceted complexes that underlie species-typical growth and patterning by operating in conjunction with other gene products that are deployed more flexibly to modulate the process (e.g., Budnik & Salinas, 2011; Davidson, 2006). The same sequence that functions to code a core protein in one tissue may even serve as a regulatory binding site for other proteins involved in the development of a different tissue (e.g., Birnbaum et al., 2012).

Moreover, transcription of RNAs from so-called noncoding sequences adds an additional dimension in the networks underlying tissue differentiation. Many kinds of RNAs act in conjunction with enzymes and other proteins to help to regulate both what will be expressed and their levels of expression, thereby fine-tuning the proliferation and specialization of tissues (Costa, 2008). Obviously then, there is no one gene "for" any trait (see also Flames & Hobert, 2009), and, given the complex interplay between the many elements involved in controlling gene expression, in response to environmental conditions, the same genotype can produce different phenotypic variants. Documented examples of such variations in response to prevailing conditions range from human physical (Kim, Shin, & White-Traut, 2003) and cognitive (Nelson, Fox, Marshall, Smyke, & Guthrie, 2007) growth as a function of environmental stimulation to remodeling of the circuitry in the mammalian central nervous system as a result of experience (Budnik & Salinas, 2011).

Indeed, learning, once considered to represent an alternative to genetically mediated individual differences, has now been shown to involve the same underlying processes. For example, changes in neural activity patterns in response to conditioning have been shown to involve enzymes that help to promote tissue differentiation via modifications of the histone proteins that "package" the chromosomes and thereby help to regulate gene expression (e.g., Gao et al., 2010; Guan et al., 2009), and the formation and retention of memory has been linked to the activities of other enzymes that mediate the attachment of methyl molecules to cytosine nucleotides in the DNA, a process also considered to be an important element in the regulation of growth and tissue differentiation (Miller et al., 2010).

In short, rigid distinctions between nature, nurture and development blur when one begins to examine the processes that underlie development and responses to environmental conditions.

2. WHAT IS INHERITED?

Moreover, as one considers what is transmitted across generations, "inheritance" includes more than just chromosomal sequences; as pointed out by Oyama, Griffiths, and Gray (2001), one's inheritance spans a spectrum from the cellular to the social and physical surroundings. Although life, as we know it, depends upon nucleic acids, it is obvious that DNA by itself will not lead to a viable organism. Genotypes have evolved to capitalize upon certain features of their terrestrial surroundings—the species' niche—to give rise to distinct life forms. This is not a unidirectional process; each life form impacts its environment to some degree so that the evolutionary process is dynamic and might be conceptualized as a spiral in time wherein, for a given genotype, phenotypes are induced in response to external conditions, and these phenotypes help to shape their offspring's environments that, in turn, exert selective pressures on the genome of the species. Insofar as habitats vary during an organism's lifetime in terms of such features as the availability of nutrients and fluctuations in the numbers of competitors and/or predators, all life forms must be able to adjust appropriately to such fluctuations (see, e.g., Rando & Verstrepen, 2007). Thus, in order to adequately understand development of any attribute, we ultimately must identify the features both within the organism and the external setting that are necessary and/or sufficient to shape the ontogeny of the trait in question.

In metazoa, at the cellular level, what is transmitted across generations includes a diverse array of components: in the nucleus, in addition to the chromosomal sequences of nucleotides in the DNA that code for enzymes and proteins—and "splice-variants" from a given protein-coding sequence (e.g., Brenet et al., 2011)—that are necessary for the differentiation of tissues and organs, the nuclear DNA includes sequences that act as binding sites for other proteins (Chakravarti & Kapoor, 2012; Georges, Benayoun, Caburet, & Veitia, 2010), and the production of RNA variants, both long sequences (Guttman & Rinn, 2012) that help to guide the attachment of proteins to certain regulatory regions of the chromosome, and short "microRNAs" (Lu & Clark, 2012) that are involved in degrading messenger RNAs and/or in blocking their access to the machinery that leads to the production of protein. Thus, the chromosomal strand encodes the basis for a complex array of interactions that determine what aspects of the genome are expressed and at what level.

Furthermore, the chromosomal strands in the nucleus are "wrapped" around histone proteins to package the chromosomes. These "nucleosomes" composed of two copies each of four histone proteins, in conjunction with additional elements that form the chromatin that surrounds segments of the DNA strand, influence the accessibility of the DNA for transcription. In addition, the histones and other regulatory proteins can be subject to various enzyme-mediated modifications such as the addition of acetyl or methyl molecules to certain amino acid elements in the histone protein bodies that alter their affinities to both DNA sequences and to other proteins, thereby facilitating or inhibiting the likelihood of gene expression within a cell line (Djupedal & Ekwall, 2008; Nahkuri, Taft, & Mattick, 2009). Moreover, as a function of certain classes of DNA-binding proteins' interactions with the DNA strands, segments of chromosomes can alter their locations within the cell nucleus thereby modifying their proximity to active sequences on the same, or even other chromosomes to facilitate or inhibit gene expression and help to maintain tissue identity (Dixon et al., 2012; Gomes & Espinosa, 2010).

In addition to the intranuclear dynamics, there is an ongoing, complementary interplay between the nucleus and the surrounding cellular cytoplasm. In rodents, transplantation of zygotic nuclei from animals of one strain to the cytoplasm of animals from a different strain has shown that the cytoplasm has a major role in regulating nuclear gene expression (e.g., Kikyo, Wade, Guschin, Ge, & Wolffe, 2000; Martin et al., 2006; Reik et al., 1993). The cytoplasm contains the cellular energy sources, mitochondria, and a range of products including proteins and RNAs produced in the nucleus and also RNAs that are products of the mitochondrial DNA, whose activities are coordinated with the nuclear DNA (Rugarti & Langer, 2012). The ribosomes, organelles responsible for the "translation" of the messenger RNA produced in the nucleus into proteins via the mitochondrial transfer RNAs, also are located in the cytoplasm. The cytoplasmic membrane receptors—encoded in the nuclear DNA—provide the gateways for transmission of extracellular signals to the nucleus, often via complex signaling pathways (Han, Mtango, Patel, Sapienza, & Latham, 2008).

As the fertilized egg divides and cells begin to differentiate, each new cell line inherits a different pattern of gene expression from its precursors (e.g., Djupedal & Ekwall, 2008; Munsky, Neuert, & van Oudenaarden, 2012), so that the products of the tissues surrounding any one cell (Stevens & George, 2005) constitute yet another dimension of inheritance, yielding signals ranging from physical tension derived from the three-dimensional conformation of the developing embryo (Engler, Sen, Sweeney, & Discher, 2006) to hormones and neurotransmitters that influence the fates of the developing tissues. Thus, even before the functional development of sensory receptors, neuronal networks and specialized signaling tissues such as endocrine glands, the developing phenotype mediates the gene–environment interplay that supports its expression (cf. Harper, 1989).

These processes unfold within—and depend upon—a broader context. Mammals can be considered to "inherit" the maternal environment wherein there is also a bidirectional interplay: not only do the fetal cells contribute to the placenta, but there is an ongoing cross-talk between the developing embryo and the maternal reproductive tract (e.g., Entringer et al., 2012). This can be seen at fertilization, at which time the oviduct plays a role in preparing the fertilized egg to divide and migrate to, and implant in, the uterus (Han et al., 2008; Lee & Yeung, 2006). The uterine environment is not only shaped by the maternal genome, as demonstrated by interstrain embryo cross-transfer and cross-fostering neonates (Gleason, Zupan, & Toth, 2011) but is likely to have been conditioned by environmental events occurring during the period when the mother's uterus was developing (e.g., Burdge et al., 2007; Gluckman, Hanson, & Beedle, 2007; Painter et al., 2008; Spencer, Hayashi, Hu, & Carpenter, 2005; Tobi et al., 2009).

In addition, the pregnant mother's responses to factors in her immediate surroundings, such as the availability of nutrients, may transmit signals to her developing offspring impacting their later metabolic responses (e.g., Hoile, Lillycrop, Thomas, Hanson, & Burdge, 2011; Ollikainen et al., 2010; Radford et al., 2012), food preferences (Mennella, Jagnow, & Beauchamp, 2001), and neural development (Mehedint, Niculescu, Craciunescu, & Zeisel, 2010). Maternal hormonal state may also "inform" the fetus with regard to maternal social standing (e.g., Dioniak, French, & Holekamp, 2006) and/ or the presence of environmental stressors (e.g., Morgan & Bale, 2011) and thereby influence offspring responsiveness to the surroundings they will encounter postnatal.

The birth canal itself is a source of inheritance. Dominguez-Bello et al. (2010) found that vaginally delivered human neonates had "bacterial communities" more similar to those of their mothers' vaginal tissues, whereas caesarian-delivered neonates' bacterial communities had more in common with their mothers' skin. While this might seem irrelevant for understanding variations in behavioral development, Heijtz et al. (2011) found that germ-free mice displayed different reactions to potentially stressful contexts when compared with mice with normal gut microbe contents. Insofar as gut flora act as intermediates for the availability of a broad range of nutrients that may include additional substrates for gene regulation and neural development such as methyl molecules (Waterland & Michels, 2007), this may be of importance.

As indicated earlier, with respect to the prenatal effects of stress, the context in which the mother resides can have an impact on offspring experience, and this may be mediated via as yet unappreciated pathways. For example, Fujita et al. (2012) found that breastfeeding mothers in Northern Kenya who were economically secure produced richer milk for sons than daughters while poor mothers produced richer milk for daughters than sons. Insofar as this effect was seen at first feeding, it was unlikely to have resulted from postpartum dietary decisions and illustrates how evolutionary models may help to identify both environmental conditions to which we respond and the nature of the responses so evoked.

Cultures clearly differ and provide relatively unambiguous evidence of transmission of lifestyles that impact individual differences (cf. Rogoff & Morelli, 1989). Nevertheless, despite very strong designs (e.g., Kovan, Chung, & Sroufe, 2009), the evidence for transgenerational effects of parenting practices in humans is still essentially correlational. On the other hand, experimental studies with rodents and primates have demonstrated unambiguously that aspects of maternal behavior can be altered by experience, and that female offspring of mothers who were more responsive tend, in turn, to be more solicitous of their young. Insofar as work with inbred strains indicates that cross-fostering offspring of low responsive animals to highly responsive foster mothers yields the same effects, the results can be interpreted as resulting from postnatal experience, per se (see, e.g., Cameron et al., 2005). This animal work also suggests that rather than interpreting these effects in terms of transmission via some pathway analogous to culture, they may be seen as results of natural selection, whereby the young were sensitized to respond to environmental "signs" (cf. Harper, 1989), often indirect, but reliable predictors of the nature of the conditions with which they will have to cope.

In sum, much more than just the DNA is inherited; there is a broad range of factors that contribute to individual similarities and/or differences across generations.

Moreover, depending on the behavior so influenced, a number of different conditions may operate relatively independently to shape any particular outcome, so that the effects of one kind of experience may be compensated or reversed by another. For example, Tang, Akers, Reeb, Romeo, and

McEwen (2006) found that male rats that were exposed to a novel setting for 3 min per day during their first 3 weeks after birth, as adults differed from control littermates in response to a variety of conditions including many that have been shown to be influenced by enhanced early maternal care. Moreover, such compensatory experience need not be concurrent. Bredy, Zhang, Grant, Diorio, and Meaney (2004) found that while a high level of maternal responsiveness during the nursing period can affect hippocampal development and facilitate rat pups' later spatial learning, the effects of low maternal responsiveness could be reversed by later, peripubertal environmental enrichment. Similar compensatory phenomena in humans have been documented by Werner and Smith (1992).

3. TRANSGENERATIONAL EPIGENETIC INHERITANCE

Clearly then, many factors can shape development across the lifespan. To date, however, particularly with respect to the ontogeny of individual differences in behavior, the emphasis has been on the direct effects of environmental conditions on the developing person. While the possibility that the effects of the experiences of previous generations could in some way be transmitted to subsequent generations via the germ line, in the absence of the conditions that evoked them, has been raised repeatedly, clear evidence for such transmission in mammals has been absent until recently.

Jablonca and Lamb (1995) compiled a large body of evidence from protozoa to vertebrates indicating that the conditions to which one generation had adjusted could shape the phenotype or responsiveness of a subsequent generation, suggesting that natural selection has favored organisms with the ability to anticipate variations in their surroundings. Subsequently, Agrawal, Laforsch, and Tollrian (1999) experimentally demonstrated that in an insect and a plant, phenotypic defenses developed in response to exposure to predators could be transmitted to subsequent generations via the germ line (eggs, seeds) and that these modifications did indeed enhance offspring survival when they were subject to predation, thereby supporting the view that natural selection would favor such a pathway of inheritance.

For mammals, as indicated earlier, the prenatal maternal environment can be a source of "information" concerning conditions to be encountered by offspring after birth. Thus, a report showing that body size and brain cell number in rat pups was reduced as a result of first-generation maternal dietary protein restriction, and that the offspring of females, but not males, so exposed also showed the same phenotype across three generations

(Zamenhof, van Marthens, & Grauel, 1971) could have been attributed to alterations in the mothers' reproductive tracts, rather than signals transmitted by the germ line.

Moreover, in view of the fact that the gametes have to be able to give rise to all the different tissues in the body, it would seem that any information relating to prior alterations in gene regulation must be "erased" from the gametes (cf. Chong & Whitelaw, 2004). However, the phenomenon of genetic "imprinting," the selective expression/silencing of alleles depending upon whether they are inherited from the mother or the father—displayed even when nuclei have been transplanted in order to clone animals (Inoue et al., 2002)—suggests that the chromosomes in the gametes can retain and transmit a relatively stable "memory" of their origin across generations. This phenomenon is of importance for understanding behavioral development because many imprinted alleles are selectively expressed in the brain (Davies, Isles, & Wilkinson, 2005) and, as indicated by the Skuse et al. (1997) study of girls with Turner syndrome (only one X chromosome), a "memory" of parent of origin of the X chromosome can impact behavior significantly.

In recent years, a number of experimental studies with rodents have supported the possibility that parental experiences may be transmitted via the mammalian germ line. Among the first of these investigations were studies addressing the question of whether early dietary conditions could impact the metabolism of subsequent generations.

Burdge et al. (2007) found differences in the expression of genes in the livers of the male offspring of female rats whose grandmothers were on a protein-restricted diet during pregnancy but on typical diets thereafter (as were the mothers). Changes in the methylation levels of two genes across the two generations were similar. Hoile et al. (2011) later examined gene expression patterns in the livers of three generations of female rats. The first generation was exposed to protein restriction during gestation and subsequent generations were on typical diets. In this latter study, patterns of gene expression differed across generations both when compared with control animals and between generations among the protein-restricted animals. Insofar as the effects were transmitted across females, as with the Zamenhof et al. (1971) experiment, Hoile et al. (2011) suggested that these outcomes might have involved interplay between changes in the maternal environments and gametic transmission.

Paternal transmission rules out the possible effects of maternal environment insofar as any relation between paternal experience and offspring phenotype can be attributed to the gametes so long as the offspring are not reared with their sires. At least two studies with rats provide such evidence: in the first such investigation, Ng et al. (2010) found that exposing male rats to high-fat diets after weaning until adulthood led to differences in a large number of pancreatic genes expressed by their female offspring as compared to the same-strain females sired by untreated, same-strain males. In the second experiment, Yasbek, Spiezio, Nadeau, and Buchner (2010) worked with male mice carrying a single mutant allele that led to reduced food intake and body weight in response to a high-fat diet. When males carrying this allele were fed a high-fat diet and subsequently mated with genetically homozygous, wild-type females, their offspring showed the same response even when they did not carry the mutant allele. Moreover, this phenotype could be transmitted via the male line for at least three generations. These studies thus indicate that the gamete, sperm, can transmit a signal relating to nutrient quality experienced by a previous generation.

Studies of the cross-generation effects of nutrient availability have indicated that comparable phenomena may occur in humans and that there may be sensitive periods for such effects. Pembrey et al. (2006) examined a sample based upon records from a Swedish province that had been exposed to occasional periods of harvest failure from the 1890s to the 1920s. They found that either maternal or paternal dietary restriction during the juvenile "slow growth period" predicted daughters' risk of early mortality. A trend suggested that mothers' access to nutrients during that period might affect their sons' life spans. When they analyzed the effects of limited juvenile nutrition on the third generation, they found that while there was no relation between limited access to food supplies during a *woman's* parents' slow growth period and (grand) offspring lifespan, in the case of males, there was an effect. When the food supply was abundant during the *paternal* grandfather's youth, grandsons had a higher risk of early mortality, and when the paternal grandmother had ample food during her early childhood, granddaughters, but not grandsons, were at higher risk. Reduced food supplies during the paternal grandparents' slow growth period was associated with a longer lifespan for grandsons. Moreover, there was a gender difference with respect to when the effects of food availability best predicted the outcomes for grand offspring. For grandmothers, the strongest association was with dietary restriction during the period from her conception to the third postnatal year; for grandfathers, the most sensitive phase seemed to be later in the slow growth period of middle childhood.

In a subsequent study, Kaati, Bygren, Pembrey, and Sjostrom (2007) replicated the work on the effects of early dietary restriction using a sample in which they could trace both the parents' and grandparents' nutritional histories and their socioeconomic backgrounds. They found that if grandparents were in the slow growth period (ages 8–11 for grandmothers, 9–12 for grandfathers) during years when there was "total crop failure," male grandchildren lived longer than those with grandfathers who had good nutrition during the same stage. The grandmother being literate and alive had demonstrable bearing on the survival of grandsons, but not granddaughters. Again, there was a patrilineal, gender-specific pattern where paternal grandfathers' food supply affected only grandsons and paternal grandmothers' access to food affected only granddaughters.

Thus, although the pathways seem complex, for both animals and humans, there is reason to believe that the effects of nutrient availability impacting a prior generation may be transmitted across generations by means that cannot be explained solely in terms of alterations in the female reproductive tract, indicating that transgenerational information can be transmitted via the gametes.

More recently, work with mice has shown that the experiences of prior generations may also affect the responses of their grand offspring to novel settings. Franklin, Linder, Russig, Thony, and Mansuy (2011) subjected male mice to randomly timed daily 3 h separations from their mothers for the first 3 weeks postpartum, at which times the mothers also were subjected body restraint or were forced to swim in cold water. Whereas, as young adults, the separated animals showed typical emotional reactions to odors of an aggressor and social defeat, when mated with nonstressed females, their offspring showed diminished reticence to behave cautiously in similar contexts. These results suggest that the fathers' early experience led to "stress resilience" in the subsequent generation.

In a second experiment, Weiss et al. (2011) examined the effects of daily 3 h physical separations (visual and olfactory contact was maintained via a wire screen) of mouse litters from their mothers during the active (dark) period for the first 2 weeks postpartum with or without additional randomly timed stressors imposed on the dam during the period of separation. The pups remained together. As adults, the males and females in the treatment group, as compared to controls, showed reduced latencies to enter an unfamiliar setting, and less cautious behavior; males spent more time in the center of an open field. The additional stress to their mothers enhanced the effects on males only. When they reached adulthood, males were mated with control females and the treatment group females were mated with control males; the litters were cross-fostered to experimental or control females. The male and female offspring of early separated sires whose dams did not experience any stress still showed less caution than did the offspring of controls. The effect was more pronounced among males—although somewhat reduced among pups reared by control females. No significant effects were found on measures of corticosterone levels or on adrenal weight, but a corticoid receptor variant was altered in the amygdala and hypothalamus of the F1 females.

In another study with mice, Morgan and Bale (2011) exposed pregnant females to stressful stimuli such as fox odors, unfamiliar objects placed in their cages, or restraint in a conical tube. While these treatments did not affect the females' food intake, gestation duration, litter size, or observed postpartum maternal behavior, it did lead to changes in the phenotype of male offspring that rendered them more similar to normal females in response to stress. When the experimental males were mated with control females, the male offspring whose grandmothers were stressed during the first week of gestation showed gene-expression levels for 13 of 17 loci in brain that were more like that of females than control males (including a receptor for a hormone involved in sexual differentiation), along with alterations in the expression levels of several microRNAs and a few proteins regulated by these microRNAs.

Insofar as the effects of early experiences of males could be transmitted across generations even when the animals were mated with control females, these experiments support the view that the effects of experiences of prior generations can be transmitted via the gametes in ways that impact the behavioral responses of subsequent generations in the absence of the precipitating conditions. They suggest that in mammals, as well as plants and insects (cf. Agrawal et al., 1999), natural selection has favored the transgenerational transmission of "preparation" for conditions that might impact the well-being of offspring and that such preparation may involve epigenetic modifications transmitted via the gametes.

4. PATHWAYS OF TRANSMISSION

This returns us to the question of "how." As indicated earlier, in order to transmit "totipotency," the ability of the fertilized egg to give rise to all the different kinds of cells in the organism, it has generally been assumed that despite the fact that the gametes are specialized cells, the "marks" on the genome that lead to tissue differentiation had to be "erased" in the germ line. To the extent that autosomal chromosomes recombine more or less at random, this would also suggest removal of the germ line marks identifying maternal and paternal origin (e.g., Hajkova et al., 2010; Reik, 2007). However, imprinted allelic expression indicates that a memory of (current) parent-of-origin is carried by the gametes and that there exist pathways by which such information is transmitted across generations.

While imprinting was once attributed to stable methylation of cytosine nucleotides in the DNA in the gametes, the fact that some imprinted patterns of expression are not expressed in tissues until after they begin to differentiate (e.g., Fan, Hagan, Kozlov, Stewart, & Muegge, 2005; Morison et al., 2005) has led to more extensive examination of the sites involved. From these investigations, it now seems that in mammals, imprinted genes tend to be found in adjacent segments in the chromosomal strand in regions characterized by repetitive elements (e.g., Koerner & Barlow, 2010). In addition to segments with cytosine–guanosine nucleotide sequences (CpGs) associated with individual genes, these "clusters" of imprinted genes also harbor so-called "control regions" characterized by high levels of CpGs, where methyl molecules may be attached to the cytosine nucleotides and thereby modify access of these regions to the enzyme complexes involved in DNA transcription. These regions are differentially methylated according to parent of origin as the gametes are differentiating (Dindot, Person, Strivens, Garcia, & Beaudet, 2009), and a stepwise progression of events leading to the maintenance of such marks as cells divide has been proposed by Koerner and Barlow (2010). At least one pathway by which such DNA methylation via methyltransferase enzymes is accomplished involves mediation by a non-protein-coding RNA (Sleutels, Zwart, & Barlow, 2002). Thus, it would seem that there are distinctive features of the genome that are related to this allele-specific memory of origin and that their expression may be facilitated by RNA-mediated processes.

Another example of transgenerational transmission of a phenotypic trait not directly attributable to the inheritance of a related allele is a phenomenon called "paramutation," originally described in maize (Brink, 1956). It is the result of interaction between two different (heterozygous) alleles of a gene that yields a phenotype that is transmitted across generations even when offspring are homozygous for the alleles not giving rise to that phenotype. Similar phenomena have been reported to occur in mammals with the most well-studied examples involving morphological traits in mice.

Rakyan et al. (2003) examined mice carrying a mutant allele that led to the development of a kinked tail, but only in a proportion of the carriers. The expression of this phenotype was correlated with differential DNA methylation at a retrotransposon, a region of the DNA within the allele which is labile, and can move to a different position in the strand during chromosomal replication under some conditions. When carriers of the mutation were mated with genetically wild-type animals, genetically wild-type offspring of both males and females showed (variable) expression of the kinked tail phenotype. Moreover, they found that methylation of the relevant locus in mature sperm of carriers was related to the allelic methylation in the same animals' somatic tissues, suggesting that methylation of a transposable element in the DNA sequence associated with a gene might be the means by which the trait was transmitted. Interestingly, the likelihood of inherited expression of the mutant phenotype varied across strains.

Another paramutation that led to obesity and a yellow (as opposed to species-typical, agouti) coat color in mice could be related to low cytosine methylation in the region of the mutant allele. Further examination of this phenomenon showed that when pregnant carrier females' diets were supplemented with methyl donors, all pups showed the species-typical agouti coat color. However, when these animals were mated, only the offspring of males whose dams were supplemented during midgestation continued to show the agouti pelage, and this effect was not seen in the third generation of mutants. Without methyl donor supplementation, the yellow phenotype was only maternally heritable (Cropley et al., 2012). Here, DNA methylation also was involved, but the transmission of the modification seemed to involve complex pathways.

The *Kit* locus in mice is another mutation that impacts melanogenesis. While lethal in dizygotes, in heterozyogtes, it leads to the appearances of white spots in the animals' coats, including the tip of the tail. When male or female heterozygotes were mated with genetically wild-type animals (including mice from another strain), their offspring, even those who were genetically wild-type, showed the white tail-tip. Across subsequent matings of genetically wild-type, but phenotypically "paramutant" animals who showed the white spots, with genetically wild-type partners, the frequency with which the offspring displayed white spots progressively declined. This phenotype could be induced when a microRNA known to interfere with the expression of *Kit* messenger RNA was injected into one-cell wild-type embryos, and it was transmitted across generations both maternally and paternally (Rassoulzadegan et al., 2006).

The latter work indicates that in addition to variations across generations in the retention of DNA methylation (cf. Rakyan et al., 2003, above),
microRNAs may also provide a pathway for gametic transmission of paramutant phenotypes. Furthermore, in a study of the effects of a gene involved
in regulating heart size, Ghanbarian, Grandjean, Cuzin, and Rassoulzadegan
(2011) found that injecting one-cell mouse embryos with factors regulating this gene, including a microRNA, could lead to cardiac hypertrophy,
and that although there was variation among littermates, the effect could
be transmitted for at least three generations when the offspring so-treated
were mated with controls. In another experiment, Grandjean et al. (2009)
injected fertilized mouse eggs with a microRNA that, when compared with
controls injected with an "irrelevant" microRNA sequence and transplanted
to the other uterine horn of the same foster mothers, led to increased body
weight, a phenotype that could be transmitted by both genders to offspring
when the carriers were mated to untreated controls. Thus, at least in mice,
microRNAs can transmit information relevant to the regulation of aspects
of body size, as well as pelage color, across generations.

The studies of the characteristics of imprinted loci raise the question of
whether there exist other distinctive chromosomal regions at which such
epigenetic modifications could be expected to endure across generations.
The work on paramutation and similar transgenerational phenomena has led
to speculation regarding the existence of "metastable epialleles" or sites that
might be particularly suited for information transmission across generations in
the absence of modifications of the DNA sequence itself. Several investigators
(e.g., Daxinger & Whtelaw, 2012) have proposed that repetitive sequences and
transposable elements—which (McClintock, 1984) had suggested could provide means for adjusting to severe conditions via alterations in gene regulation—might provide loci for transgenerational alterations in gene expression.

There is evidence for epigenetic modifications of both transposable
elements and repetitive sequences in sperm. Popp et al. (2010) found that
retrotransposons were the sites of high levels of DNA methylation in
sperm, and Flanagan et al. (2005) found high intraindividual variation in
the levels of DNA methylation within human sperm at repetitive elements.
Consistent with the view that repetitive elements could be "marked" to
convey a memory across generations, Koerner and Barlow (2010) noted
that some imprinting control elements included such sequences. Thus,
insofar as these elements are common in the human genome (see also
Costa, 2008), phenomena comparable to paramutation might be found in
homo as well as rodents.

On the basis of examinations of differences in DNA methylation in human monozygotic twins, Waterland et al. (2010) reported that there were a number of loci that showed apparently stochastic methylation patterns, suggesting that they could represent sites that were particularly sensitive to environmental conditions. This possibility gained support when they analyzed buccal samples from the offspring of a group of "subsistence farmers" in Gambia. This analysis revealed that, when compared with those who were conceived during food abundance and low work demands, individuals who were conceived during periods of reduced food availability and high work loads showed higher DNA methylation patterns in the sensitive sites they had identified.

The later findings thus lend further credence to the possibility that the gametes could provide an avenue by which modifications in gene expression made to environmental demands by humans could be transmitted to their offspring. This leads to questions regarding what is known about the contents of ova and sperm and how and when they might be modified.

4.1. The Gametes

The egg is a highly differentiated cell; the egg cytoplasm transmits a broad range of "resources" for the development of the zygote. Along with the mitochondria, there is an abundance of messenger RNAs, proteins, and nutrients that are necessary for embryonic development (Watson, 2007). As demonstrated by the successful cloning of whole animals by transplanting the nuclei of differentiated cells into the cytoplasm of oocytes (Kikyo et al., 2000), in addition to transmitting nutrient supplies and RNAs, the oocyte contains factors critical for regulating gene expression. For example, Nakamura et al. (2012) found that in the eggs of mice, there was a protein required for the maintenance of DNA methylation in early embryogenesis and that the same protein also protected the methylation of sperm histones.

As noted earlier, in most cells, histone proteins form clusters of two copies each of four "canonical" variants (H2, H3A, H3B and H4) to form the nucleosomes around which the DNA strands wrap. While carrying relatively more canonical histones than sperm, the egg nucleus also harbors at least one unique variant that is involved in linking nucleosomes for the formation of heterochromatin (thought to modulate access of the DNA region for transcription). In general, nucleosome histones in oocytes display relatively more histone acetylation, which is thought to render the nucleosome more easily displaced from the chromosomal strand for transcription (Kimmins & Sassone-Corsi, 2005). Likewise, relative to sperm, at least in mice, oocyte DNA is relatively unmethylated, and when methylated CpG sites were found, they were located more frequently at promoter regions (the locus at which the transcriptional machinery assembles when a gene sequence is being transcribed). In addition, cytosine nucleotides other than CpGs were more frequently methylated in mouse oocytes. However, according to Chotalla et al. (2009), imprinted regions were more methylated in oocytes than sperm.

In mouse oocytes, Watanabe, Totoki, Toyoda, and et al. (2008) also observed substantial levels of double-stranded RNAs, which are targets for enzymatic modification to produce the microRNAs that regulate protein-coding genes and retrotransposons. According to Tang et al. (2007), relative to sperm, the mouse maternal microRNAs contribute more to early zygotic regulation. Thus, the oocyte clearly transmits factors that can determine the plasticity of the developing zygote (see also Watson, 2007).

Obviously, sperm cells also represent highly specialized cell forms. Sperm chromatin is largely compacted by a number of proteins "protamines" not found in somatic cells; however, along with a sperm-specific histone variant, a few canonical histone variants also are found in sperm, leading Kimmins and Sassone-Corsi (2005) to propose that, to the degree that histones play a role in gene regulation, these loci at which histones are found have the potential to transmit information across generations. Subsequent investigations indicated that the canonical nucleosomes in sperm are nonrandomly located, being more frequently positioned at segments of the strands that code for protein products (Nahkuri et al., 2009) and that these nucleosomes tended to be found at sites that were involved in embryonic development, including loci from which microRNAs were transcribed in human sperm (Hammoud et al., 2009). Vavouri and Lehner (2011) also found that the nucleosomes in human sperm were most likely to bind to cytosine–guanosine-rich DNA sequences that are often implicated as regions suited for epigenetic regulation via the addition of methyl molecules to the cytosine nucleotides.

Amino acid segments of the histone proteins can be modified most often by the addition of methyl or acetyl molecules that, respectively, increase or decrease the affinities of the histones to DNA segments, and to other elements involved in the transcription process, thereby regulating access of the DNA to the enzyme clusters responsible for the production of messenger RNAs. Brykczynska et al. (2010) reported that in mouse and human sperm, such methyl modifications of histone H3 are retained at regulatory sequences involved in early embryonic development. Thus, histone modifications could be a pathway for transmission of epigenetic information (see also Vavouri & Lehner, 2011).

As noted earlier, small RNAs are now considered to also have an important role in controlling gene expression, and when involved in degrading mRNAs, do so when bound to a family of proteins, one of which the "Piwi" family, is largely restricted to human germ cells (Kaya & Doudna, 2012). Insofar as, at least in mice, RNA content and the chromatin in sperm could be influenced by diet (Carone et al., 2010), RNA-mediated transgenerational modifications might also be transmitted across generations via sperm.

While much remains to be determined, the gametes transmit not only chromosomes but also elements that regulate their expression during development. This further supports the possibility that they carry information regarding adjustments made by prior generation(s)—which leads to the question of when such signals might be encoded.

4.2. Timing

Insofar as some traits affected by transgenerational epigenetic inheritance seem to be gender-specific, an examination of current knowledge regarding the chronology of the development of sperm and oocytes may help in the search for as yet unidentified examples.

The precursor cells for both gametes begin to differentiate in the early prenatal period, and for sperm, it is generally understood that their subsequent maturation is mediated via the Sertoli cells in the gonad. The Sertoli cells are evident in fetal gonads, and their proliferation decreases postnatally, ending at puberty at which time they become fully responsive to androgens produced by the gonadal Leydig cells (Petersen & Soder, 2006; Rey, Musse, Venara, & Chenes, 2009). A recent report by Morgan, Dennis, Ruffman, Bilkey, and McLennan (2011) indicated that the Sertoli cells in young boys secrete "Mullerian inhibiting substance" and "inhibin B" hormones that may act to restrict physical growth (height) during childhood. In view of the fact that Pembrey et al. (2006) and Kaati et al. (2007) found that the "slow growth period" was the phase during which males seemed to be most sensitive to nutrient availability as a signal for transmission to their offspring, it would seem that the Sertoli cells may mediate this process insofar as they are thought to be the primary contributors to the development of mature sperm. Moreover, given the fact that growth can be regulated by nutrient availability, the Sertoli cells would seem to be influenced by the conditions faced by the developing child. Thus, middle childhood would seem to be a particularly important phase to examine closely for the possibility of epigenetic transmission via the male germ line.

In the search for maternal transmission of epigenetic modifications via the oocyte, there is a potential confound presented by the fact that the uterine environment also may be a pathway of transmission. Although there is much more to be learned, it seems that tissues in the uterus are subject to environmentally induced modifications both pre- and postnatally (Spencer et al., 2005).

As the studies of the consequences of reduced food supplies on humans indicated that the effects of restricted diet of grandmothers could be conveyed by sons to granddaughters (Kaati et al., 2007; Pembrey et al., 2006), implying something possibly related to imprinting, transmission via the oocyte is indicated. Thus, although there is controversy regarding whether or not the development of oocytes is limited to the prepubertal period (e.g., Bukovsky, 2011), there is consensus that ovarian tissues develop during the perinatal period and that the granulosa cells are particularly important in determining the fates of the developing eggs (Sarraj & Drummond, 2012). This suggests that the period when the granulosa cells are developing might be an especially important time for modification of the regulatory-related content of oocytes. There is also evidence that modifications involved in imprinting in oocytes occur postnatally as the eggs arrest in meiosis (Chotalla et al., 2009). Thus, it seems that for maternal transmission, both the perinatal and early slow-growth periods would be worthy of close scrutiny when looking for evidence of maternal transgenerational transmission of epigenetic modifications.

5. CONDITIONS FAVORING TRANSGENERATIONAL INHERITANCE

As indicated in the introductory discussion of what is inherited from conception, beginning with cross-talk between the zygote and tissues in the maternal reproductive tract, there is the potential for offspring development to be moderated in response to prevailing environmental conditions. So we must address the question of "why" there should be yet another pathway moderating phenotypic plasticity. Current views suggest that evolutionary pressure favoring the transmission across generations of anticipatory adjustments that, to some degree, limit plasticity in response to external demands would be greatest under a limited set of circumstances.

Insofar as most examples of transgenerational epigenetic inheritance involve some sort of constraint on the heirs' developmental range of response to external conditions, the cost of making an appropriate adjustment after the fact would seem to be a major consideration. For example, although limiting one's size and storing energy in the form of fat can have deleterious effects in the long run under conditions of adequate food supplies, allocation of energy supplies during development to growth in size, particularly if also minimizing the potential for lipid storage, is likely to be even more disadvantageous—even fatal—if one then encounters a situation in which there is a prolonged reduction in available nutrients, especially if high expenditures of physical energy are required for viability. In short, selection should favor an adjustment, that perhaps suboptimal under good conditions, protects the individual against the full impact of adverse circumstances.

This leads to the question of the kinds of circumstances that would be expected to cause selection to favor the transgenerational transmission of phenotypic modifications even if those conditions are not currently present. Given that natural selection operates across generations, favoring those who are most able to thrive under the conditions that their species is likely to encounter, a key feature would be the probability that a set of circumstances demanding costly, relatively long-term, adjustments will occur within the span of two generations or more. That is, while not predictable within any one lifespan, when conditions that place enduring, high-cost demands for adjustment on a species occur regularly over longer periods of time, selection would favor those individuals whose offspring were prepared to deal with them in advance (Tollrian & Harvell, 1999).

As suggested by the results of the studies of the effects of periodic crop failures in Sweden (Kaati et al., 2007; Pembrey et al., 2006), periods of famine represent one condition likely to have shaped transgenerational inheritance. Although the Swedish studies examined the effects of relatively brief periods of limited food supplies, cyclic climate fluctuations that can lead to long-term effects on nutrient availability also occur over much longer time spans. For example, the declines of civilizations in the Midwestern United States (Kloor, 2007) in Latin America (Medina-Elizalde & Rohling, 2012), and middle Asia (Lawler, 2007) have been attributed to prolonged periods of drought. Thus, the conditions favorable for transgenerational transmission of metabolic adjustments to limited food supplies would apply in this case.

Another event occurring cyclically in human history is the occurrence of migration and/or invasion, often leading to the enslavement of the vanquished peoples for multiple generations. While there exist no data on transmission in humans, given the recent results of studies of the trans-generational transmission of responses to stress on rodent behavior (e.g., Franklin et al., 2011; Weiss et al., 2011), it would seem likely that humans might be likely to "prepare" their progeny for conditions that would favor the display of "cautious" behavior in some contexts.

In addition, although speculative, it is possible that the tendency to "think outside the box" could be constrained. For example, under conditions in which access to resources is limited and there exist few, if any, alternatives to currently practiced strategies for successfully extracting them, to the extent that members of a population remain in such a setting, there could be transmission of a tendency to be less open to speculation regarding "what might happen" and attempts at innovative thinking (Harper, 2010). Moreover, to the extent that subjugated populations are under strict surveillance by their oppressors, and when deviance is likely to result in severe punishment or even death, it is possible that a constraint on the tendency to entertain behavioral options could be evoked by the conditions facing previous generations.

Finally, insofar as humans are a cosmopolitan species inhabiting a wide range of settings presenting different adaptive challenges (e.g., altitude, climate variation) to the extent that only certain (sub) populations have inhabited such settings over relatively long periods of time, it is likely that additional dimensions of flexibility in response to cyclic exigencies have been subject to constraint via epigenetic inheritance.

6. IMPLICATIONS FOR RESEARCH AND POLICY

In order to determine the degree to which individual differences that are shared across generations can be attributed to epigenetic transmission, in addition to assessing epigenetic "marks" that differ across individuals and families, such as DNA methylation or histone modifications in serum or buccal tissue samples (cf. Ollikainen et al., 2010; Waterland et al., 2010), along with assessments of the characteristic in question, as much information as possible should be acquired about conditions faced by the lineage of each subject across at least three generations. This is particularly important when assessing similarities and differences possibly attributable to experiences encountered within the matriline, given that alterations in the uterine environment could mimic germ line effects (e.g., Spencer et al., 2005; Tobi et al., 2009). Thus, this information should include both details related to the exigencies faced by progenitors and also the point(s) in their lifetimes at which such events occurred.

Moreover, as indicated by studies of the effects not only of diet on human longevity (e.g., Pembrey et al., 2006) and offspring size (see, e.g., Tobi et al., 2009) but also the effects of stress on later behavior in mice (Franklin et al., 2011; Morgan & Bale, 2011), one may expect gender differences both in terms of when progenitors are likely to be affected and in terms of the quality and/or degree of response to a specific kind of experience that may be epigenetically transmitted. Likewise, in view of the fact that transgenerational responses to diet may be mediated via adjustments to the expression of imprinted genes (e.g., Pembrey et al., 2006), apparently indirect pathways may exist, suggesting that a broad range of relationships need to be examined.

Another consideration relates to the results of work with animals indicating the existence of pleiotropic effects (Bendesky & Bargmann, 2011)—including strain differences in the patterns of expression of imprinted genes in the brain (Gregg, Zhang, Butler, Haig, & Dulac, 2010)—and in the degree to which a particular response to a given set of conditions may lead to transmission via the gametes (e.g., Rakyan et al., 2003). This is of particular significance in view of the fact that there is ample evidence that human (sub)populations have resided for many generations in habitats that have subjected them to different selective pressures. For example, Tibetans living at high altitude differ from Han Chinese and Danes with respect to genes involved in responses to hypoxia (Yi et al., 2010). Likewise, lifestyle seems to have differentially shaped other aspects of the human genome. For example, an examination of the remains of 5000-year-old humans yielded DNA differences between neolithic hunter-gatherers and farmers living in relatively close proximity. These allelic differences were reflected to some degree in modern populations in Northern Europe and in the Mediterranean region (Skoglund et al., 2012). The impact of differences in subsistence strategies are still relevant as can be seen from the fact that members of a Canadian native American hunter-gatherer group whose diets had previously been relatively low in carbohydrates, upon recently moving to urban settings in northern Canada, displayed extremely high rates of diabetes mellitus due to a variant in a gene regulating glucose tolerance (Millar & Dean, 2012).

Thus, there may be ethnic differences in the likelihood of transgenerational epigenetic transmission via the germ line. That such inter population differences are likely to be widespread is supported by the findings of Stranger et al. (2012) indicating the existence of substantial genetic variation across some eight human ethnic groups in alleles associated with gene regulation and Lu and Clark (2012) who linked such differences to population differences in variations of microRNAs—elements identified earlier as potential vehicles for transmission of epigenetic modulation of gene expression.

There are probable interactions between gender and population differences in the likelihood of transmission as a result of group and/or cultural differences in the potential impact of such enduring conditions as resource availability. Presumably, some gender-related variation could result from stable cultural differences in the degree to which inheritance or residence is matrilineal or patrilineal. Obviously, inequality of status according to gender within a population is another potential domain of interest.

Thus, one should not necessarily expect all groups to display the same or comparable anticipatory adjustments to variations in setting. A consideration of the diversity of habitats humans occupy and the sociocultural histories of human (sub)populations can provide useful cues relevant to identifying the kinds of physical and social environments—and combinations thereof—that might act as cues for the restriction or facilitation of particular adaptive strategies.

This latter point has implications that need to be taken seriously when shaping policies related to health and welfare. Given the substantial range of variability within and across human (sub) groups when devising means to overcome risks or other potential disadvantages, we cannot expect that the same ameliorative strategies will apply equally well to everyone. Moreover, insofar as the effects of transgenerational epigenetic inheritance endure across generations, environmentally induced constraints on development may take two or more generations to be fully ameliorated. Thus, in the case of what might appear to be a minimally effective remediation, it will be necessary to perform an examination of the life histories of the participants' ancestors before making final decisions relating to the effectiveness of an intervention.

REFERENCES

Agrawal, A. A., Laforsch, C., & Tollrian, R. (1999). Transgenerational induction of defenses in animals and plants. *Nature, 401*, 60–63.

Bendesky, A., & Bargmann, C. I. (2011). Genetic contributions to behavioural diversity at the gene-environment interface. *Nature Reviews Genetics, 12*, 809–820.

Birnbaum, R. Y., Clowney, E. J., Agamy, O., Kim, M. J., Zhao, J., Yamada, T., et al. (2012). Coding exons function as tissue-specific enhancers of nearby genes. *Genome Research.* http://dx.doi.org/10.1101/gr. 133546.111.

Bredy, T. W., Zhang, T. Y., Grant, R. J., Diorio, J., & Meaney, M. J. (2004). Peripubertal environmental enrichment reverses the effects of maternal care on hippocampal development and glutamate receptor subunit expression. *European Journal of Neuroscience, 20*, 1355–1362.

Brenet, F., Moh, M., Funk, P., Fierstein, E., Viale, A. J., Socci, N. D., et al. (2011). DNA methylation of the first exon is tightly linked to transcription silencing. *PLoS ONE*, *6*(1), e14524.

Brink, R. A. (1956). A genetic change associated with the R locus in maize which is directed and potentially reversible. *Genetics*, *41*, 872–889.

Brykczynska, U., Hisano, M., Erkek, S., Ramos, L., Oakeley, E. J., Roloff, T. C., et al. (2010). Repressive and active histone methylation mark distinct promoters in human and mouse spermatozoa. *Nature Structural and Molecular Biology*, *17*, 679–687.

Budnik, V., & Salinas, P. C. (2011). Wnt signaling during synaptic development and plasticity. *Current Opinion in Neurobiology*, *21*, 151–159.

Bukovsky, A. (2011). Ovarian stem cell niche and follicular renewal in mammals. *Anatomical Record*, *294*, 1284–1306.

Burdge, G. C., Slater-Jefferies, J., Torrens, C., Phillips, E. S., Hanson, M. A., & Lillicrop, K. A. (2007). Dietary protein restriction of pregnant rats in the Fo generation induces altered methylation of hepatic gene promoters in the adult male offspring in the F1 and F2 generations. *British Journal of Nutrition*, *97*, 435–439.

Cameron, N. M., Champagne, F. A., Parent, C., Fish, E. W., Ozaki-Kuroda, K., & Meaney, M. J. (2005). The programming of individual differences in defensive responses and reproductive strategies in the rat through variations in maternal care. *Neuroscience and Biobehavioral Reviews*, *29*, 843–865.

Carone, B. R., Fauquier, L., Habib, N., Shea, J., Hart, C. E., Li, R., et al. (2010). Paternally induced transgenerational environmental reprogramming of metabolic gene expression in mammals. *Cell*, *143*, 1084–1096.

Chakravarti, A., & Kapoor, A. (2012). Mendelian puzzles. *Science*, *335*, 930–931.

Chong, S., & Whitelaw, E. (2004). Epigenetic germline inheritance. *Current Opinion in Genetics and Development*, *14*, 692–696.

Chotalla, M., Smallwood, S. A., Ruf, N., Dawson, C., Lucifero, D., Frontera, M., et al. (2009). Transcription is required for establishment of germline methylation marks at imprinted genes. *Genes and Development*, *23*, 105–117.

Costa, F. F. (2008). Non-coding RNAs, epigenetics and complexity. *Gene*, *410*, 9–17.

Cropley, J., Dang, T. H. Y., Martin, D. I. K., & Suter, C. M. (2012). The penetrance of an epigenetic trait in mice is progressively yet reversibly increased by selection and environment. *Proceedings of the Royal Society B*. http://dx.doi.org/10.1098/rspb.2011.2646.

Davidson, E. H. (2006). *The regulatory genome*. NY: Academic Press.

Davies, W., Isles, A. R., & Wilkinson, L. S. (2005). Imprinted gene expression in the brain. *Neuroscience and Biobehavioral Reviews*, *29*, 421–430.

Daxinger, L., & Whtelaw, E. (2012). Understanding transgenerational epigenetic inheritance via the gametes in mammals. *Nature Reviews Genetics*, *13*, 153–161.

Dindot, S. V., Person, R., Strivens, M., Garcia, R., & Beaudet, A. L. (2009). Epigenetic profiling at mouse imprinted gene clusters reveals novel epigenetic and genetic features at differentially methylated regions. *Genome Research*, *19*, 1374–1383.

Dioniak, S. M., French, J. A., & Holekamp, K. E. (2006). Rank-related maternal effects of androgens on behaviour in wild spotted hyenas. *Nature*, *440*, 1190–1193.

Dixon, J. R., Selvaraj, S., Yue, F., Kim, A., Li, Y., Shen, Y., et al. (2012). Topological domains in mammalian genomes identified by analysis of chromatin interactions. *Nature*, *485*, 376–380.

Djupedal, I., & Ekwall, K. (2008). The paradox of silent heterochromatin. *Science*, *320*, 624–625.

Dominguez-Bello, M. G., Costello, E. K., Contreras, M., Magris, M., Hidalgo, G., Fierer, N., et al. (2010). Delivery mode shapes the acquisition and structure of the initial microbiota across multiple body habitats in newborns. *Proceedings of the National Academy of Sciences of the United States of America*, *107*, 11971–11975.

Eggan, K., Baldwin, K., Tackett, M., Osborne, J., Gogos, J., Chess, A., et al. (2004). Mice cloned from olfactory sensory neurons. *Nature*, *428*, 44–49.

Engler, A. J., Sen, S., Sweeney, H. L., & Discher, D. E. (2006). Matrix elasticity directs stem cell lineage specification. *Cell, 126,* 677–689.

Entringer, S., Buss, C., Swanson, J. M., Cooper, D. M., Wing, D. A., Waffarn, F., et al. (2012). Fetal programming of body composition, obesity, and metabolic function: the role of intrauterine stress and stress biology. *Journal of Nutrition and Metabolism.* http://dx.doi.org/10.1155/2012.632548. Article ID 532548.

Fan, T., Hagan, J. P., Kozlov, S. V., Stewart, C. L., & Muegge, K. (2005). Lsh controls silencing of the imprinted *Cdkn1c* gene. *Development, 132,* 635–644.

Flames, N., & Hobert, O. (2009). Gene regulatory logic of dopamine neuron differentiation. *Nature, 458,* 885–889.

Flanagan, J. F., Mi, L.-Z., Chruszcz, M., Cymborowski, M., Clines, K. L., Kim, Y., et al. (2005). Double chromodomains cooperate to recognize the methylated histone H3 tail. *Nature, 438,* 1181–1185.

Franklin, T. B., Linder, N., Russig, H., Thony, B., & Mansuy, I. (2011). Influence of early stress on social abilities and serontonergic functions across generations. *PLoS ONE, 6*(7), e21842.

Fujita, M., Roth, E., Lo, Y.-J., Hurst, C., Vollner, J., & Kendell, A. (2012). In poor families, milk is richer in daughters than sons: a test of Trivers–Willard hypothesis in agropastoral settlements in northern Kenya. *American Journal of Physical Anthropology.* http://dx.doi.org/10.1002/ajpa.22092.

Gao, J., Wang, W.-Y., Mao, Y.-W., Graff, J., Guan, J.-S., Pan, L., et al. (2010). A novel pathway regulates memory and plasticity via SIRT1 and miR-134. *Nature, 466,* 1105–1109.

Georges, A. B., Benayoun, B. A., Caburet, S., & Veitia, R. A. (2010). Generic binding sites, generic DNA-binding domains: where does specific promoter recognition come from? *FASEB Journal, 24,* 346–356.

Ghanbarian, H., Grandjean, V., Cuzin, F., & Rassoulzadegan, R. (2011). A network of regulations by small non-coding RNAs: the P-TEFb kinase in development and pathology. *Frontiers in Genetics, 2.* Article 95.

Gleason, G., Zupan, B., & Toth, M. (2011). Maternal genetic mutations as gestational and early life influences in producing psychiatric disease-like phenotypes in mice. *Frontiers in Psychiatry, 2.* Article 25.

Gluckman, P. D., Hanson, M. A., & Beedle, A. S. (2007). Non-genomic transgenerational inheritance of disease risk. *BioEssays, 29,* 145–154.

Gomes, N. P., & Espinosa, J. M. (2010). Gene-specific repression of the p53 target gene PUMA via intragenic CTCF-Cohesin binding. *Genes & Development, 24,* 1022–1034.

Gottlieb, G. (2002). Developmental-behavioral initiation of evolutionary change. *Psychological Review, 109,* 211–218.

Grandjean, V., Gounon, P., Wagner, N., Martin, L., Wagner, K. D., Bernex, F., et al. (2009). The *miR-124-Sox9* paramutation: RNA-mediated epigenetic control of embryonic and adult growth. *Development, 136,* 3647–3655.

Gregg, C., Zhang, J., Butler, J. E., Haig, D., & Dulac, C. (2010). Sex-specific parent-of-origin allelic expression in the mouse brain. *Science, 329,* 682–685.

Guan, J.-S., Haggarty, S. J., Giacometti, E., Dannenberg, J.-H., Joseph, N., Gao, J., et al. (2009). HDAC2 negatively regulates memory formation and synaptic plasticity. *Nature, 459,* 55–60.

Guttman, M., & Rinn, J. L. (2012). Modular regulatory principles of large non-coding RNAs. *Nature, 482,* 339–346.

Hajkova, P., Jeffries, S. J., Lee, C., Miller, N., Jackson, S. P., & Surani, M. A. (2010). Genome-wide reprogramming in the mouse germ line entails the base excision repair pathway. *Science, 329,* 78–82.

Hammoud, S. S., Nix, D. A., Zhang, H., Purwar, J., Carrell, D. T., & Cairns, B. R. (2009). Distinctive chromatin in human sperm packages genes for embryo development. *Nature, 460,* 473–479.

Han, Z., Mtango, N. R., Patel, B. G., Sapienza, C., & Latham, K. E. (2008). Hybrid vigor and transgenerational epigenetic effects on early mouse embryo phenotype. *Biology of Reproduction, 79*, 638–648.

Harper, L.V. (1989). *The nurture of human behavior.* Norwood, NJ: Ablex.

Harper, L.V. (2010). Epigenetic inheritance. In B. Mesquita, L. F. Barrett & E. R. Smith (Eds.), *The mind in context* (pp. 25–41). NY: Guilford Press.

Heijtz, R. D., Wang, S., Anuar, F., Qian, Y., Bjorkholm, B., Samuelsson, A., et al. (2011). Normal gut microbiota modulates brain development and behavior. *Proceedings of the National Academy of Sciences of the United States of America, 108*, 3047–3052.

Hoile, S. P., Lillycrop, K. A., Thomas, N. A., Hanson, M. A., & Burdge, G. C. (2011). Dietary protein restriction during Fo pregnancy in rats induced transgenerational changes in the hepatic transcriptome in female offspring. *PLoS ONE, 6*(7), e21668.

Inoue, K., Khoda, T., Lee, J., Ogonuki, N., Mochida, K., Noguchi, Y., et al. (2002). Faithful expression of imprinted genes in cloned mice. *Science, 295*, 297.

Jablonca, E., & Lamb, M. J. (1995). *Epigenetic inheritance and evolution.* Oxford: Oxford University Press.

Johnson, L. J., & Tricker, P. J. (2010). Epigenetic plasticity within populations: its evolutionary significance and potential. *Heredity, 105*, 113–121.

Kaati, G., Bygren, L. O., Pembrey, M., & Sjostrom, M. (2007). Transgenerational response to nutrition, early life circumstances and longevity. *European Journal of Human Genetics, 15*, 784–790.

Kaya, E., & Doudna, J. A. (2012). Guided tour to the heart of RISC. *Science, 336*, 985–986.

Kikyo, N., Wade, D., Guschin, J., Ge, H., & Wolffe, A. P. (2000). Active remodeling of somatic nuclei in egg cytoplasm by the nucleosomal ATPase ISWI. *Science, 289*, 2360–2362.

Kimmins, S., & Sassone-Corsi, P. (2005). Chromatin remodelling and epigenetic features of germ cells. *Nature, 434*, 583–589.

Kim, T. I., Shin, Y. H., & White-Traut, R. C. (2003). Multisensory intervention improves physical growth and illness rates in Korean orphaned newborn children. *Research in Nursing & Health, 26*, 424–433.

Kloor, K. (2007). The vanishing Fremont. *Science, 318*, 1540–1543.

Koerner, M.V., & Barlow, D. P. (2010). Genomic imprinting—an epigenetic gene-regulatory model. *Current Opinion in Genetics and Development, 20*, 164–170.

Kovan, N. M., Chung, A. L., & Sroufe, L. A. (2009). The intergenerational continuity of observed early parenting: a prospective longitudinal study. *Developmental Psychology, 45*, 1205–1213.

Laland, K. N., Sterelny, K., Odling-Smee, J., Hoppitt, w., & Uller, T. (2011). Cause and effect in biology revisited: is Mayr's proximate-ultimate dichotomy still useful? *Science, 334*, 1512–1516.

Lawler, A. (2007). Middle Asia takes center stage. *Science, 317*, 586–590.

Lee, K.-F., & Yeung, W. S. B. (2006). Gamete/embryo-oviduct interactions: implications on *in vitro* culture. *Human Fertility, 9*, 137–143.

Lu, J., & Clark, A. G. (2012). Impact of microRNA regulation on variation in human gene expression. *Genome Research.* http://dx.doi.org/10.1101/gr.132514.111.

Martin, C., Beaujean, N., Brochard, V., Audouard, V., Zink, D., & Debey, P. (2006). Genome restructuring in mouse embryos during reprogramming and early development. *Developmental Biology, 292*, 317–332.

McClintock, B. (1984). The significance of responses of the genome to challenge. *Science, 226*, 792–801.

McGowan, P. O., Sasaki, A., D'Alessio, A. C., Dymov, S., Labonte, B., Szyf, M., et al. (2009). Epigenetic regulation of the glucocorticoid receptor in human brain associates with childhood abuse. *Nature Neuroscience, 12*, 342–348.

Medina-Elizalde, M., & Rohling, E. J. (2012). Collapse of classic Maya civilization related to modest reduction in precipitation. *Science*, *335*, 956–959.

Mehedint, M. G., Niculescu, M. D., Craciunescu, C. N., & Zeisel, S. H. (2010). Choline deficiency alters global histone methylation and epigenetic marking at the Re1 site of the calbindin 1 gene. *FASEB Journal*, *24*, 184–195.

Mennella, J. A., Jagnow, C. P., & Beauchamp, G. K. (2001). Prenatal and postnatal flavor learning by human infants. *Pediatrics*, *107*, E 88.

Millar, K., & Dean, H. (2012). Developmental origins of type 2 diabetes in aboriginal youth in Canada: It is more than diet and exercise. *Journal of Nutrition and Metabolism*. Article ID 127452.

Miller, C. A., Gavin, C. F., White, J. A., Parrish, R. R., Honasoge, A., Yancey, C. R., et al. (2010). Cortical DNA methylation maintains remote memory. *Nature Neuroscience*, *13*, 664–666.

Morgan, C. P., & Bale, T. L. (2011). Early prenatal stress epigenetically programs dysmasculinization in second-generation offspring via the paternal lineage. *The Journal of Neuroscience*, *31*, 11748–11755.

Morgan, K., Dennis, N., Ruffman, T., Bilkey, D. K., & McLennan, I. S. (2011). The stature of boys is inversely correlated to the levels of their Sertoli cell hormones: do the testes restrain the maturation of boys? *PLoS ONE*, *6*(6), e20533.

Morison, I. M., Ramsay, J. P., & Spencer, H. G. (2005). A census of mammalian imprinting. *Trends in Genetics*, *21*, 457–465.

Munsky, B., Neuert, G., & van Oudenaarden, A. (2012). Using gene expression noise to understand gene regulation. *Science*, *336*, 183–187.

Nahkuri, S., Taft, R. J., & Mattick, J. S. (2009). Nucleosomes are preferentially positioned at exons in somatic and sperm cells. *Cell Cycle*, *8*, 3420–3424.

Nakamura, T., Liu, Y.-J., Nakashima, H., Umehara, H., Inoue, K., Matoba, S., et al. (2012). PGC7 binds histone H3K9me2 to protect against conversion of 5mC to 5hmC in early embryos. *Nature*, *486*, 415–419.

Nelson, C. A., Fox, N. A., Marshall, P. J., Smyke, A. T., & Guthrie, D. (2007). Cognitive recovery in socially deprived children: the Bucharest early intervention project. *Science*, *318*, 1937–1940.

Ng, S.-F., Lin, R. C.Y., Laybutt, D. R., Barres, R., Owens, J. A., & Morris, M. J. (2010). Chronic high-fat diet in fathers programs beta-cell dysfunction in female rat offspring. *Nature*, *467*, 963–967.

Ollikainen, M., Smith, K. R., Joo, E. J.-H., Ng, H. K., Andronikos, R., Novakovic, B., et al. (2010). DNA methylation analysis of multiple tissues from newborn twins reveals both genetic and intrauterine components to variations in the human neonatal epigenome. *Human Molecular Genetics*, *19*, 4176–4188.

Oyama, S., Griffiths, P. E., & Gray, R. D. (2001). *Cycles of contingency: Developmental systems and evolution*. Cambridge, MA: MIT Press.

Painter, R. C., Osmond, C., Gluckman, P., Hanson, M., Phillips, D. I. W., & Rosenboom, T. J. (2008). Transgenerational effects of prenatal exposure to the Dutch famine on neonatal adiposity and health in later life. *British Journal of Obstetrics and Gynecology*, *115*, 1243–1249.

Pembrey, M., Bygren, L. O., Kaati, G., Edvinsson, S., Northstone, K., Sjostrom, M., et al. (2006). Sex-specific, male-line transgenerational responses in humans. *European Journal of Human Genetics*, *14*, 159–166.

Petersen, C., & Soder, O. (2006). The Sertoli cell—a hormonal target and 'super' nurse for germ cells that determines testicular size. *Hormone Research*, *66*, 153–161.

Popp, C., Dean, W., Feng, S., Cokus, S. J., Andrews, S., Pellegrini, M., et al. (2010). Genome-wide erasure of DNA methylation in mouse primordial germ cells is affected by AID deficiency. *Nature*, *463*, 1101–1105.

Radford, E. J., Isganaitis, E., Jimenez-Chillaron, J., Schroeder, J., Molla, M., Andrews, S., et al. (2012). An unbiased assessment of the role of imprinted genes in an intergenerational model of developmental programming. *PLoS Genetics, 8*(4), e1002605.

Rakyan, V. K., Chong, S., Champ, M. E., Cuthbert, P. C., Morgan, H. D., Luu, K.V.K., et al. (2003). Transgenerational inheritance of epigenetic states at the murine *AxinFu* allele occurs after maternal and paternal transmission. *Proceedings of the National Academy of Sciences of the United States of America, 100,* 2538–2543.

Rando, O. J., & Verstrepen, K. J. (2007). Timescales of genetic and epigenetic inheritance. *Cell, 128,* 655–668.

Rassoulzadegan, M., Grandjean, V., Gounon, P., Vincent, S., Gillot, I., & Cuzin, F. (2006). RNA-mediated non-Mendelian inheritance of an epigenetic change in the mouse. *Nature, 441,* 469–474.

Reik, W. (2007). Stability and flexibility of epigenetic gene regulation in mammalian development. *Nature, 447,* 425–432.

Reik, W., Romer, I., Barton, S. C., Surani, M. A., Howlett, S. K., & Close, J. (1993). Adult phenotype in the mouse can be affected by epigenetic events in the early embryo. *Development, 119,* 933–942.

Rey, R. A., Musse, M., Venara, M., & Chenes, H. E. (2009). Ontogeny of the androgen receptor expression in the fetal and postnatal testis: Its relevance to Sertoli cell maturation and the onset of adult spermatogenesis. *Microscopy Research and Technique, 72,* 787–795.

Rogoff, B., & Morelli, G. (1989). Perspectives on children's development from cultural psychology. *American Psychologist, 44,* 343–348.

Rugarti, E. I., & Langer, T. (2012). Mitochondrial quality control: a matter of life and death for neurons. *EMBO Journal, 3,* 1336–1349.

Rutter, M. (2002). Nature, nurture and development: from evangelism through science toward policy and practice. *Child Development, 73,* 1–21.

Sarraj, M. A., & Drummond, A. E. (2012). Mammalian foetal ovarian development: consequences for health and disease. *Reproduction, 143,* 151–163.

Skoglund, P., Malmstrom, H., Raghavan, M., Stora, J., Hall, P., Willerslev, E., et al. (2012). Origins and genetic legacy of Neolithic farmers and hunter-gatherers in Europe. *Science, 336,* 466–469.

Skuse, D. H., James, R. S., Bishop, D. V. M., Coppin, B., Dalton, P., Aanodt-Leeper, G., et al. (1997). Evidence from Turner's syndrome of an imprinted, X-linked locus affecting cognitive function. *Nature, 387,* 705–708.

Sleutels, F., Zwart, R., & Barlow, D. P. (2002). The non-coding *Air*RNA is required for silencing autosomal imprinted genes. *Nature, 415,* 810–813.

Spencer, T. E., Hayashi, K., Hu, J., & Carpenter, K. D. (2005). Comparative developmental biology of the mammalian uterus. *Current Topics in Developmental Biology, 68,* 85–122.

Stevens, M. M., & George, J. H. (2005). Exploring and engineering the cell surface interface. *Science, 310,* 1135–1138.

Stranger, B. E., Montgomery, S. B., Dimas, A. S., Parts, L., Stegle, O., Ingle, C. E., et al. (2012). Patterns of *cis* regulatory variation in diverse human populations. *PLoS Genetics, 8*(4), e1002639.

Tang, A. C., Akers, K. G., Reeb, B. C., Romeo, R. D., & McEwen, B. S. (2006). Programming social, cognitive and neuroendocrine development by early exposure to novelty. *Proceedings of the National Academy of Sciences of the United States of America, 103,* 15716–15721.

Tang, F., Kaneda, M., O'Carroll, D., Hajkova, P., Barton, S. C., Sun, Y. A., et al. (2007). Maternal microRNAs are essential for mouse zygotic development. *Genes and Development, 21,* 644–648.

Tanner, J. M. (1990). *Fetus into man* (Revised ed.). Cambridge, MA: Harvard University Press.

Tobi, E. W., Lumey, L. H., Talens, R. P., Kremer, D., Putter, H., Stein, A. D., et al. (2009). DNA methylation differences after exposure to prenatal famine are common and timing- and sex-specific. *Human Molecular Genetics, 18,* 4046–4053.

Tollrian, R., & Harvell, C. D. (Eds.), (1999). *The ecology and evolution of inducible defenses.* Princeton, NJ: Princeton University Press.

Vavouri, T., & Lehner, B. (2011). Chromatin organization in sperm may be the major functional consequence of base composition variation in the human genome. *PLoS Genetics,* 7(4), e1002036.

Watanabe, T., Totoki, Y., Toyoda, A., Kaneda, M., Kuramochi-Miyagawa, S., Obata, Y. et al. (2008). Endogenous siRNAs from naturally formed ds RNAs regulate transcripts in mouse oocytes. *Nature, 453,* 539–543.

Waterland, R. A., Kellermayer, R., Laritsky, E., Rayco-Solon, P., Harris, R. A., Travisano, M., et al. (2010). Season of conception in rural Gambia affects DNA methylation at putative human metastable epialleles. *PLoS Genetics,* 6(12), e1001252.

Waterland, R. A., & Michels, K. B. (2007). Epigenetic epidemiology of the developmental origins hypothesis. *Annual Review of Nutrition, 27,* 363–388.

Watson, A. J. (2007). Oocyte cytoplasmic maturation: a key mediator of oocyte and embryo developmental competence. *Journal of Animal Science, 85*(Suppl. E.), E1–E3.

Weber, B. H., & Depew, D. J. (2003). *Evolution and learning. The Baldwin effect reconsidered.* Cambridge, MA: MIT Press.

Weiss, I. C., Franklin, T. B., Vizi, S., & Mansuy, I. M. (February, 2011). Inheritable effect of unpredictable maternal separation on behavioral responses in mice. *Frontiers in Behavioral Neuroscience, 5.* Article 3.

Werner, E. E., & Smith, R. S. (1992). *Overcoming the odds. High risk children from birth to adulthood.* Ithica, NY: Cornell University Press.

West-Eberhard, M. J. (2003). *Developmental plasticity and evolution.* Oxford: Oxford University Press.

Yasbek, S. N., Spiezio, S. H., Nadeau, J. H., & Buchner, D. A. (2010). Ancestral paternal genotype controls body weight and food intake for multiple generations. *Human Molecular Genetics, 19,* 4134–4144.

Yi, X., Liang, Y., Huerta-Sanchez, E., Jin, Z., Cuo, Z. X.P., Pool, J. E., et al. (2010). Sequencing of 50 human exomes reveals adaptation to high altitude. *Science, 329,* 75–78.

Zamenhof, S., van Marthens, E., & Grauel, L. (1971). DNA (cell number) in neonatal brain: second generation (F2) alteration by maternal (Fo) dietary protein restriction. *Science, 172,* 850–851.

SUBJECT INDEX

AUTHOR INDEX

CONTENTS OF PREVIOUS VOLUMES